Dynamic Models for Structural Plasticity

Springer
London
Berlin
Heidelberg
New York
Barcelona
Budapest
Hong Kong
Milan
Paris
Santa Clara
Singapore
Tokyo

W.J. Stronge and T.X. Yu

Dynamic Models for Structural Plasticity

With 175 Figures

Springer

William James Stronge, PhD, MS, BS
Department of Engineering, University of Cambridge, Trumpington Street, Cambridge CB2 1PZ, UK

Tongxi Yu, PhD, ScD
Department of Mechanical Engineering, Hong Kong University of Science and Technology, Clear Water Bay, Kowloon, Hong Kong

ISBN 3-540-76013-X Springer-Verlag Berlin Heidelberg New York

British Library Cataloguing in Publication Data
A catalogue record for this book is available from the British Library

Printed in Great Britain
Corrected 2nd printing, 1995

Typeset by Fox Design, Bramley, Guildford, Surrey
Printed at the Athenæum Press Ltd., Gateshead
69/3830-54321 Printed on acid-free paper

Preface

Our topic is irreversible or plastic deformation of structural elements composed of relatively thin ductile materials. These deformations are commonly used in sheet metal forming operations to produce lightweight parts of any particular shape. In another context, this type of plastic deformation is described as impact damage in the case of structural components involved in collision. Here we are concerned with mechanics of both static and dynamic deformation processes. The purpose is to use typical material properties and structural characteristics to calculate the deformation for certain types of load; in particular to find the final deflection and shape of the deformed structure and to illustrate how the development of this final shape depends on the constitutive model used to represent the material behavior. The major issue to be addressed is which structural and constitutive properties are important for calculating response to either static or brief but intense dynamic loads. Furthermore, how do the results of various constitutive models compare with observed behavior.

Static analysis of plastic deformation is concerned with quantitative assessment of structural safety and function; in practice either live loads or foundation settlement can cause small sections of a structure to experience relatively large plastic deformation. The margin between the initial yielding load and the similar load for fully plastic collapse is a measure of the ability of a structure to occasionally withstand deformations that exceed the elastic limit. Static analysis of plastic deformation is used also for sheet metal forming operations to find the press forces and die shapes necessary to achieve a particular final shape.

Dynamic plastic deformation of structures is applicable to blast or impact loading; i.e. intense transient loads where the forces are substantially more than the plastic collapse force but these forces are applied for only a very brief period. This type of loading is used in high-rate forming operations and it is the cause of structural damage in collisions. Calculations of dynamic structural deformation are useful for design of soft crash barriers that minimize damage to occupants of a colliding vehicle; similar calculations are useful for investigating structural integrity of the vehicle during all stages of a collision.

This book develops calculations of irreversible or plastic deformation for an elementary structural component on the basis of *simple constitutive models*. These models represent various aspects of material behavior by means of equations developed from a theory; e.g. elasticity, plasticity or viscoplasticity. The constitutive models are formulated by combining the material representation with an approximation for the distribution of local deformation; this approximation primarily represents flexure of structures that have a thickness which is small in comparison with the length. This restriction applies since the constitutive equations are all based on a deformation approximation which assumes that plane sections remain plane — the Euler–Bernoulli approximation. So although the effect of in-plane and shear forces are considered, our attention is mainly on the static or dynamic transverse deflection of slender ductile structural components. At this point it is worth noting that in this structural application, the term *constitutive model* generally implies both a mathematical representation of material properties and the approximation that plane sections remain plane.

Here our approach is to present various constitutive models and determine the effect of each model on structural response to generic classes of loading. These models involve material properties that are representative of different forms of real structural behavior and they can be used to analyze complex arrangements of loads. The book begins with elastic behavior of slender members in flexure and finds limits on flexure that are determined by the onset of plasticity. This occurs if curvature κ at a section is larger than the elastic limit κ_Y so that the section has an elastic core while near at least one surface the strain exceeds the elastic limit. Since the stress–strain relation for plastically deforming material is irreversible, the moment–curvature relation for a plastically deforming section is also irreversible; i.e. if $\kappa > \kappa_Y$ the moment–curvature relation for loading is different from that for unloading. To determine structural response, often it is convenient to assume that the stress resultants on a section are related to the deformation by a *rigid–perfectly plastic* constitutive equation. The rigid–perfectly plastic structural deformation model is the fundamental constitutive equation considered in this monograph. For material that is actually elastic–plastic with negligible strain hardening, the rigid–perfectly plastic constitutive relation is the asymptotic limit for stress resultants as curvature becomes indefinitely large; consequently it is appropriate for large deformations of rate–independent materials. We will investigate the accuracy and range of validity of the rigid–perfectly plastic model by examining differences between the deformations predicted by this elementary constitutive model and more complete representations of real materials.

Of course the true test of accuracy for any theory is comparison with experimental measurements. Some effects such as strain hardening and strain-rate dependence are due to material behavior such as dislocation multiplication and dislocation drag, while other effects such as the influence of large deflections or complex loads are independent of the material; thus validation of these theories for application to any particular class of materials requires many tests over a range of load intensities. Where experimental data are currently available, the consequences of different constitutive and theoretical models developed in this book are compared with these data. Some of the data used for validation are new, but most have been compiled from the literature. It will become obvious that there is not sufficient data available to validate all of the different effects; some material effects have not been tested and some tests lack the necessary documentation to permit us to calculate the material parameters. Experimental work in dynamic plasticity is especially difficult but nonetheless should be strongly encouraged if there is to be continuing progress in adequately modeling impact damage.

This book compares effects of constitutive modeling by determining structural response of an elementary structure to a range of different loads. The slender cantilever is the structural element used as a basis of comparison. In the earliest development of this subject the cantilever was a useful element for identifying characteristics of dynamic, as opposed to static, behavior. For a heavy particle striking the tip of a uniform cantilever, Parkes calculated the dynamic response of a rigid–perfectly plastic cantilever and showed that the final deformed shape depends on the ratio of the mass of the colliding particle to the mass of the cantilever. The qualitative behavior predicted by this elementary theory was verified by experiments that used both light and heavy missiles striking the tip of cantilevers in a transverse direction. The change in final shape of the cantilever as a function of mass ratio was impressive and stimulated much subsequent work. While Parkes was not the first to examine the effect of inertia on impact response of plastically deforming structures, his analysis was influential because it demonstrated that complex behavior could be represented by a rigid–perfectly plastic constitutive model — this simple constitutive approximation yielded results that appeared to be qualitatively correct. Subsequent papers on dynamic plasticity have discussed the necessary conditions where this approximation is valid; e.g. Symonds found that elastic effects are negligible if the impact (or input) energy is large in comparison with the limiting strain energy for elastic deformation of the cantilever. This energy ratio R still provides a guide to applicability although it must be augmented by other conditions in some cases. Other investigations by Bodner and Symonds introduced modifications such as strain-rate effects that brought the analysis and experimental results for some materials into better quantitative agreement. So historically, the analysis of a transversely loaded cantilever has been important for developing understanding of phenomena associated with dynamic plastic deformation.

The cantilever is a particularly attractive structural element for comparison of constitutive models because it can undergo large, mainly flexural deformations wherein the effects of resultant forces and bending moment at a section are only coupled through the associated plastic flow law. In the case of the rigid–perfectly plastic constitutive model, pure bending results in regions of plastic deformation that are localized at discrete plastic hinges. For any analysis this substantially simplifies the admissible fields of plastic deformation. The effect of large deflections on some other structures such as clamped beams or plates is necessarily linked with stretching that soon dominates the development of plastic deformation. For a cantilever, however, the effects of large deflection are introduced only through the equations of motion. Consequently, any coupling between flexure and stretching or flexure and shear is obviously a consequence of the plastic flow law and not of large deflections. The cantilever also has advantages for any experimental work; it exhibits large, easily measured deflections for rather modest deformations. Moreover, the specified boundary conditions are easy to achieve in practice because the degree of clamping affects neither the stress resultants nor the inertia.

While the fundamental rigid–perfectly plastic structural response approximation can represent many aspects of deformation for rate-independent elastic–plastic materials, there are other materials and deformation conditions where this basic constitutive model is not adequate. These conditions will be exposed, then methods are presented to analyze deformation of structural elements where second-order effects significantly modify the response. This book emphasizes qualitative rather than quantitative differences that become obvious in comparisons of deformations calculated with various constitutive models. In almost all cases the consideration of second-order effects increases the complexity of the analysis; nonetheless, our aim is to reduce these complexities and provide the simplest formulation which contains the essence of a material and deformation effect. There are many numerical analysis methods which can provide more precise detail on development of deformation at every point for any particular structure and load. The codes that implement these methods are necessary to analyze deformation fields that depend on more than one spatial variable; they are not suited, however, to identifying common patterns of structural response. In the wealth of analytical detail provided by these programs one may lose sight of general phenomena or not identify the most important parameters for representing a particular class of materials, loads or structures — we lose sight of the forest because of the trees. Hence the approach adopted here is to employ analytical methods to obtain both static and dynamic behavior of a simple structural element — the cantilever. Using these methods the results of a variety of constitutive models are examined by comparing the deformations developed in cantilevers subjected to similar loads. This approach is useful for segregating effects of constitutive modeling in part because deformation of the cantilever involves only a single spatial coordinate. Also, it explicitly identifies the essential parameters for representing an effect from among the many possibilities presented by dimensional analysis.

The organization of this monograph has been designed to assist in learning to recognize patterns of structural behavior associated with effects of different material properties. Thus we begin with development of a range of constitutive models using a strength of materials type approximation for the distribution of deformation at any section. These models are employed to analyze a series of static loads that result in increasingly complex deformation fields. Most of the book is concerned, however, with dynamic response to brief but intense loads that result from impact. In this case the final displacement field is the culmination of a deformation process in which the shape of the structure is continually evolving.

The framework for this monograph has been used for teaching at the graduate level and many of the details were developed for courses on Dynamic Plasticity and Impact Response of Structures taught by the authors in Cambridge and Peking Universities. The book is also aimed at research workers in nonlinear structural dynamics who can employ the analytical methods in further developments. For practicing engineers who are concerned with design of high-rate forming processes, impact damage prediction or design and evaluation of structurally crashworthy vehicles, we offer a compilation of data on different materials that will assist in applying these methods of structural analysis. This will be handy in using the methods presented here to perform any particular calculation.

Here we wish to acknowledge our gratitude to our friend, Professor Bill Johnson, who brought us together and raised many of the questions addressed in this book. His enthusiasm and insight into problems of metal forming have been sources of inspiration. Tongxi Yu is grateful to The Royal Society of London who provided a visiting fellowship that permitted him to devote full time to this project in the initial stage of writing. Also, we greatly appreciate the support and encouragement of our students, former students and colleagues who have critically read and painstakingly commented on various sections of the book; in particular we thank Norman Jones, Dongquan Liu, Steve Reid, Victor Shim and Tieguang Zhang. Their criticisms have helped to clarify some points and catch some errors. Nevertheless, the authors are responsible for any remaining blunders; we will appreciate being informed of any errors that the reader detects. Finally we wish to thank our wives, Katerina and Shiying, who did most of the typing and a lot of preliminary editing; their careful attention smoothed some of our rough edges.

6 March 1993 W.J. Stronge
Cambridge, U.K. T.X. Yu

Contents

List of Symbols

A	cross-sectional area
a	radius of circular cross-section; half side-length of square block (Sect. 5.3)
b	width of rectangular cross-section
C	elastic spring coefficient
c	wall thickness of thin-walled section
$\bar{c}$	characteristic flaw size (Sect. 6.8)
D	total energy dissipation due to plastic deformation
D_m	energy dissipation due to bending
D_q	energy dissipation due to shear (Sect. 5.3)
$\dot{D}$	energy dissipation rate due to plastic deformation
d	$= D/M_p$, nondimensional energy dissipation
d_m	$= D_m/M_p$, nondimensional energy dissipation due to bending
d_q	$= D_q/M_p$, nondimensional energy dissipation due to shear (Sect. 5.3)
E	Young's modulus
E_{in}	input energy
E_t	tangent modulus for plastic flow
e_{in}	$= E_{in}/M_p$, nondimensional input energy
e_0	$= K_0/M_p$, impact energy ratio
e_r	nondimensional rupture energy (Sect. 5.3)
$\mathbf{e}_i$	unit vectors fixed in undeformed configuration
F	transverse force
F_c	static plastic collapse force
F_i	components of external traction
f	$= F/F_c = FL/M_p$, nondimensional force
G	mass of colliding particle; elastic shear modulus

g	distributed force per unit length
h	depth of doubly symmetric cross-section in plane of deformation
I	second moment of area about transverse axis through centroid
I_{ox}, I_{oy}, I_{oz}	second moment of area about X-, Y- and Z-axis, respectively
I_ξ, I_η, I_H	first and second moments of mass about plastic hinge (Sect. 5.4)
J	Lee's functional (Sect. 2.5, 5.6); a functional related to Tamuzh's principle (Sect. 6.3); J-integral (Sect. 6.7)
J_o	$= I_{ox}$, polar moment of inertia per unit length
K	kinetic energy
K_o	$= GV_o^2/2$, impact energy
k	$= \kappa L$, nondimensional curvature; Boltzmann's constant
k_Y	$= \kappa_Y L$, nondimensional curvature at yield
$\bar{k}$	$= \kappa/\kappa_Y$, curvature ratio
$\bar{k}_f$	$= \kappa_f/\kappa_Y$, curvature ratio after unloading
$\bar{k}_*$	$= \kappa_*/\kappa_Y$, largest curvature ratio before unloading
L	length of beam or slender bar
L_h	effective length of plastic hinge (Sects 5.1, 5.2)
M	bending moment
M_p	fully plastic bending moment
M_Y	elastic limit moment
m	$= M/M_p$, fully plastic bending moment ratio (Chaps 4,5)
$\tilde{m}$	$= M/M_p$, fully plastic bending moment ratio (Chaps 1-3,6)
$\bar{m}$	$= M/M_Y$, elastic bending moment ratio
$\bar{m}_*$	$= M_*/M_Y$, largest bending moment ratio before unloading
N	axial force
N_p	fully plastic axial force
N_Y	elastic limit axial force
n	$= NL/M_p$, nondimensional axial force; number of segments (Sect. 5.5)
$\tilde{n}$	$= N/N_p$, fully plastic axial force ratio
$\bar{n}$	$= N/N_Y = \tilde{n}$, elastic axial force ratio
$\bar{n}_*$	$= N_*/N_Y$, largest axial force ratio before unloading
P	impulse of a loading pulse
P_{eff}	effective total impulse
P_f	total impulse

p	$= PL/M_pT_0$, nondimensional impulse of a loading pulse
p_f	$= P_fL/M_pT_0$, nondimensional total impulse
p_0	$= \gamma v_0$, nondimensional initial impulse
Q	shear force
Q_j	generalized stresses; nodal shear force (Sect. 5.5)
Q_p	fully plastic shear force
Q_Y	elastic limit shear force
q	$= QL/M_p$, nondimensional shear force
q_p	$= Q_pL/M_p$, nondimensional fully plastic shear force
$\dot{q}_j$	generalized strain-rate vector
$\tilde{q}$	$= Q/Q_p$, fully plastic shear force ratio
$\bar{q}$	$= Q/Q_Y$, elastic shear force ratio
R	energy ratio E_{in}/U_e^{max} (Sects 5.5, 5.6); fracture toughness (Sect. 5.3); initial radius of curved bar (Sects 6.3, 6.5)
r	strain-rate index; radial coordinate
r_g	radius of gyration of colliding block (Sect. 5.3)
S	arc length along axis
s	$= S/L$, nondimensional arc length along axis
T	torque
T_p	fully plastic torque
T_Y	elastic limit torque
T_0	characteristic time, $L\sqrt{\rho L/M_p}$
T_1	fundamental period of elastic vibration
$\bar{T}$	tearing modulus (Sect. 6.7)
t	time
t_d	duration of a pulse
$\tilde{t}$	$= T/T_p$, fully plastic torque ratio
$\bar{t}$	$= T/T_Y$, elastic torque ratio
U	axial displacement
U_e	elastic deformation energy
U_e^{max}	maximum elastic flexural deformation energy for structure
V	transverse velocity of colliding mass; volume (Chap 2)
V_0	initial transverse velocity of colliding mass
V_x	axial velocity of colliding mass in oblique impact (Sect. 6.3)
V_z	transverse velocity of colliding mass in oblique impact (Sect. 6.3)
v	$= VT_0/L$, nondimensional transverse velocity
v_0	$= V_0T_0/L$, nondimensional initial transverse velocity of colliding mass

v_x	$= V_x T_o/L$, nondimensional axial velocity of colliding mass
v_z	$= V_z T_o/L$, nondimensional transverse velocity of colliding mass
W	transverse deflection of centroid
w	$= W/L$, nondimensional deflection
X	axial coordinate in undeformed configuration
x	$= X/L$, nondimensional coordinate
Y	yield stress; transverse coordinate normal to plane of loading (and symmetry) in undeformed configuration
Z	transverse coordinate in plane of loading (and symmetry) measured from centroid in undeformed configuration
$\tilde{Z}$	transverse distance from neutral surface
$\hat{Z}$	distance between centroid and neutral surface

α	$= E_t/E$, coefficient of strain hardening
α_m	coefficient of strain hardening for bending moment (Sect. 5.2)
$\tilde{\alpha}$	$\alpha_m L/M_p$ (Sect. 5.2)
β	material or section constant (Chap. 1); coefficient showing the variation in beam width (Sect. 6.2); angle subtended by hinge position in circular cantilever (Sects 6.4, 6.5); bend angle of a bent cantilever (Sect. 6.6)
Γ	area where surface tractions applied
γ	$= G/\rho L$, mass ratio
γ_q	transverse shear strain (shear angle)
Δ	transverse deflection at the tip
Δ_q	relative transverse displacement due to shearing (Sect. 5.3)
δ	$= \Delta/L$, nondimensional transverse deflection at the tip
δ_q	$= \Delta_q/L$, nondimensional relative transverse displacement due to shear
ε	axial strain
ε_o	material constant
$\tilde{\varepsilon}_o$	rate dependent material constant
ε_{fr}	true tensile strain at fracture (Sect. 5.3)
ε_{ij}	strain
ε_Y	axial yield strain
ζ	nondimensional distance between centroid and neutral

	surface (Chap. 1); velocity ratio (Sect. 5.6); ratio $4L/h$ (Sect. 6.3); coefficient of imperfection (Sect. 6.7)
η	constant defined by (5.3) in (Sect. 5.1); coordinate with origin at hinge H (Sect. 5.4); ratio T_p/M_p (Sect. 6.5)
η_L	$= L_2/L_1$, ratio of lengths for stepped cantilever/Sect. 6.6)
η_M	$= M_{p2}/M_{p1}$, ratio of fully plastic bending moments for stepped cantilever
η_ρ	$= \rho_2/\rho_1$, ratio of densities per unit length for stepped cantilever (Sect. 6.6)
ϑ	$= \mathrm{d}\theta_t/\mathrm{d}X$
ϑ_Y	value of ϑ at yield
θ	rotation angle, inclination
$\dot{\theta}$	rotation rate related to bending
θ_t	rotation of section about centroidal axis
κ	curvature
κ_f	residual curvature after unloading
κ_Y	maximum elastic curvature
Λ	coordinate of hinge location H
λ	$= \Lambda/L$, nondimensional coordinate of hinge
μ	flexibility parameter (Sect, 3.1); ratio of peak force to yield force (Sect. 5.6); ratio $\tilde{t}_{\mathrm{H}}/\tilde{m}_{\mathrm{H}}$ (Sect. 6.5)
ξ	nondimensional elastic part of half depth (Chap. 1); coordinate with origin at hinge H (Sect. 5.4)
ρ	$= \rho_{\mathrm{v}}A$, mass per unit length of cantilever
ρ_{c}	radius of curvature of deformed centroid
ρ_{v}	density of material
σ	normal stress on cross-section
σ_f	final residual stress
σ_{ij}	stress
σ_u	ultimate stress (Sect. 5.3)
σ_Y	$= Y$, uniaxial yield stress
τ	$= t/T_0$, nondimensional time; shear stress on cross-section
τ_Y	yield stress in shear
ϕ	nondimensional group $\tilde{\alpha}e_0$ (Sect. 5.2); angle (Sects 6.4, 6.5)
ϕ_j	shape factor vector
ϕ_m	shape factor for bending
ϕ_n	shape factor for axial force
ϕ_q	shape factor for shear
ϕ_t	shape factor for torsion

$\phi_i{}^*$	components of mode-shape
Ψ	yield function
Ψ_e	elastic limit surface
Ψ_p	fully plastic limit surface
ψ	relative rotation angle (Sect. 5.5); angular coordinate for circular cantilever (Sects 6.3, 6.5)
ψ_Y	relative rotation angle at yield (Sect. 5.5)
Ω	angular velocity
ω	nondimensional angular velocity, ΩT_0

Superscripts

c	kinematically admissible
d	dynamically admissible
e	elastic
p	plastic
$*$	modal
$'$	differentiation with respect to coordinate x
$\cdot$	differentiation with respect to time variable
$+$	positive side
$-$	negative side

Subscripts

A	tip of cantilever
B	root of cantilever
C	section where cross-section varies or bends
f	final
H	plastic hinge
in	input
m	bending
q	shearing
t	torsion
Y	yield
0	initial
1	phase 1
2	phase 2

Chapter 1

Elastoplastic and Viscoplastic Constitutive Relations

1.1 Stress Resultants and Generalized Stress – Deformations and Generalized Strain

In Chapter 1 we develop constitutive modeling for slender structural elements. Together with equations of motion, the constitutive equations will be used to relate nonlinear structural deformations to the applied forces or loads that cause these deformations. Structural deformations depend on load, structural configuration and properties of the materials of which the structure is composed.

Material properties are obtained from tests that produce a uniform state of stress, temperature and rate of deformation in a specimen; e.g. a uniaxial tensile test. For a homogeneous material, the results of these tests are approximated by *constitutive relations*; i.e. mathematical expressions that represent the material response to stress. For an arbitrary material particle these expressions relate the current state of stress $\boldsymbol{\sigma}$ to the current state of strain $\boldsymbol{\varepsilon}$, the strain rate $\dot{\boldsymbol{\varepsilon}}$ and the history of strain. The constitutive relations describe different types of material behavior such as elastic, elastic–plastic or viscoplastic; for any particular material, the appropriate constitutive relation can depend on the current strain, history of strain and temperature of the material particle.

For slender structural members such as bars, beams or plates the normal stress component through the depth is negligible so it is convenient to express constitutive relations in terms of generalized measures of stress and deformation on a cross-section. These stress and deformation variables are related by means of structural properties for a typical cross-section; e.g. the bending or elongation stiffnesses. For a bar or beam with axial coordinate X and cross-sectional area A, the components of stress σ_{ij} on any cross-section give the following stress resultants for axial force N, shear force Q, axial torque T and bending moment M:

$$N = \int_A \sigma_{xx}\, \mathrm{d}A\,, \qquad Q = \int_A \sigma_{xz}\, \mathrm{d}A$$

$$T = \int_A (Y\sigma_{xz} - Z\sigma_{xy})\, \mathrm{d}A\,, \qquad M = -\int_A Z\sigma_{xx}\, \mathrm{d}A$$

where Y, Z are transverse coordinates measured from the axis through the centroid of every section. These *generalized stresses* are counterparts of stress; for beams or plates it is convenient to develop constitutive equations relating generalized stresses $\{N, Q, T, M\}$ to deformation variables termed *generalized strains*.

For slender members, out-of-plane warping of cross-sections is negligible since the depth is small in comparison with the length; consequently, plane sections remain plane. This kinematic constraint means that deformations are completely described by stretching ε and shear γ, axial twist ϑ and curvature κ of the centroidal axis for the undeformed section. Hence relations between stress, strain and strain-rate translate to relations between generalized stresses $\{N, Q, T, M\}$, generalized strains $\{\varepsilon, \gamma, \vartheta, \kappa\}$ and the strain-rates $\{\dot{\varepsilon}, \dot{\gamma}, \dot{\vartheta}, \dot{\kappa}\}$ where differentiation with respect to time t is denoted by $(\dot{\ }) \equiv \mathrm{d}(\ )/\mathrm{d}t$. A convenient notation defines a vector of generalized stresses $Q_\alpha \equiv \{Q_1, Q_2, Q_3, Q_4\} = \{N, Q, T, M\}$ and corresponding vectors of conjugate strains $q_\alpha \equiv \{q_1, q_2, q_3, q_4\} = \{\varepsilon, \gamma, \vartheta, \kappa\}$ and strain-rates $\dot{q}_\alpha \equiv \{\dot{q}_1, \dot{q}_2, \dot{q}_3, \dot{q}_4\} = \{\dot{\varepsilon}, \dot{\gamma}, \dot{\vartheta}, \dot{\kappa}\}$. The *conjugate* strains q_α for a set of generalized stresses Q_α give the rate of energy dissipation $\dot{D}$ in a member of length L as the sum of contributions from the components of generalized stress that are incorporated in the yield criterion

$$\dot{D} = \int_0^L Q_\alpha \dot{q}_\alpha \, \mathrm{d}X$$

Materials that exhibit reversible deformations during loading and unloading are termed *elastic*. In the range of small strain, most structural metals are represented by a linear elastic relation (Hooke's law). The range of strain for elastic behavior is limited by yielding and the onset of *plastic* deformation. Plastic deformation is irreversible and it can be either rate–independent or rate–dependent depending on the material, temperature and state of strain. The choice of an appropriate constitutive model that represents these effects often depends more on the application of the results and less on a precise material description. In general, analytical solutions for structural deformation due to impact are attained only for the simplest material descriptions, so this chapter focuses on elementary constitutive models.

1.2 Pure Bending of Rate–Independent Bar

1.2.1 Kinematics of Deformation

Finding the relationship between stress resultants (or generalized stresses) and deformations at any section of a uniform beam is facilitated by starting from a description of the deformations and then deriving the corresponding stress resultants for any particular constitutive equation. Here these relationships are found for a straight and uniform beam that has cross-sections with a common plane of symmetry. The ends of the beam are loaded by equal but opposed couples that act in directions perpendicular to the plane of symmetry. Thus the beam bends in the plane of symmetry and does not twist. Deformation in response to these couples is known as pure bending; in a uniform segment the axial fibers that were initially straight are deformed to circular arcs. Moreover, cross-sections perpendicular to the axis of the undeformed beam are deformed into plane surfaces perpendicular to the axis of the deformed beam; i.e. *plane sections remain plane*. This kinematic condition is exact for pure bending if cross-sections have in-plane deformations that are negligible; this condition can be shown to follow from considerations of symmetry and compatibility of displacements for adjacent elements. The result follows also from the theory of elasticity (e.g. Timoshenko and Goodier [1971], p 248).

Deformations are defined relative to a set of coordinates X, Z with fixed orientation in relation to the initial undeformed configuration. The X axis is the line passing through the centroid of each cross-section of the beam while the Z axis is transverse to the axis of the beam and in the plane of symmetry. In addition, an axial coordinate S is measured along the centroid in the deformed configuration. Axial and transverse displacements are denoted by $U(X)$ and $W(X)$, respectively. If forces are applied to the beam it deforms as the load increases; the axis becomes inclined at an angle $\theta_m(X)$ that is given by $\theta_m(X) = \arctan(\mathrm{d}W(X)/\mathrm{d}X) \approx \mathrm{d}W/\mathrm{d}X$ where the approximation is accurate if $|\theta_m| << 1$. In pure bending the cross-sections rotate with the axis, so every longitudinal fiber is bent into a circular arc. If the centroid has a radius of curvature $\rho_c(X)$, the arc length $\mathrm{d}S = \rho_c\,\mathrm{d}\theta_m$ where ρ_c depends on the rate of change of inclination along the beam. Since $\mathrm{d}S = (1+(\mathrm{d}W/\mathrm{d}X)^2)^{1/2}\,\mathrm{d}X$, the curvature $\kappa(X)$ of the centroid is given by

$$\kappa = \rho_c^{-1} = \frac{\mathrm{d}^2W/\mathrm{d}X^2}{\left[1+(\mathrm{d}W/\mathrm{d}X)^2\right]^{3/2}} \approx \frac{\mathrm{d}^2W}{\mathrm{d}X^2} \tag{1.1}$$

Once more, the approximation applies to small inclinations $|\theta_m| << 1$. Notice that with the convention shown in Fig. 1.1, positive curvature is associated with 'hogging'.

During pure bending, longitudinal fibers are stretched but not sheared. Near the convex surface of the beam longitudinal fibers are elongated while near the concave surface they are compressed. However, somewhere in the cross-section there is a layer with zero elongation; this is known as the *neutral surface*. In planar deformation the neutral surface is represented by the *neutral axis*. In order

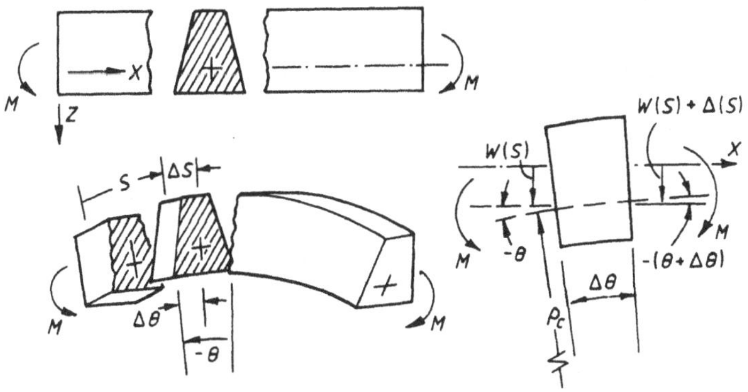

Fig. 1.1 Pure bending of symmetric prismatic bar by moment M acting about axis normal to plane of symmetry. The transverse deflection of the centroid $W(S) \approx W(X)$ is measured from the initial undeformed configuration.

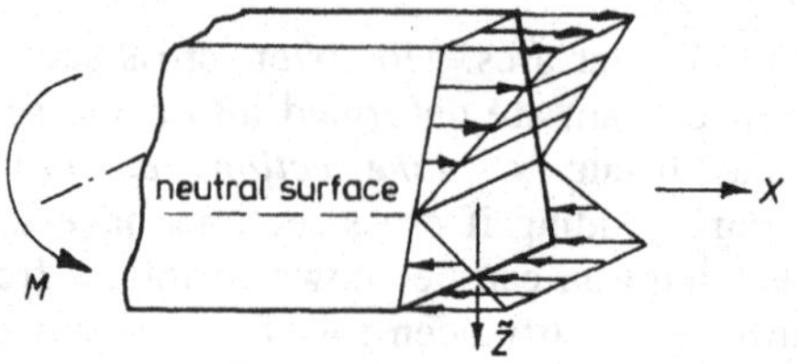

Fig. 1.2 Normal stress distribution on cross-section for elastic flexure.

to satisfy the condition that plane sections remain plane the strain (elongation per unit length) normal to the cross-section must vary in proportion to the distance from the neutral axis. Let $\tilde{Z}$ be the distance from the neutral axis (positive in the same sense as Z). Then the longitudinal strain is $\varepsilon \equiv \partial U/\partial X = -\tilde{Z}\kappa$; in any section the strain has a linear variation with depth.

1.2.2 Elastic Constitutive Equation

For elastic materials Hooke's law provides a unique relation between uniaxial stress σ and longitudinal strain ε for deformation from an unstressed initial state:

$$\sigma = E\varepsilon \tag{1.2}$$

where E is Young's modulus. This elastic modulus is just the slope of the stress–strain curve in the region of small strain. For pure bending the normal stress increases linearly with distance from the neutral surface; i.e. $\sigma = -E\tilde{Z}\kappa$ as shown in Fig. 1.2.

1.2.3 Stress Resultants (Axial Force and Bending Moment)

The resultant *axial force* N is the integrated effect of the normal stresses acting on a cross-section. For pure bending the resultant force vanishes. Thus for cross-sections with area A, the axial force obtained from the stress distribution for pure bending is given by

$$N = \int_A \sigma \, \mathrm{d}A = -E\kappa \int_A \tilde{Z} \, \mathrm{d}A = 0 \tag{1.3}$$

The second integral in this expression is known as the first moment of area about the neutral surface. Vanishing of the first moment of area implies that the neutral axis is coincident with the centroid of the section since the distance $\hat{Z}$ from the neutral axis to the centroid is defined as $\hat{Z} = A^{-1}\int_A \tilde{Z} \, \mathrm{d}A$.[1]

The *bending moment* $M(X)$ at a section is the first moment of normal stress on the cross-section about the centroid,

[1]For a beam with substantial curvature in the unstressed state, the neutral axis for pure bending is not coincident with the centroid of cross-sections. Bending of a curved beam results in plane sections that remain plane, but the strain distribution is no longer linear since longitudinal fibers in a differential element have undeformed length that varies across the section.

$$M \equiv -\int_A \tilde{Z}\sigma \, dA = +EI_0\kappa \tag{1.4}$$

where $I_0 \equiv \int_A \tilde{Z}^2 \, dA$ is the second moment of area about the transverse axis through the centroid (frequently this is termed the moment of inertia). From equilibrium we obtain that the moment M is equal in magnitude to the couple at the end of the beam. Thus Eq. (1.4) describes the moment–curvature relation for pure bending of elastic beams. Although this expression is exact only for pure bending, the expression is still useful for a beam that carries both shear force and a bending moment since the deformation due to shear is usually negligible. This is the case if the beam is slender, $L/h >> 1$; in this case, only a negligibly small part of the strain energy of deformation is caused by shear.

The previous expressions are valid for bending of elastic bars; i.e. if throughout the section the strains are less than the yield strain ε_Y. This limiting strain is related to the uniaxial yield stress Y through $\varepsilon_Y = Y/E$. If the neutral surface is also a plane of symmetry for cross-sections, the largest strains are at the top and bottom surface — a distance $h/2$ from neutral axis. Thus the largest elastic curvature is $\kappa_Y = \pm 2\varepsilon_Y/h = \pm 2Y/Eh$, while the corresponding *bending moment at yield* M_Y is given by

$$M_Y = \pm EI_0\kappa_Y = \pm 2I_0 Y/h \tag{1.5}$$

If the bending moment is larger than this elastic limit, the fibers of the cross-section furthest from the neutral surface are strained beyond the elastic limit ε_Y.

1.2.4 Elastic–Plastic Constitutive Equations

Plastic deformation is related to dislocation motion on slip planes in crystalline materials. In both polycrystalline materials and crystals with second phase particles, dislocation motion only occurs if the shear stress is sufficiently large on slip planes. The yield strain ε_Y is the manifestation of this limit; in a polycrystalline solid it is the deformation where slip is activated on a large number of slip planes. The yield stress Y is the corresponding limit for slip activation. These parameters are limits for the elastic range; i.e. the range wherein deformations are reversible. After plastic deformation begins it is assumed that the strain ε can be separated into elastic and plastic parts, $\varepsilon = \varepsilon^e + \varepsilon^p$; only the elastic part of the strain is reversible so after plastic deformation starts to develop, unloading of the body leaves a state of residual plastic deformation. Ordinarily the unloading modulus is identical with the elastic modulus E so in the unstressed state the residual strain is ε^p.

In many solids the stress required to continue plastic deformation increases with the plastic strain; i.e. the material strain hardens. This increase in stress that is required to continue dislocation motion on slip planes has been associated with the stress to overcome dislocation entanglements. Part of this increase is offset, however, by an increase in the number of active slip planes as stress increases.

A simple phenomenological model for an elastic–plastic material with *strain hardening* presumes that stress increases in proportion to elastic strain if $\varepsilon < \varepsilon_Y$ while if $\varepsilon > \varepsilon_Y$ the stress increases in proportion to total strain. Thus plastic deformation develops if $\varepsilon > \varepsilon_Y$; in this range if strain is increasing, $d\varepsilon/dt > 0$, the stress and total strain are related by a linear strain hardening modulus E_t.

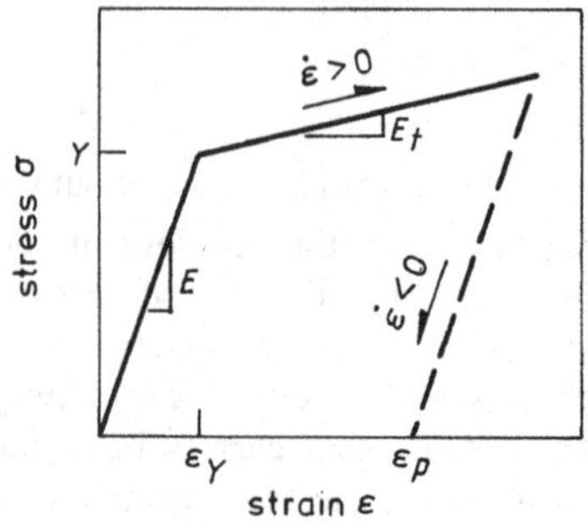

Fig. 1.3 Elastic–plastic material with linear strain hardening and elastic unloading.

This counterpart of the elastic modulus E is shown in Fig. 1.3

Let the strain hardening and elastic moduli have a ratio $\alpha = E_t / E$, where $\alpha < 1$. For pure bending of rate–independent materials, the elastic and plastic constitutive relations result in a unique moment–curvature relationship while the bending moment increases,

$$\begin{aligned} \sigma &= -E\tilde{Z}\kappa, && |\kappa| \le \kappa_Y \\ \sigma &= -\alpha E\tilde{Z}\kappa - Y(1-\alpha)\,\mathrm{sgn}(\kappa)\,\mathrm{sgn}(\tilde{Z}), && |\kappa| > \kappa_Y, \quad \kappa\dot{\kappa} > 0 \end{aligned} \tag{1.6}$$

The second of the expressions applies only if the location of the neutral axis is independent of curvature; i.e. if the cross-section is doubly symmetric. The bending moment for the cross-section is obtained by integrating the first moment of the stress as in Eq. (1.4); the moment is taken about the transverse axis through the centroid. Thus, for a *rectangular cross-section* where $M_Y = Ybh^2/6$,

$$\begin{aligned} M/M_Y &= \kappa/\kappa_Y, && |\kappa| \le \kappa_Y \\ M/M_Y &= \alpha(\kappa/\kappa_Y) + 0.5(3 - \kappa_Y^2/\kappa^2)(1-\alpha)\mathrm{sgn}(\kappa), && |\kappa| > \kappa_Y, \quad \kappa\dot{\kappa} > 0 \end{aligned} \tag{1.7}$$

It is often convenient to express this relationship in terms of the curvature resulting from some applied bending moment. For an elastic–plastic material this results in a cubic equation that has only one real root,

$$0 = (\kappa/\kappa_Y)^3 - (\overline{m}/\alpha)(\kappa/\kappa_Y)^2 + 0.5(1 - 1/\alpha)[1 - 3(\kappa/\kappa_Y)^2], \quad |\overline{m}| > 1 \tag{1.8}$$

where $\overline{m} = M/M_Y$. The elastic–plastic relations (1.7) and (1.8) apply also to strain softening materials $(\alpha < 0)$ if the moment–curvature relation is single valued. This condition requires $\mathrm{sgn}(M) = \mathrm{sgn}(\kappa)$ which is satisfied if $|\kappa|/\kappa_Y < \sqrt{1 - 1/\alpha}$.

If the stress–strain idealization is *elastic–perfectly plastic* the moment–curvature relation for a rectangular cross-section is obtained directly from Eq. (1.7) by considering the limiting case as $\alpha \to 0$. Thus

$$\begin{aligned} M/M_Y &= \kappa/\kappa_Y, && |\kappa| \le \kappa_Y \\ M/M_Y &= 0.5(3 - \kappa_Y^2/\kappa^2)\mathrm{sgn}(\kappa), && |\kappa| > \kappa_Y \end{aligned} \tag{1.9}$$

or

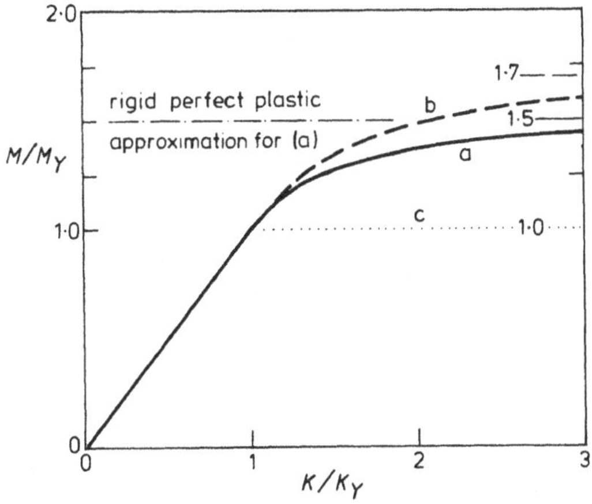

Fig. 1.4 Elastic–perfectly plastic moment–curvature relations for three cross-sections: (**a**) rectangular, (**b**) circular and (**c**) an ideal I-beam. The asymptotic values for bending moment at large curvature are indicated.

$$\kappa/\kappa_Y = [3 - 2|M|/M_Y]^{-1/2}\operatorname{sgn}(M), \qquad |M| > M_Y$$

The moment–curvature relation for pure bending depends on the shape of the cross-section. For a *circular cross-section* with radius a the limiting elastic curvature is $\kappa_Y = Y/Ea$ while the limiting elastic moment $M_Y = EI\kappa_Y = \pi Ya^3/4$. Thus for a circular cross-section the elastic–perfectly plastic stress–strain idealization gives a moment–curvature relation

$$\frac{|M|}{M_Y} = \frac{2}{3\pi}\left[\left(5 - \frac{2\kappa_Y^2}{\kappa^2}\right)\sqrt{1 - \frac{\kappa_Y^2}{\kappa^2}} + \frac{3\kappa}{\kappa_Y}\sin^{-1}\left(\frac{\kappa_Y}{\kappa}\right)\right], \qquad \kappa > \kappa_Y \tag{1.10}$$

For elastic–perfectly plastic materials, the moment–curvature relations for pure bending of rectangular and circular cross-sections are compared in Fig. 1.4. The *fully plastic bending moment* M_p is the asymptotic limit of the moment–curvature relation for a perfectly plastic material as curvature κ becomes indefinitely large. Thus $M_p/M_Y = 1.5$ and 1.7 for rectangular and circular cross-sections, respectively. This ratio of the fully plastic bending moment M_p to the elastic limit M_Y is known as the *shape factor* for bending ϕ_m

$$\phi_m \equiv M_p/M_Y \tag{1.11}$$

The shape factor is defined only for perfectly plastic materials since with strain hardening, there is no asymptotic limit for bending moment as the curvature becomes large without bound. It depends on the distribution of stress in the cross-section; i.e. on both the cross-section shape and the stress resultant. Table 1.1 gives this structural property for separate bending, shear and torsion loads acting on typical cross-sections. Tension N has not been included since in all cases the elastic and fully plastic limit forces are identical, $N_Y \equiv YA = N_p$; consequently, the shape factor for axial force is $\phi_n = 1$

Table 1.1 Yield and fully plastic limits for bending, shear and torsion
(Tresca yield $k = Y/2$, von Mises yield $k = Y/\sqrt{3}$)

Cross-section	Area	2nd moment									
	A	I_o	$\frac{M_Y}{YA}$	$\frac{M_p}{YA}$	ϕ_m	$\frac{Q_Y}{kA}$	$\frac{Q_p}{kA}$	ϕ_q	$\frac{T_Y}{kA}$	$\frac{T_p}{kA}$	ϕ_t
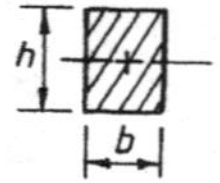	bh	$Ah^2/12$	$h/6$	$h/4$	1.5	2/3	1	1.5	$\beta_2 b$[†]	$\frac{b(3-b/h)}{6}$[*]	$\frac{3-b/h}{6\beta_2}$
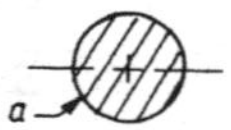	πa^2	$Aa^2/4$	$a/4$	$4a/3\pi$	1.7	3/4	1	1.33	$a/2$	$2a/3$	1.33
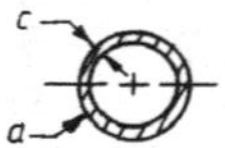	$2\pi ac$	$Aa^2/2$	$a/2$	$2a/\pi$	1.27	1/2	1	2.0	a	a	1.0
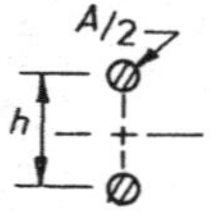	A	$Ah^2/4$	$h/2$	$h/2$	1.0	1	1	1.0	h	h	1.0

[†]For narrow rectangles, Timoshenko and Goodier [1971] give values of $\beta_2(h/b)$: $\beta_2(1.0) = 0.208$, $\beta_2(1.5) = 0.231$, $\beta_2(2.0) = 0.246$.
*Expression for $b/h < 1$.

1.2.5 Elastic–Power Law Hardening Constitutive Equations

The linear strain hardening examined previously is not the behavior commonly observed in materials stretched to a large extent. Over a large range of strain, a power law hardening relation may be a better fit to measured properties. For plane stress, the stress–strain relation for *elastic–power law hardening* materials with yield stress Y can be expressed as

$$\begin{aligned} \sigma &= E\varepsilon, & |\varepsilon| < \varepsilon_Y = Y/E \\ \sigma &= Y(E\varepsilon/Y)^{\beta}, & |\varepsilon| > \varepsilon_Y \end{aligned} \tag{1.12}$$

where β is a strain hardening parameter. For a rectangular cross-section deformed by pure bending this stress–strain idealization results in a moment–curvature relation,

$$\frac{M}{M_Y} = \frac{1}{2+\beta}\left\{3\left|\frac{\kappa}{\kappa_Y}\right|^{\beta} - (1-\beta)\left(\frac{\kappa_Y}{\kappa}\right)^2\right\}\mathrm{sgn}(\kappa), \quad |\kappa| > \kappa_Y \tag{1.13}$$

Again if the hardening coefficient is positive, $\beta > 0$, the bending moment has no asymptotic limit as curvature becomes indefinitely large. In a strain hardening material the moment always increases with curvature; this tends to spread or diffuse the regions with large curvature. Equation (1.13) has an asymptotic limit as $\beta \rightarrow 0$ that is identical with the moment–curvature relation for vanishing strain hardening obtained from Eq. (1.6); i.e. $M/M_Y = 0.5[3-(\kappa_Y/\kappa)^2]$.

The inverse of (1.13) gives curvature if the bending moment M is specified. In a power law hardening material this curvature is obtained as a root of the nonlinear equation

$$0 = \left|\frac{\kappa}{\kappa_Y}\right|^{\beta} - \frac{1-\beta}{3}\left(\frac{\kappa_Y}{\kappa}\right)^2 - \frac{2+\beta}{3}\frac{M}{M_Y}\mathrm{sgn}(\kappa), \quad |\kappa| > \kappa_Y \tag{1.14}$$

1.3 Pure Bending of Rate–Dependent Bar

1.3.1 Strain-Rate Dependent Constitutive Equations

Material rate–dependence can be of two kinds. Materials such as polymers at temperatures above the glass transition temperature Θ_g exhibit viscoplastic behavior that manifests itself as creep; this behavior is related to viscous flow as polymer chains slip past one another. In polycrystalline materials (e.g. metals), viscous drag due to dislocation motion around barriers exhibits similar behavior that has been termed rate sensitivity of strain hardening. Effects of viscoplastic constitutive relations on deformation of impulsively loaded structures have been considered by Ting [1964] and Ting and Symonds [1962]. In many metals, however, rate–dependence at temperatures substantially less than melting is related to an additional stress required to generate and accelerate dislocations; i.e. to initiate changes in the rate of plastic flow. The latter effect is most prevalent in metals that have a distinct and easily defined yield stress; e.g. BCC metals and precipitate hardened alloys (Hartley and Duffy, [1984]). Low carbon steel is notoriously rate sensitive in this respect; at strain-rates that commonly occur during impact of hard bodies (e.g. 10^3 s^{-1}) the yield stress of mild steel can increase by more than a factor of two over the static value.

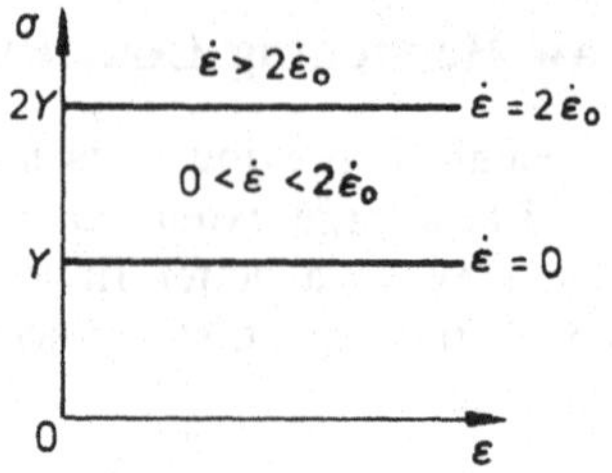

Fig. 1.5 Rate–dependent flow stress Y^d for nonhardening material.

Microstructural models of rate effects are based on thermal activation of dislocation motion; thus the effective plastic strain-rate is given by an Arrhenius equation,

$$\dot{\varepsilon}^p / \dot{\varepsilon}_0^p = \exp(-\Delta U / k\Theta)$$

where ΔU is an activation energy which depends on the barriers opposing dislocation motion, k is Boltzmann's constant, Θ is absolute temperature and $\dot{\varepsilon}_0^p$ is a characteristic strain-rate for the material. The activation energy decreases with increasing stress. This model for rate effects results in a stress that depends on only the current strain-rate and not on strain history.[2] A unidirectional constitutive relation representing these effects is known as the Cowper–Symonds [1957] relation.

The Cowper–Symonds relation represents a rigid–perfectly plastic material with dynamic yield or flow stress $\sigma(\dot{\varepsilon})$ that depends on the strain-rate $\dot{\varepsilon}$. (In this model all strain is plastic so it is convenient to omit the superscript p.) Thus the ratio of dynamic to static yield stress Y^d/Y is

$$\frac{Y^d}{Y} = 1 + \left(\frac{\dot{\varepsilon}}{\dot{\varepsilon}_0}\right)^{1/r}, \qquad \dot{\varepsilon} > 0 \tag{1.15}$$

where $\dot{\varepsilon}_0$ and r are material constants. The characteristic rate $\dot{\varepsilon}_0$ is the strain-rate at which $Y^d = 2Y$ as shown in Fig. 1.5, while the material constant r is a measure of sensitivity to strain-rate.

For pure bending of beams or other slender members the relationship between bending moment and rate of curvature is obtained from the condition that plane sections remain plane. Therefore, at a distance $\tilde{Z}$ from the neutral axis

$$\dot{\varepsilon} = -\tilde{Z}\dot{\kappa} \tag{1.16}$$

where the rate of change of curvature at a section is $\dot{\kappa} \equiv d\kappa/dt$. Thus for pure bending of a rigid–perfectly plastic rate–dependent section where $\sigma = Y^d$, Perzyna [1962] has given the stress for a linear strain–rate distribution (1.16) as

$$\frac{\sigma}{Y} = 1 + \left(\frac{-\tilde{Z}\dot{\kappa}}{\dot{\varepsilon}_0}\right)^{1/r}$$

This stress distribution is illustrated in Fig. 1.6. The normal stresses on a cross-section of area A yield a resultant couple

---

[2]Hartley and Duffy [1984] observed that strain-rate history effects are more pronounced in FCC metals, whereas strain-rate and temperature sensitivity is stronger in BCC metals.

Table 1.2 Material parameters for rate effect

Material	$\dot{\varepsilon}_0(s^{-1})$	r	$\dot{\varepsilon}_{0r}(s^{-1})$	Reference
Mild steel	40.4	5	65	Forrestal and Wesenberg [1977]
Stainless steel	100	10	160	Forrestal and Sagartz [1978]
Titanium (Ti 50A)	120	9	195	
Aluminum 6061-T6	1.70×10^6	4	2.72×10^6	Symonds [1965]
Aluminum 3003-H14	0.27×10^6	8	0.44×10^6	Bodner and Speirs [1963]

$$M_p^d = -Y\int_A \left[1+\left(\frac{-\tilde{Z}\dot{\kappa}}{\dot{\varepsilon}_0}\right)^{1/r}\right]\tilde{Z}\,dA$$

while for a rectangular cross-section with depth h,

$$\frac{M_p^d}{M_p} = 1 + \frac{2r}{2r+1}\left(\frac{h\dot{\kappa}}{2\dot{\varepsilon}_0}\right)^{1/r}, \qquad \dot{\kappa} > 0 \tag{1.17}$$

where $M_p^d = Ybh^2/4 = 3M_Y/2$ is the static fully plastic moment for the rectangular section. The corresponding relation for rate of change of curvature $\dot{\kappa}$ is

$$\frac{h\dot{\kappa}}{2\dot{\varepsilon}_{0r}} = \left(\frac{M_p^d}{M_p} - 1\right)^r, \qquad M_p^d > M_p \tag{1.18}$$

where $\dot{\varepsilon}_{0r} \equiv \dot{\varepsilon}_0(1+1/2r)^r$.

Some representative values of material parameters for use in the Cowper–Symonds relation are given in Table 1.2. These values were obtained for strain $\varepsilon \approx 0.05$; they may not be accurate for strains that are either very large or very small in comparison with the elastic limit.

The methods described thus far have focused on rate effects and ignored strain hardening. While this is acceptable for small strains, hardening effects can be more important than strain-rate at large strains. Perrone [1970] has suggested a constitutive relation that includes both effects; his relation has the advantage of being separable and thus more easily evaluated.

$$\frac{\sigma}{Y} = \left[1 + \left(\frac{\dot{\varepsilon}}{\dot{\varepsilon}_0}\right)^{1/r}\right]\left(1+\frac{\varepsilon}{\varepsilon_0}\right) \tag{1.19}$$

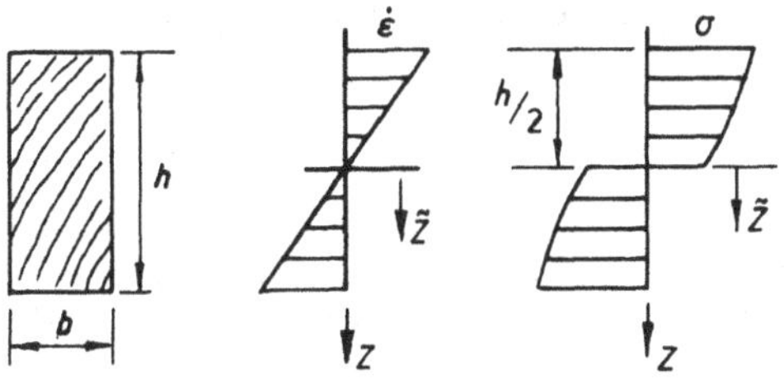

Fig. 1.6 Strain-rate and stress distributions in a rate–dependent deforming section where $\dot{\kappa} > 0$.

1.4 Interaction Yield Functions and Associated Plastic Flow

1.4.1 Elastic Limit for Bending and Tension

In general, slender structures are subjected to a combination of simultaneous loads rather than pure bending; in addition to a couple, the beam may support axial and shear forces as well as an axial torque. For elastic deformations the state of stress associated with each of these stress resultants is linearly related to the strain so the principle of superposition is applicable; each component of stress and strain is just the sum of contributions from the individual stress resultants. Thus a beam with axial force N in addition to bending moment M has normal stresses σ on a cross-section

$$\sigma = \frac{N}{A} - \frac{MZ}{I_0}, \qquad |\sigma| < Y$$

where I_0 is the second moment of area about the transverse axis for a cross-section of area A and Z is the transverse distance from the centroidal axis. Initial yield for each of these stress resultants acting independently is denoted by N_Y and M_Y where, for a doubly symmetric section of depth h,

$$M_Y = EI_0\kappa_Y = 2YI_0/h, \qquad N_Y = YA \tag{1.20}$$

In terms of these separate limits for N, M the elastic range is given by the *yield condition*

$$\frac{\sigma}{Y} = \frac{|N|}{N_Y} + \frac{|M|}{M_Y} < 1 \tag{1.21}$$

The stress distribution varies linearly through the depth of a section; it has tension on one side of the neutral surface and compression on the other side. For pure bending the neutral axis is coincident with the centroid but if there is axial force N in addition to a couple, the neutral axis is located some distance $\zeta h/2$ from the centroid. Hence yield occurs first at either the top or bottom fibers of the beam; thereafter as M and N increase an increasing part of the cross-section is stressed beyond yield. Around the neutral surface there is a core of elastically deforming material but elastic–plastic deformation is present beyond some distance ε_Y/κ from the neutral surface. These plastic deformations occur for only a small range of stress resultants if the material is elastic–perfectly plastic. For any stress resultant Q_α acting independently, the ratio of the fully plastic to the elastic limit is given by the shape factor ϕ_α.

1.4.2 Fully Plastic Limit Surface for Bending and Tension in Elastic–Perfectly Plastic Bar

The stress resultants for tension N, shear Q, torsion T and bending M in the theory of plasticity for slender structural members are analogous to stress components in a continuum; hence they are sometimes referred to as generalized stresses Q_α where $Q_1 = N$, $Q_2 = Q$, $Q_3 = T$ and $Q_4 = M$. For sections that are partly plastic, in the plastically deforming region the stresses are related by a yield criterion; e.g. the von Mises or Tresca yield conditions. The yield condition

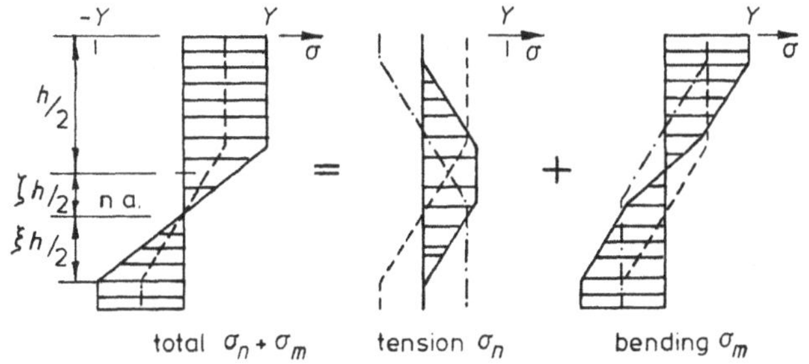

Fig. 1.7 Normal stress distribution for elastic–perfectly plastic bending plus tension. For doubly symmetric cross-sections the stress distribution separates into a symmetric part (tension) and an antisymmetric part (bending). The chain line (– – –) is at half the total stress. The chain–dot line is the reflection of the chain line about the centroidal axis. The sum of tension and bending stresses under the chain–dot line vanishes.

for stress can be used to form a *yield function* $\Psi_e(Q_\alpha)$ for stress resultants; e.g. for a section subjected to simultaneous axial tension N and bending M the yield function is

$$\Psi_e = \frac{|M|}{M_Y} + \frac{|N|}{N_Y} - 1 \tag{1.22}$$

This function is a measure of the state of generalized stress in the section. The *elastic limit* is given by a *yield condition* $\Psi_e = 0$, that provides an upper bound on generalized stresses corresponding to elastic (reversible) strains at every point in the cross-section; thus the stresses are elastic if $\Psi < \Psi_e$ while $\Psi > \Psi_e$ implies that some parts of the cross-section are strained beyond the elastic limit. In stress space the *elastic limit* $\Psi_e = 0$ is a surface that relates the generalized stresses.

For a material that is elastic–perfectly plastic, there are ultimate values for stress resultants; these asymptotic limits for large deformations are termed the fully plastic axial force N_p and the fully plastic bending moment M_p. The independent fully plastic stress resultants for generalized stresses are respectively

$$M_p = \phi_m M_Y, \qquad N_p = N_Y = YA \tag{1.23}$$

A state of stress that results in part of the cross-section being strained beyond the elastic limit gives $\Psi_e > 0$. Further proportional increases in the stress resultants increase Ψ_e and the deformation; the increases in stress asymptotically approach a limiting or fully plastic state of stress as the deformation becomes indefinitely large. A *fully plastic limit function* $\Psi_p(Q_\alpha)$ can be defined that gives a relationship between stress resultants in the fully plastic state, $\Psi_p = 0$. This can be obtained from the stress resultants expressed as a function of the location of the neutral surface. For example, Fig. 1.7 illustrates the normal stress distribution on a section that carries both bending moment M and axial force N; the cross-section is rectangular with depth h and width b. The deformation is such that the stress distribution is elastic–plastic near the top and bottom surfaces; the stresses are tensile above and compressive below the neutral surface located a distance $\zeta h/2$ from the centroid. The perfectly plastic stress distribution for a doubly symmetric section is readily separated into a part that is antisymmetric with respect to the centroid and a remainder of one sign. The stress resultants for these separate parts are the bending moment and axial force respectively

$$\frac{|M|}{M_Y} = \frac{1}{2}\left(3 - \frac{\kappa_Y^2}{\kappa^2} - 3\zeta^2\right), \qquad \frac{N}{N_Y} = \zeta \tag{1.24}$$

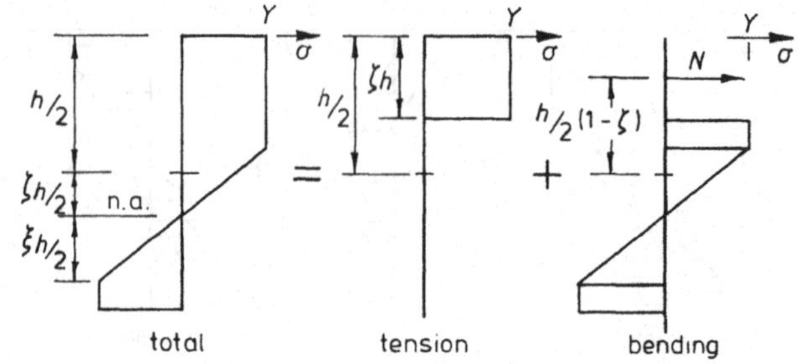

Fig. 1.8 Elastoplastic tension N and bending moment M for centroid obtained from alternative separation of normal stress distribution. This method of separation depends on the cross-section; it divides the section into a part that is antisymmetric about the neutral axis and the remainder. $N/N_Y = \zeta$, $M/M_Y = \xi^2 + 3\zeta(1-\zeta) + 3\left[(1-\zeta)^2 - \xi^2\right]/2$.

where the elastic core has half depth $\xi h/2 = h\kappa_Y/\kappa$. These expressions can also be obtained from the alternative separation of the stress distribution illustrated in Fig. 1.8; for a rectangular cross-section the tension force N is obtained from the part of the normal stress field that is not antisymmetric about the *neutral axis.* The bending moment M is the moment about the centroid of the normal stresses on the entire section. For combined forces on a section the curvature κ is obtained by eliminating ζ from the expressions above,

$$\frac{\kappa}{\kappa_Y} = \left[3 - \frac{2|M|}{M_Y} - \frac{3N^2}{N_Y^2}\right]^{-1/2}, \qquad |\kappa| > \kappa_Y \tag{1.25}$$

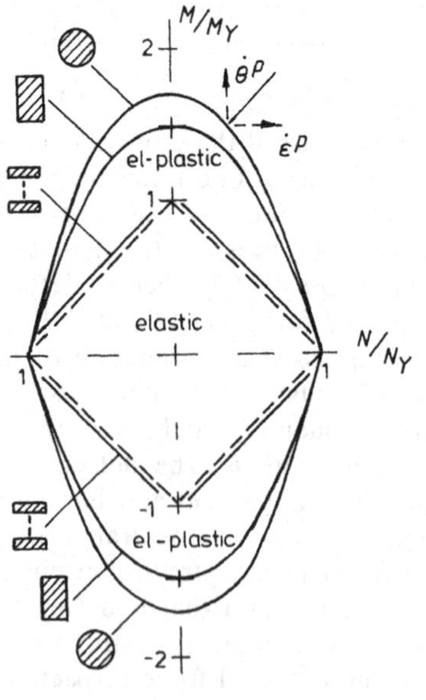

Fig. 1.9 Axial force N and bending moment M for stress states on fully plastic (——) and yield (– – –) surfaces of beams with circular, rectangular or ideal I-beam cross-sections. The associated flow law for the fully plastic state relates the rates of stretching and relative rotation.

For a cross-section composed of an elastic–perfectly plastic material, a fully plastic state of stress is approached in the limit as curvature becomes indefinitely large. The *fully plastic limit function* Ψ_p for generalized stresses in a rectangular cross-section is given by

$$\Psi_p = \frac{2|M|}{3M_Y} + \frac{N^2}{N_Y^2} - 1$$

The *fully plastic limit* is $\Psi_p = 0$ while in partly elastic sections, the stress resultants satisfy $\Psi_p < 0$. Figure 1.9 compares the elastic and fully plastic limits of rectangular and circular cross-sections. In general, the state of stress at yield is circumscribed by the fully plastic limit.

For application to *rigid–perfectly plastic* materials it is convenient to express the fully plastic limit function in terms of independent fully plastic stress resultants M_p, N_p instead of the independent yield limits. Defining nondimensional fully plastic stress resultants $\tilde{m}$, $\tilde{n}$,

$$\tilde{m} \equiv \frac{|M|}{M_p} = \frac{|M|}{\phi_m M_Y} \quad \text{and} \quad \tilde{n} \equiv \frac{|N|}{N_p} = \frac{|N|}{N_Y} \tag{1.26}$$

gives the plastic limit function Ψ_p for stress resultants that correspond to fully plastic stresses in a rectangular cross-section,

$$\Psi_p = |\tilde{m}| + \tilde{n}^2 - 1 \tag{1.27}$$

Stress states that give $\Psi_p = 0$ are fully plastic while sections with a stress state $\Psi_p < 0$ are at least partly elastic. (For a rigid–perfectly plastic material there is no deformation if $\Psi_p < 0$.)

The fully plastic limit function for other cross-sections can also be obtained by integrating over the cross-section the stresses and the first moment of stresses about the centroid. Thus for a beam with thin–walled circular cross-section, the stress resultants for perfectly plastic stresses are bounded by a fully plastic upper limit $\Psi_p = 0$ where

$$\Psi_p = |\tilde{m}| - \cos\left(\frac{\pi\tilde{n}}{2}\right) \tag{1.28a}$$

For a beam with solid circular cross-section the fully plastic limit function is

$$\Psi_p = |\tilde{m}| - \left[\cos\left(\frac{\pi\tilde{n}}{2} - \tilde{m}^{1/3}\sqrt{1-\tilde{m}^{2/3}}\right)\right]^3 \tag{1.28b}$$

Also, the fully plastic limit for an ideal sandwich or I-beam is shown in Fig. 1.9; this is an I-beam that carries stress only in thin flanges. The fully plastic limit is obtained by considering an I-beam with finite thickness flanges and taking the limit as flange thickness decreases. This gives $M_p = M_Y$ and $N_p = N_Y$ so the elastic and fully plastic limits are identical; thus

$$\Psi_p = |\tilde{m}| + |\tilde{n}| - 1 \tag{1.29}$$

Interaction limit functions for stress resultants in perfectly plastic plates and shells are discussed by Sawczuk [1989, p 80] and Save and Massonet [1972].

1.4.3 Yield and Fully Plastic Stress Condition

The state of stress required to initiate plastic deformation can be expressed as a *yield condition* that relates components of stress to a limiting condition for stress or strain energy; e.g. Tresca's criterion corresponds to a maximum shear stress while the von Mises criterion corresponds to a maximum deviatoric strain energy. The yield condition is both an upper bound on the state of stress for elastic response and the stress at which plastic deformation commences. Usually the yield condition $\Psi_e(Q_\alpha)=0$ is an equation that relates the largest magnitude of various components of generalized stress for elastic (reversible) behavior. Elastic stresses throughout a section give $\Psi_e(Q_\alpha)<0$, while a section that is partly elastic and partly plastic has $\Psi_e(Q_\alpha)>0$. If the material is elastic–perfectly plastic (i.e. no strain hardening), generalized stresses asymptotically approach a limiting or *fully plastic stress condition*, $\Psi_p(Q_\alpha)=0$. It is worth noting that yield and fully plastic stress conditions are almost always expressed in terms of a selected set of generalized stresses

$$\Psi_e(Q_1, Q_2, \cdots, Q_\eta) = 0, \quad \Psi_p(Q_1, Q_2, \cdots, Q_\eta) = 0, \qquad \text{where} \quad \eta \le 6$$

Only generalized stresses for the more important sources of energy dissipation need to be included in the yield condition for a particular problem.[3] The analysis is simplified by neglecting the effect on yield of small components of stress; hence, $\eta \le 6$.

Yield and fully plastic stress conditions are limiting surfaces in stress space — they are continuous and piecewise differentiable surfaces surrounding the unstressed state. For perfectly plastic materials where strain hardening and strain-rate effects are negligible, these surfaces are upper bounds for elastic and elastic–plastic behavior, respectively. Yield and fully plastic stress conditions can be used as functions indicating the state of stress in a section.

$$\begin{array}{lll} \Psi_e < 0 & & \text{elastic stresses} \\ \Psi_e > 0, & \Psi_p < 0 & \text{elastic–plastic stresses} \\ \Psi_e > 0, & \Psi_p = 0 & \text{fully plastic stresses} \end{array} \tag{1.30}$$

The fully plastic limit is an upper bound for stress states that satisfy yield in any part of the cross-section. These limit surfaces are path independent; i.e. independent of the history of stress in the section. For *rigid–perfectly plastic materials, the yield and fully plastic stress conditions are identical* so these terms are synonymous.

1.4.4 Associated Flow Rule for Plastic Deformations

The fully plastic limit is a convex surface in generalized stress space. This surface is a differentiable function of the stress resultants except at isolated corners such as $\tilde{m}=0$, $\tilde{n}=\pm 1$ in Eq. (1.27). Convexity of the locus of fully plastic stresses implies that if one stress resultant increases the others cannot simultaneously increase.

Both the yield and fully plastic loci of generalized stresses must be convex if the rate of work done by stresses is to be positive for any plastic deformation. Consequently, for any state of stress on the yield (or fully plastic) surface, energy is dissipated by internal plastic work during an incremental change of stress to

[3] In developing a constitutive model for an isotropic continuum, we require the yield function $\Psi_e(\sigma_{ij})$ to satisfy a principle of objectivity; i.e. that yield is invarient with respect to orientation of the coordinate system. This principle does not apply to the yield function $\Psi_e(Q_\alpha)$ for a slender member since the cross-sectional properties are not isotropic.

any different state on the yield (fully plastic) surface. Plastic deformation always dissipates energy. Thus if σ_{ij} is a state of stress on the fully plastic locus and σ_{ij}^s is any state of stress inside this locus, an increment of plastic strain $d\varepsilon_{ij}^p$ for σ_{ij} satisfies

$$(\sigma_{ij} - \sigma_{ij}^s)\, d\varepsilon_{ij}^p \geq 0 \qquad \text{or} \qquad (Q_\alpha - Q_\alpha^s)\, dq_\alpha^p \geq 0 \tag{1.31}$$

where q_α^p are measures of deformation or generalized strains corresponding to the generalized stresses included in the fully plastic stress function Ψ_p. The proposition (1.31) expresses a *postulate of nonnegative dissipation by plastic deformation* — this was given independently by Taylor [1947] and Hill [1948]. Here this postulate has also been expressed in terms of the analogous generalized stresses Q_α and the conjugate generalized plastic strains q_α^p. A restricted form of this proposition is known as *Drucker's postulate* for a stable, dissipative material [Drucker 1959]. It immediately gives a positive increment of plastic work $\sigma_{ij}\, d\varepsilon_{ij}^p \geq 0$ when we consider $\sigma_{ij}^s = 0$. For a smooth fully plastic locus Drucker's postulate requires a positive rate of work for plastic stresses. This is satisfied if the plastic strain increment $d\varepsilon_{ij}^p$ has a direction normal to the locus of fully plastic stresses. Thus for a state of stress where the convex surface is smooth, the following plastic strain-rate increments correspond to $\Psi_p = 0$:

$$\dot{\varepsilon}_{ij}^p \equiv \frac{d\varepsilon_{ij}^p}{dt} = \nu \frac{\partial \Psi_p}{\partial \sigma_{ij}} \qquad \text{or} \qquad dq_\alpha^p = \nu \frac{\partial \Psi_p}{\partial Q_\alpha} \tag{1.32}$$

where ν is a positive scalar constant. This flow rule satisfies the *normality* condition relating increments of deformation to the state of stress; consequently, it is termed an *associated flow rule.*

In elastic–perfectly plastic beams and slender members the deformations are stretch and curvature along the centroidal axis. The states of stress for these deformations are located between the yield and fully plastic loci. For elastic–perfectly plastic stress resultants that approach the fully plastic limit surface, the curvature becomes large without bound. An indefinitely large curvature corresponds to relative rotation at some point along the axis of sections on either side of the point. Hence for stresses on the fully plastic surface, the correct measure of flexural rate of deformation is rate of rotation $\dot{\theta}_m(X)$ rather than rate of change of curvature. In particular, for static or quasistatic deformation the rigid–perfectly plastic material idealization always results in discontinuities in inclination at stationary 'plastic hinges'.

In a rigid–perfectly plastic beam or slender member the increment of deformation (stretch and rotation) can be related to a fully plastic state of stress $\Psi_p = 0$ by an associated flow rule. From the fully plastic limit surface for a rectangular cross-section, $|\bar{m}| + \bar{n}^2 - 1 = 0$, one obtains generalized plastic strain-rates $\dot{\varepsilon}^p = 2\nu N/N_p^2$ and $\dot{\theta}_m^p = \nu/M_p$. Hence, in a plastically deforming section the stretch and relative flexural rotation are related by

$$\frac{\dot{\varepsilon}^p}{\dot{\theta}_m^p} = 2\frac{M_p N}{N_p^2} \qquad \text{for} \qquad |\dot{\varepsilon}^p| \leq 2\left|\frac{M_p}{N_p}\dot{\theta}_m^p\right| \tag{1.33}$$

This relation applies to states of stress on smooth parts of the fully plastic limit surface where the limit function is differentiable; it does not apply at corners. Similar relations can be obtained also between any other generalized plastic strains that are coupled through the fully plastic limit surface for the conjugate generalized stresses.

1.4.5 Separated Yield Functions and Separated Plastic Flow

A consistent model formulation requires a deformation field with a component of deformation for each component of generalized stress in the yield function; i.e. if axial forces significantly influence the bending moment in a plastically deforming beam, the equations of motion must consider stretching as well as flexural deformations. Thus interaction fully plastic limit surfaces, such as those illustrated in Fig. 1.9, introduce several complications for analyses of structural deformations. These are primarily associated with satisfying the flow rule in plastically deforming regions while simultaneously satisfying the equations of equilibrium or equations of motion for the system. These complications are avoided by an additional constitutive approximation which circumvents the flow rule.

The curved surfaces for fully plastic generalized stress in Fig. 1.9 can be approximated by a plastic limit function that provides square limit surface $\Psi_p = 0$. For sections subjected to simultaneous bending $\tilde{m}$ and axial force $\tilde{n}$ this square plastic limit function is given by

$$\Psi_p = \tilde{m} - 1 \qquad |\tilde{n}| \leq 1$$

$$\Psi_p = \tilde{n} - 1 \qquad |\tilde{m}| \leq 1 \tag{1.34}$$

This fully plastic locus circumscribes the true fully plastic surface for an elastic–plastic material as shown in Fig. 1.10. Hence it overestimates the stiffness during plastic deformation; i.e. the reduction in stiffness due to axial force has been neglected. While the circumscribing square gives an upper bound on fully plastic generalized stresses, a lower bound can be obtained from an inscribing square. The main benefits derived from these approximations arise from the associated flow rule which provides the following relations between generalized strains for $\Psi_p = 0$,

$$\dot{\varepsilon} = 0, \qquad \dot{\theta}_m > 0 \qquad \text{for} \quad |\tilde{m}| = 1, \quad |\tilde{n}| < 1$$

$$\dot{\varepsilon} > 0, \qquad \dot{\theta}_m = 0 \qquad \text{for} \quad |\tilde{m}| < 1, \quad |\tilde{n}| = 1$$

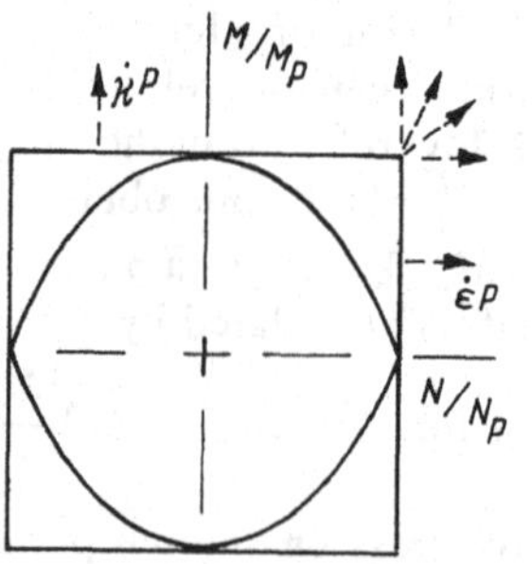

Fig. 1.10 Separated limiting stress approximation circumscribing the fully plastic stress locus. The normality condition provides the ratio between components of deformation rate for any state of stress on the fully plastic surface.

At corners where $|\tilde{m}| = |\tilde{n}| = 1$, a range of ratios $\dot{\varepsilon}/\dot{\theta}_m$ for deformation rates satisfies the flow rule; at these points the only thing that can be said is that $\text{sgn}(\dot{\varepsilon}) = \text{sgn}(\tilde{n})$ and $\text{sgn}(\dot{\theta}_m) = \text{sgn}(\tilde{m})$.

1.5 Interaction Yield Surfaces Including Shear

If a bar or beam carries shear force or torsion in addition to either axial force or bending moment, both normal and shear stress components act on the cross-sections. In parts of a section that are plastically deforming the yield criterion involves both these stress components. For thick walled or solid cross-sections the distribution of individual stress components in the plastically deforming parts of the cross-section can change with deformation. Here these changes are ignored. For a perfectly plastic material an *upper bound* for the yield condition can be obtained with stresses that are conjugate to a *kinematically admissible velocity field*; i.e. a continuous velocity field that satisfies the displacement constraints. This bound for the fully plastic state will be based on an assumption for the distribution of normal stress; namely, that in the fully plastic stress state the normal stress distribution with shear is identical to that without shear. This approach was used by Hill and Siebel [1953]; the present development uses their method but obtains yield functions for rectangular cross-sections that carry tension, shear and bending and for circular cross-sections that carry tension, torsion and bending. Necessary conditions for application of the bound theorems are described in Chap. 2.

1.5.1 Tension, Shear and Bending in Rectangular Cross-Section

Tension and Shear Let σ and τ be normal and shear stress components on a section that is plastically deforming. If the stresses are uniformly distributed on a cross-section of area A they have stress resultants for tension N and shear Q,

$$N = \sigma A, \qquad Q = \tau A \tag{1.35}$$

Notice that this stress field corresponds to a uniform shear stress imposed on a pretensioned bar; thus the shear stresses in the cross-section do not vanish at the top and bottom surfaces. For boundary conditions of negligible shear stress on transverse surfaces, this stress distribution is not in equilibrium at the surfaces; nevertheless, it is useful for obtaining an upper bound of the fully plastic state of stress.

For a perfectly plastic material the yield limit for this combined state of stress (1.35) is given by the von Mises yield criterion

$$\sigma^2 + 3\tau^2 = Y^2 \tag{1.36}$$

In contrast if either tension or shear vanishes the fully plastic stress resultants are $N_p = YA$ and $Q_p = YA/\sqrt{3}$. After dividing Eq. (1.36) by Y^2, the yield criterion can be expressed in terms of nondimensional generalized stresses $\tilde{n} = N/N_p$ and $\tilde{q} = Q/Q_p$,

$$\tilde{n}^2 + \tilde{q}^2 = 1 \tag{1.37}$$

Hence, the fully plastic axial stress σ_Y is reduced by shear

$$\sigma_Y = \frac{Y}{\sqrt{1-\tilde{q}^2}} \tag{1.38}$$

Tension, Shear and Bending The effect of shear is incorporated in the fully plastic limit function for tension and bending by reducing the normal stresses by the factor found in (1.38); i.e. for a rectangular cross-section the limit function (1.27) with shear gives

$$\Psi_p = \frac{\tilde{m}}{\sqrt{1-\tilde{q}^2}} + \frac{\tilde{n}^2}{1-\tilde{q}^2} - 1 \tag{1.39a}$$

or alternatively

$$\Psi_p = \tilde{m}\sqrt{1-\tilde{q}^2} + \tilde{n}^2 + \tilde{q}^2 - 1 \tag{1.39b}$$

If $\Psi_p = 0$ the cross-section is plastically deforming, while if $\Psi_p < 0$ the section is rigid or undeforming. Although this surface is an upper bound for the fully plastic limit surface, it is very slightly smaller than a different upper bound given by Hodge [1957] for the case of negligible tension.

1.5.2 Tension, Torsion and Bending in Circular Cross-Section

Torsion and Bending For a solid circular cross-section of radius a, the cross-section area is $A = \pi a^2$, the second moment of area about the neutral axis is $I_o = \pi a^4/4$, while the polar second moment of area for the centroidal axis is $J_o = 2I_o$. On an elastic cross-section, the bending moment M gives normal stress σ in a fiber at distance Z from the neutral axis; this stress varies with curvature κ at the section. Similarly, a torque T gives a shear stress τ in a fiber at distance r from the centroidal axis so in the elastic range torque varies with axial twist $\vartheta = \mathrm{d}\theta_t/\mathrm{d}X$ where θ_t is the rotation of the section about the longitudinal axis. Thus for an elastic bar,

$$\frac{\sigma}{Z} = -\frac{M}{I_o} = -E\kappa\,, \qquad \frac{\tau}{r} = \frac{T}{J_o} = G\vartheta \tag{1.40}$$

where $G = E/2(1+\nu)$ is the shear modulus and ν is Poisson's ratio. The largest stresses occur in the outer fibers at distance a from both the neutral axis for bending and the centroidal axis for twist. For each stress resultant in isolation, yield in these outer fibers occurs for a bending moment $M_Y = \pm YI_o/a$ or a torque $T_Y = \pm YJ_o/a\sqrt{3}$, where the yield condition for pure shear $\tau_Y = Y/\sqrt{3}$. Thus for combined bending and shear the von Mises condition (1.36) gives a criterion $\Psi_e = 0$ for the stress resultants that cause yield in the outer fibers,

$$\Psi_e = \left(\frac{M}{M_Y}\right)^2 + \left(\frac{T}{T_Y}\right)^2 - 1 \tag{1.41}$$

The section is elastic for $\Psi_e < 0$. If the stress resultants are on the yield surface then $\Psi_e = 0$, while if $\Psi_e > 0$ and $\Psi_p < 0$ the plastically deforming region covers an increasingly large part of the cross-section with increasing Ψ_e. Because the elastic stress distributions due to bending moment and torque are both nonuniform and different, the boundary between the elastic and plastic regions is a non-circular surface; plastic deformation is present in a lens shaped region at the top and bottom of the cross-section. At each point in the plastically deforming segments, the stress components change in accord with the flow law as deformation increases. If either one of these stress resultants vanishes, the other resultant for elastoplastic stresses in a circular cross-section is given by

$$\frac{|M|}{M_Y} = \frac{2}{3\pi}\left\{\left(5 - \frac{2\kappa_Y^2}{\kappa^2}\right)\sqrt{1 - \frac{\kappa_Y^2}{\kappa^2}} + \frac{3\kappa}{\kappa_Y}\sin^{-1}\left(\frac{\kappa_Y}{\kappa}\right)\right\}, \quad \kappa > \kappa_Y, \ T = 0$$

$$\frac{|T|}{T_Y} = \frac{4}{3}\left\{1 - \frac{1}{4}\left(\frac{\vartheta_Y}{\vartheta}\right)^3\right\} \qquad \vartheta > \vartheta_Y, \ M = 0 \tag{1.42}$$

For a combined stress state where both M and T are nonzero, in the limit as curvature or twist become indefinitely large either the bending moment or torque approaches the fully plastic state. The shape factors relating fully plastic and yield stress resultants for the circular cross-section are obtained by taking the limit of Eqs (1.10) and (1.42) as curvature and twist become indefinitely large; i.e. $M_p/M_Y = 16/3\pi \approx 1.7$ and $T_p/T_Y = 4/3$. A *lower bound* on the limit function Ψ_p for fully plastic stress resultants is obtained from Eq. (1.36) with the *assumption* that in the elongated and in the compressed halves of the cross-section both σ *and* τ *are uniform*. Notice that for any cross-section the combination of the independent fully plastic stress distributions for torsion and bending also satisfies the boundary conditions; this is a necessary condition for a lower bound. Hence if $\tilde{m} = M/M_p$ and $\tilde{t} = T/T_p$, the fully plastic limit is given by $\Psi_p = 0$, where

$$\Psi_p = \tilde{m}^2 + \tilde{t}^2 - 1, \qquad \tilde{m} < 1, \qquad \tilde{t} < 1 \tag{1.43}$$

Yield and fully plastic limit surfaces for a circular cross-section subjected to simultaneous bending and torsion are shown in Fig. 1.11.

Tension and Torsion Consider a circular cross-section loaded by simultaneous tension and torsion. In an elastic section the axial and shear stresses σ, τ yield tension and torque stress resultants, N and T respectively. The axial stress is uniform while shear stress increases in proportion with radius;

$$\sigma = \frac{N}{A} = E\varepsilon, \qquad \frac{\tau}{r} = \frac{T}{J_o} = G\vartheta \tag{1.44}$$

where the deformations (i.e. axial strain ε and twist ϑ) are related to the stress resultants through elastic moduli E, G and section properties A, I_o defined in Sect. 1.2. At the elastic limit the stress components (1.44) in the outer fibers, $r = a$, satisfy the yield criterion (1.36); thus at initial yield the tension and torque are on the yield surface $\Psi_e = 0$ where the yield function is

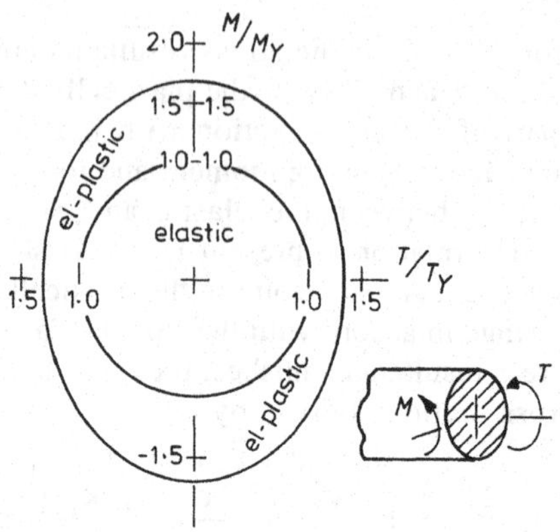

Fig. 1.11 Yield and fully plastic limits for combined bending and torsion in bar with circular cross-section.

$$\Psi_e = \left(\frac{N}{N_Y}\right)^2 + \left(\frac{T}{T_Y}\right)^2 - 1 \tag{1.45}$$

In this expression the tensile yield force is $N_Y = \pm YA$ and the yield torque is $T_Y = \pm YJ_o / a\sqrt{3}$. (Recall that for a circular or square cross-section, the polar second moment of area J_o is given by $J_o \equiv I_{oy} + I_{oz} = 2I_o$.)

A lower bound on the fully plastic limit surface can be obtained with any statically admissible stress field; i.e. a field that satisfies equilibrium throughout the cross-section. First suppose the bar is loaded by a tensile force N and torque T that satisfy $\Psi_e = 0$. Consider axial and shear stresses with constant magnitudes across the entire section; this results in uniform tensile stress $\sigma = N/A$ and hence from Eq. (1.36), uniform shear stress $\tau = aT/J_o$. This stress field is a lower bound on the actual stresses for this deformation process. Thus for tension and torque in a circular section the fully plastic limit function is

$$\Psi_p \approx \tilde{n}^2 + \tilde{t}^2 - 1 \tag{1.46}$$

where $\tilde{n} = N/N_p$ and $\tilde{t} = T/T_p$.

The actual stresses may not be uniform so this is a lower bound for the fully plastic limit surface. Note that the shear stress due to torsion reduces the fully plastic axial stress σ_Y; i.e.

$$\sigma_Y = \frac{Y}{\sqrt{1-\tilde{t}^2}} \tag{1.47}$$

Tension, Torsion and Bending The effect of shear stress due to torsion is to reduce the fully plastic axial stresses and hence the stress resultants for tension N and bending M. Thus the suggestion of Hill and Siebel [1953] given in (1.47), when substituted into the fully plastic limit function for a circular cross-section (1.28b), yields

$$\Psi_p = \frac{\tilde{m}}{\sqrt{1-\tilde{t}^2}} - \cos^3\left\{\frac{\pi\tilde{n}}{2\sqrt{1-\tilde{t}^2}} - \left(\frac{\tilde{m}}{\sqrt{1-\tilde{t}^2}}\right)^{1/3}\left[1 - \left(\frac{\tilde{m}}{\sqrt{1-\tilde{t}^2}}\right)^{2/3}\right]^{1/2}\right\}$$

For comparison, the yield limit $\Psi_e = 0$ obtained from the von Mises criterion can be expressed as

$$\Psi_e = (|\bar{n}| + |\bar{m}|)^2 + \bar{t}^2 - 1$$

where $\bar{m} = M/M_Y$, $\bar{n} = N/N_Y$ and $\bar{t} = T/T_Y$.

1.6 Elastic Springback

When forces are removed from any section of a structure, the section *springs back* as the stresses are relieved; the deformation decreases somewhat from the largest values. For most materials, unloading occurs elastically irrespective of the current state of strain. Figure 1.12 depicts the uniaxial stress–strain behavior of a linear elastic material that exhibits linear strain hardening if the strain exceeds the yield strain ε_Y. This material unloads elastically along the dashed line. If the material is stretched to some largest strain ε_* and then the stress is gradually removed, during unloading the stress is continually in balance with the applied load. When the applied stress vanishes the material has a residual state of plastic deformation with strain ε^p as a consequence of the history of stress for that location. In this section springback relations are presented for both linear strain hardening and perfectly plastic material characterizations; these are developed for bars in pure bending and combined bending and stretching.

1.6.1 Pure Bending

During bending of a bar, beam or plate, the stresses vary across a section so the stress history depends on distance from the neutral axis $\tilde{Z}$. Flexure and stretching result in a linear distribution of normal strain on a section as a function of $\tilde{Z}$ but the normal stress distribution is not linear if part of the section is plastically deforming. For pure bending, the neutral axis is at the centroid $\tilde{Z} = Z$, so the

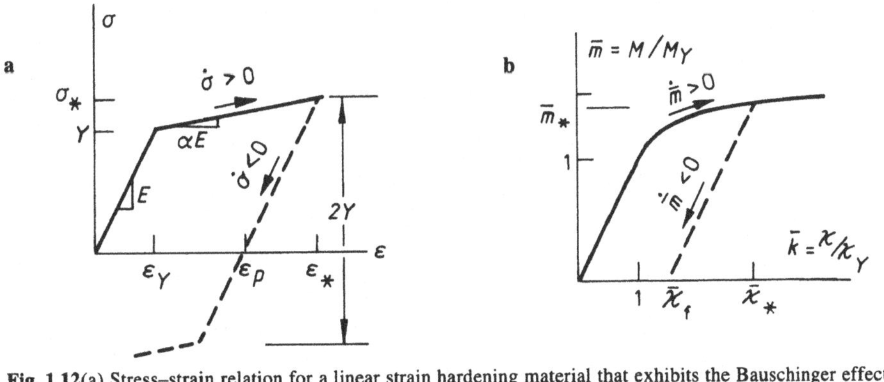

Fig. 1.12(a) Stress–strain relation for a linear strain hardening material that exhibits the Bauschinger effect upon unloading; (b) the corresponding moment–curvature relation for loading $\dot{\bar{m}} > 0$ and unloading $\dot{\bar{m}} < 0$ of an elastic–plastic section.

kinematic condition that plane sections remain plane gives

$$\frac{\sigma(Z)}{Y} = \frac{2Z\kappa}{h\kappa_Y}, \qquad \frac{Z}{h} \le \frac{\kappa_Y}{2\kappa}$$
$$\frac{\sigma(Z)}{Y} = 1-(1-\alpha)\left(1-\frac{2Z\kappa}{h\kappa_Y}\right) \qquad \frac{Z}{h} \ge \frac{\kappa_Y}{2\kappa} \tag{1.48}$$

where $\alpha = E_t/E$, as given in Sect. 1.2.4. After taking the first moment of stress about the neutral axis and integrating this over the cross-sectional area, the following nondimensional moment–curvature relations for loading a rectangular cross-section are obtained

$$\bar{m} = \bar{k}, \qquad \bar{k} \le 1$$
$$\bar{m} = \frac{1}{2}(1-\alpha)(3-\bar{k}^{-2})+\alpha\bar{k}, \qquad \bar{k} > 1 \tag{1.49}$$

where the bending moment $\bar{m} = M/M_Y$ and the curvature $\bar{k} = \kappa/\kappa_Y$. At any section loading terminates at some maximum bending moment $\bar{m}_*$ and curvature $\bar{k}_*$. Thereafter the stress resultants are removed so the section unloads completely to the final bending moment $\bar{m}_f = 0$. During elastic unloading there is a change in curvature equal to $-\bar{m}_*$, so finally

$$\bar{k}_f = \bar{k}_* - \bar{m}_* \tag{1.50}$$

Unloading occurs elastically across the entire section as long as there is no reverse yielding. Reverse yielding first occurs in the outer fibers; for *kinematic strain hardening* there can be no reverse yielding if $\bar{m}_* < 2$. (Kinematic hardening gives a yield surface $\Psi_e(Q_\alpha) = 0$ which expands in a self-similar pattern with increasing plastic strain; this results in the Bauschinger effect shown in Fig. 1.12.) For loading followed by unloading of a beam in bending, the distribution of stress in a rectangular cross-section is illustrated in Fig. 1.13.

In an elastic–plastic material, as the stress resultant is unloaded the stresses decrease in proportion to their distance from the neutral axis. The residual stresses that remain after the stress resultants vanish are locked in by the variation in plastic strain ε^p across the section.

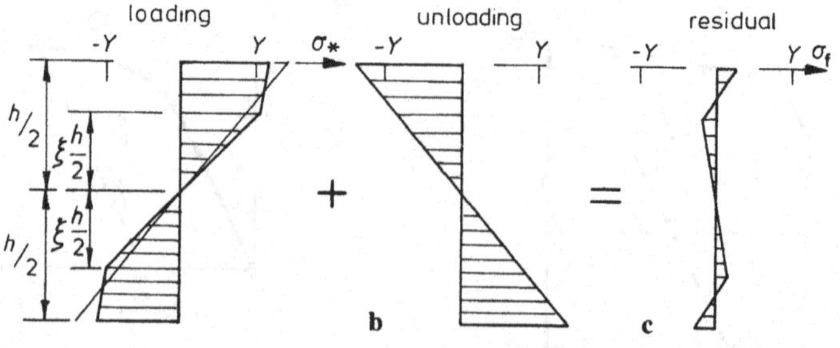

Fig. 1.13 Stress distributions for pure bending in a doubly symmetric linear strain hardening bar during (a) loading in the elastic–plastic range $1 < \bar{m} < \phi_m \bar{m}$, (b) elastic unloading and (c) the residual stress when finally $\bar{m} = 0$ (sum of **a** plus **b**).

An early analysis of springback was developed by Gardiner [1957] for pure bending of elastic–perfectly plastic sheet material in plane stress. The analysis was extended to include the effect of linear strain hardening by Woo and Marshall [1959]. Elastic springback relations for linear and power law hardening in both plane stress and plane strain were obtained by Johnson and Yu [1981]; they calculated final curvatures in rectangular and circular plates subjected to biaxial bending.

1.6.2 Bending and Tension

In many forming problems tension is applied to a sheet or plate to control wrinkling and reduce the required bending moment during forming processes; this is termed *stretch–bending*. Another source of normal in-plane forces that reduce the fully plastic bending moment are axial loads induced by impact at an angle of incidence measured from the transverse direction. The in-plane forces result in a neutral axis at some distance from the centroid and this also affects springback. For an elastic–perfectly plastic material in plane stress, Yu and Johnson [1982] obtained ratios for the change in one dimensional curvature during loading.

Figure 1.14 shows the stress distribution through the thickness of an elastic–perfectly plastic bar loaded in combined tension and bending. The tension has moved the neutral surface a distance $\zeta h/2$ from the centroid. This bar has an elastic region that extends a distance $\xi h/2$ from the neutral surface; the remainder of the section is fully plastic at tensile yield stress Y.

For combined loading, an elastic–plastic bar has several different patterns of stress distribution depending on the tension N and bending moment M. These patterns characteristically appear in specific regions relative to the elastic yield surface (3) and the fully plastic limit surface (1) illustrated in Fig. 1.15. This figure depicts the yield and fully plastic surfaces in generalized stress space,

$$\overline{m} = M/M_Y, \qquad \overline{n} = N/N_Y$$

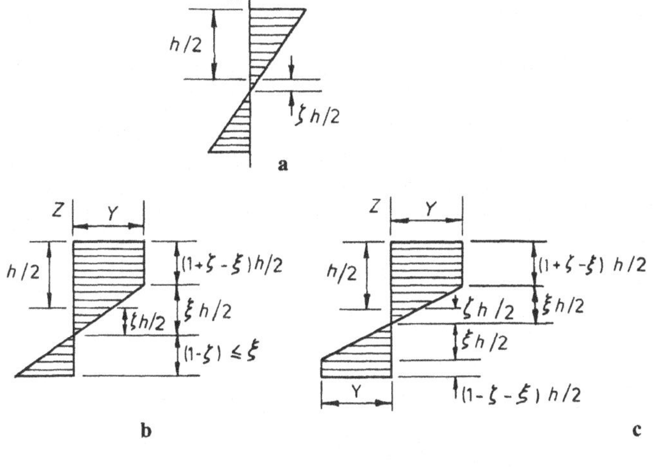

Fig. 1.14 Stress distributions in elastic–perfectly plastic section with both bending moment M and axial force N. The curves correspond to (a) elastic stresses (b) primary plastic stress distribution (c) secondary plastic stress distribution.

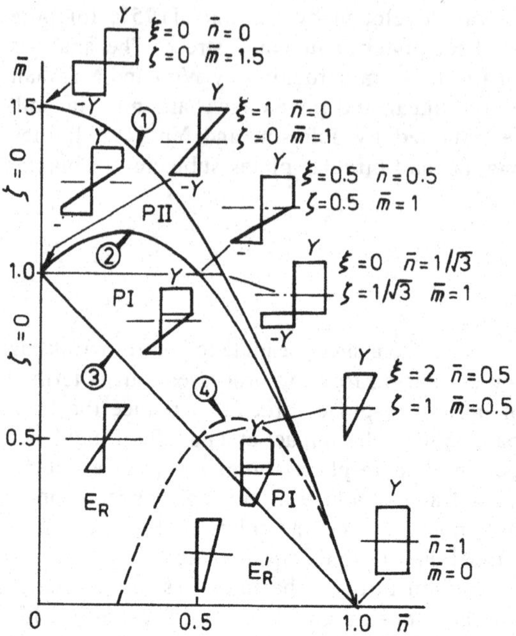

Fig. 1.15 Patterns of stress distribution in elastic–perfectly plastic bar with rectangular cross-section depend on the magnitudes of nondimensional bending moment $\overline{m}$ and axial force $\overline{n}$.

where the stress resultants in a bar with rectangular cross-section have independent elastic limits $M_Y = Ybh^2/6$, $N_Y = Ybh$. Also, the curvature κ can be related to the elastic limit $\kappa_Y = 2Y/Eh$ for a rectangular cross-section; i.e. $\overline{k} = \kappa/\kappa_Y$. Thus during loading of a rectangular cross-section, the curvature is related to the stress resultants by

$$\overline{k} = \begin{cases} \overline{m}, & 0 \le \overline{m} \le 1-\overline{n} \\ 4(1-\overline{n})^3(3-\overline{m}-3\overline{n})^{-2}, & 1-\overline{n} \le \overline{m} \le 1+\overline{n}-2\overline{n}^2 \\ [3(1-\overline{n}^2)-2\overline{m}]^{-1/2}, & 1+\overline{n}-2\overline{n}^2 \le \overline{m} \le 3(1-\overline{n}^2)/2 \end{cases} \tag{1.51}$$

These equations refer to the elastic, primary plastic and secondary plastic regions, respectively.

If maximum values of the loads $\overline{m}_*$, $\overline{n}_*$ exceed the yield limit, $\Psi(\overline{m}_*, \overline{n}_*) > 0$, after unloading there will be residual stress σ_f in the cross-section and a residual curvature κ_f. The residual curvature is related to the curvature at yield κ_Y by $\overline{k}_f = \kappa_f/\kappa_Y$. This residual curvature is given by $\overline{k}_f = \overline{k}_* - \overline{m}_*$. Often the final curvature is expressed as a springback ratio $\overline{k}_f/\overline{k}_*$; i.e. the part of the largest curvature which is recovered on unloading

$$\frac{\overline{k}_f}{\overline{k}_*} = \begin{cases} 0, & 0 \le \overline{m}_* \le 1-\overline{n}_* \\ 1-3\overline{k}_*^{-1}(1-\overline{n}_*^2)/2+\overline{k}_*^{-3}/2, & 1-\overline{n}_* \le \overline{m}_* \le 1+\overline{n}_*-2\overline{n}_*^2 \\ 1-3\overline{k}_*^{-1}(1-\overline{n}_*)+2[\overline{k}_*^{-1}(1-\overline{n}_*)]^{3/2}, & 1+\overline{n}_*-2\overline{n}_*^2 \le \overline{m}_* \le 3(1-\overline{n}_*^2)/2 \end{cases} \tag{1.52}$$

Figure 1.16 illustrates the springback ratio for a rectangular cross-section; this

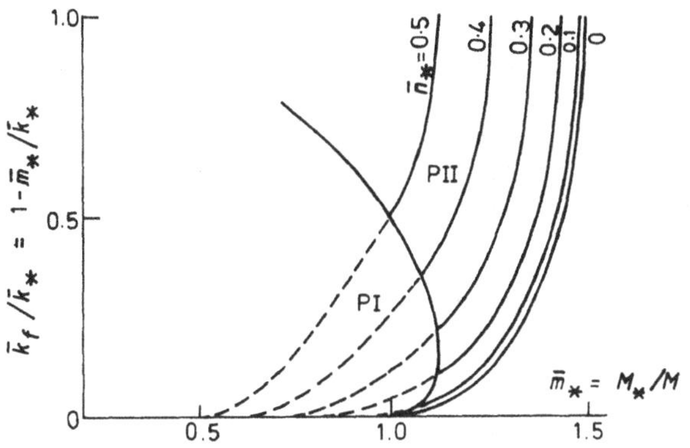

Fig. 1.16 Springback ratio in elastic–perfectly plastic beam with rectangular cross-section subjected to maximum bending moment $\bar{m}_*$ and axial force $\bar{n}_*$.

indicates that stretch–bending increases the springback ratio in comparison with that for pure bending.

References

Bodner, S.R. and Speirs, W.G. [1963]. Dynamic plasticity experiments on aluminum cantilever beams at elevated temperature. *J. Mech. Phys. Solids* **11**, 65–77.

Cowper, G.R. and Symonds, P.S. [1957]. Strain–hardening and strain-rate effects in the impact loading of cantilever beams. *Brown Univ. Dept. Appl. Math. T.R.* 28.

Drucker, D.C. [1959]. A definition of a stable inelastic material. *J. Appl. Mech.* **26**, 101–106.

Forrestal, M.J. and Sagartz, M.J. [1978]. Elastic–plastic response of 304 stainless steel beams to impulse loads. *J. Appl. Mech.* **45**, 685–687.

Forrestal, M.J. and Wesenberg, D.L. [1977]. Elastic–plastic response of simply supported 1018 steel beams to impulse loads. *J. Appl. Mech.* **44**, 779–780.

Gardiner, F.J. [1957]. The spring back of metals. *Trans. ASME* **79**, 1–7.

Hartley, K.A. and Duffy, J. [1984]. Strain-rate and temperature history effects during deformation of FCC and BCC metals. *Mechanical Properties at High Rates of Strain* (ed. J. Harding). Inst. of Physics, London, 21–30.

Hill, R. [1948]. A variational principle of maximum plastic work in classical plasticity. *Q. J. Mech. Appl. Math.* **1**, 18.

Hill, R. and Siebel, M.P.L. [1953]. On the plastic distortion of solid bars by combined bending and twisting. *J. Mech. Phys. Solids* **1**, 207–214.

Hodge, P.G., Jr [1957]. Interaction curves for shear and bending of plastic beams. *J. Appl. Mech.* **24**, 453.

Johnson, W. and Yu, T.X. [1981]. On springback after the pure bending of beams and plates of elastic work-hardening materials – III. *Int. J. Mech. Sci.* **23**(11), 687–695.

Perrone, N.J. [1970]. Impulsively loaded strain hardening rate-sensitive rings and tubes. *Int. J. Solids Structures* **6**, 1119–1132.

Perzyna, P. [1962]. The constitutive equation for rate sensitive plastic materials. *Q. Appl. Math.* **20**, 321–332.

Save, M.S. and Massonet, C. [1972]. *Plastic Analysis and Design of Plates, Shells and Disks*. North-Holland, Amsterdam.

Sawczuk, A. [1989]. *Mechanics and Plasticity of Structures*. Ellis Horwood, Chichester.

Symonds, P.S. [1965]. Viscoplastic behavior in response of structures to dynamic loading. *Behavior of Materials under Dynamic Loading* (ed. N. Huffington). ASME, 106–124.

Taylor, G.I. [1947]. Connection between the criterion of yield and the strain ratio relationship in plastic solids. *Proc. Roy. Soc. Lond.* **A191**, 441.

Timoshenko, S. and Goodier, J.N. [1971]. *Theory of Elasticity*. McGraw-Hill, New York.

Ting, T.C.T. [1964]. The plastic deformation of a cantilever beam with strain rate sensitivity under impulsive loading. *J. Appl. Mech.* **31**, 38–42.

Ting, T.C.T. and Symonds, P.S. [1962]. Impact on a cantilever beam with strain rate sensitivity. *Proc. Fourth US Natl. Cong. Appl. Mech. ASME*, 1153–1165.

Woo, D.M. and Marshall, J. [1959]. Spring-back and stretch-forming of sheet metal. *The Engineer*, 135–136.

Yu, T.X. and Johnson, W. [1982]. Influence of axial force on the elastic–plastic bending and springback of a beam. *J. Mech. Working Tech.* **6**, 5–21.

Chapter 2

Principles of Mechanics

The response of bodies to slowly varying loads is *static* or *quasistatic*. For slowly varying loads, the sum of all forces acting on any segment of a body are in balance since there are no accelerations; i.e. for any segment of the body, the resultant of tractions on the surface of the segment is equal in magnitude but opposite in direction to any external force acting on the segment. On the other hand, rapid changes in load cause the body to accelerate; in this case the resultants of stresses acting on any part of the body are related to the product of acceleration and inertia by the laws of motion. For these two broad classes of loading the differential equations for variations in stress (or stress resultant) across an arbitrary segment of the body are obtained from either equilibrium equations or the laws of motion. These laws form the basis of several useful principles. In this chapter special forms of these principles will be developed that apply to small deflection theory for planar displacements of slender bodies.

2.1 Kinematics

Begin by fixing the directions of a triad of mutually perpendicular unit vectors $\mathbf{e}_j$, $j = 1, 2, 3$ in the initial undeformed configuration of a body at rest. These vectors form the basis of a material or *Lagrangian reference frame* where the current position of a particle is measured relative to that in the undeformed configuration using spatial coordinates X_j fixed in the undeformed configuration. The displacement of each material particle from its initial position is a vector $\mathbf{W}(\mathbf{X}) = W_i(X_j)$. In this Lagrangian description the velocity field $\dot{\mathbf{W}}(\mathbf{X}) = \dot{W}_i(X_j) \equiv \mathrm{d}W_i(X_j)/\mathrm{d}t$ is the rate of change for the position vector of a particle that was initially located at X_j. Relative displacement of neighboring points in a body is the sum of deformation of the material and rotation of the set of unit vectors. The infinitesimal deformation in any differential element is given by the tensorial strain ε_{ij}, while components of rotation about $\mathbf{e}_i \times \mathbf{e}_j$ are denoted by ω_{ij}:[1]

[1]In this index notation a repeated subscript indicates a sum of components; e.g. $\varepsilon_{ii} \equiv \varepsilon_{11} + \varepsilon_{22} + \varepsilon_{33}$.

$$\varepsilon_{ij} = \frac{1}{2}\left(\frac{\partial W_j}{\partial X_i} + \frac{\partial W_i}{\partial X_j}\right), \qquad \omega_{ij} = \frac{1}{2}\left(\frac{\partial W_j}{\partial X_i} - \frac{\partial W_i}{\partial X_j}\right) \tag{2.1}$$

The deformation is related to stress σ_{ij} by constitutive relations such as Hooke's law for a linear elastic material.

This monograph focuses on slender prismatic bars or beams where the cross-section has a plane of symmetry. Applied forces, axial torques and initial velocities all act in the plane of symmetry so there is no out-of-plane flexure; i.e. the axis passing through the centroid of every section suffers solely in-plane displacements. Many of our examples consider uniform cross-sections, but it is only necessary that any changes in section are gradual. For slender members, let $\mathbf{e}_1$ be the axial direction through the centroid of every cross-section and $\mathbf{e}_2$ be the normal to the plane of loading. Thus if loading is applied in a plane of symmetry for the cross-section the displacement field $\mathbf{W}(\mathbf{X})$ has only axial and transverse displacements of the centroidal axis, $\mathbf{W} = (W_1, 0, W_3)$ as shown in Fig. 2.1. Likewise the rotation $\boldsymbol{\theta}$ of a cross-section has components $\boldsymbol{\theta} = (\theta_t, -\theta_m, 0)$ that are termed *axial* and *flexural rotation*; in general $\theta_i = e_{ijk}\omega_{jk}$.[2] Notice that the flexural component of rotation $\theta_m = -\theta_2$ has a sign that is adopted from beam convention. For a slender beam that is simultaneously stretched, twisted and bent, the axial and transverse displacements (velocities) vary linearly over the cross-section.

For a slender structural member, the equations of motion relate axial variations of stress resultants for cross-sections to accelerations in the member. In Chap. 1 these stress resultants were termed generalized stresses; there is a conjugate set of *generalized strains* for the cross-section. If the section has an outward normal, $\mathbf{n} = n_i\mathbf{e}_i$ in direction $\mathbf{e}_1$, the generalized strain vector for the section is obtained from the strain tensor ε_{ij} and rotation vector θ_i evaluated at the centroid $\{\varepsilon_{11}, \varepsilon_{21}, \varepsilon_{31}, \theta_1, \theta_2, \theta_3\}$. For planar deflections, however, $\varepsilon_{21} = 0$ and $\omega_{12} = \theta_3 = 0$ so that engineering strains and rotations are defined as

$$\varepsilon \equiv \varepsilon_{11}, \qquad 0 \equiv \varepsilon_{21}, \qquad \gamma_3 \equiv 2\,\varepsilon_{31},$$

$$\theta_t \equiv \omega_{23}, \qquad \theta_m \equiv -\omega_{31}, \qquad 0 \equiv \omega_{12}$$

where axial twisting θ_t and in-plane rotation θ_m have a positive sense as indicated in Fig. 2.1. Usually these variables are expressed as a reduced set of deformations or generalized strains q_i for deflections in the $\mathbf{e}_1$, $\mathbf{e}_3$ plane

$$q_\alpha = \{\varepsilon, \gamma_3, \theta_t, \theta_m\}$$

Generalized strains (or strain-rates) exist at every point in the body if the displacement (velocity) field is at least piecewise continuous so that it is differentiable. A piecewise continuous displacement field W_i^c (or velocity field $\dot{W}_i^c$) that satisfies all displacement and velocity constraints is said to be *kinematically admissible*; i.e. it satisfies *compatibility*. Kinematically admissible fields yield generalized strains q_i^c and strain rates $\dot{q}_i^c$.

---

[2]The permutation tensor e_{ijk} is defined as $e_{ijk} \equiv 1$ if i, j, k are in cyclic order; $e_{ijk} \equiv 0$ if $i = j$, $j = k$ or $k = i$; and $e_{ijk} \equiv -1$ if i, j, k are in anticyclic order. Thus scalar and vector multiplications of vectors $\mathbf{u} = u_i\mathbf{e}_i$ and $\mathbf{v} = v_i\mathbf{e}_i$ can be expressed as $\mathbf{u}\cdot\mathbf{v} = u_i\mathbf{e}_i \cdot \mathbf{e}_j v_j = u_i v_i$ and $\mathbf{u}\times\mathbf{v} = \mathbf{e}_i e_{ijk} u_j v_k$, respectively.

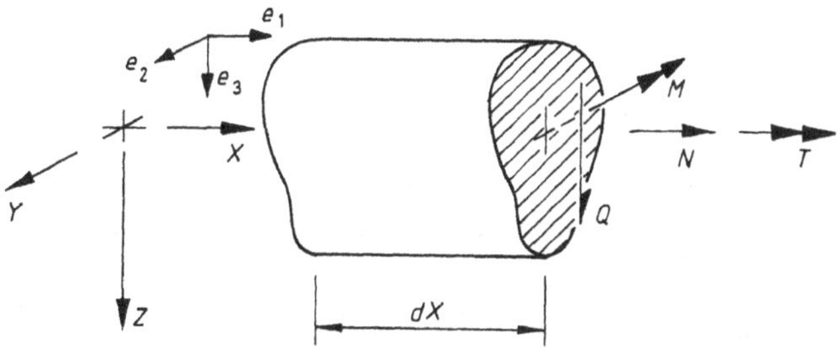

Fig. 2.1 Stress resultants on cross-section of beam that give in-plane deflections of centroidal axis.

2.1.1 Inertia Properties of Cross-Section

For a uniform bar or beam, consider a differential element taken at any point along the length. The differential element is a cross-sectional lamina with a normal $\mathbf{n} = \mathbf{e}_1$; this has an area equal to the cross-sectional area A and length $dS = \mathbf{n} \cdot d\mathbf{X} = n_j \, dX_j$. If the bar is composed of a homogeneous material with density ρ_v, the differential element has mass ρdS and moments of inertia $\rho_v I_i \, dS$ about the centroid:

$$\rho = \int \rho_v \, dA \,, \qquad \rho_v I_i = \int \rho_v \overline{\mathbf{X}} \times (\mathbf{e}_i \times \overline{\mathbf{X}}) \, dA \tag{2.2}$$

where ρ is the mass per unit length and $\overline{\mathbf{X}} = \mathbf{X} - \mathbf{X} \cdot \mathbf{nn}$ lies in the plane of the lamina. With these definitions, the moments of inertia are directly related to the second moments of area I_i for the section. Thus for a homogeneous bar with axial coordinate $X_1 = X$, the cross-section has second moments of area for principal axes through the centroid,

$$I_x = \int \left(Y^2 + Z^2\right) dA \equiv J_o, \qquad I_y = \int Z^2 \, dA\,, \qquad I_z = \int Y^2 \, dA \tag{2.3}$$

2.2 Balance of Forces

2.2.1 Stress Resultants and Generalized Stresses

In a homogeneous body the stress $\boldsymbol{\sigma}$ is a continuous function of the spatial coordinates $\mathbf{X}$. At every point the components of stress σ_{ij} are related to directions given by the set of mutually perpendicular unit vectors $\mathbf{e}_i$; thus the second order tensor $\boldsymbol{\sigma}$ can be expressed as a dyadic $\boldsymbol{\sigma} = \sigma_{ij} \, \mathbf{e}_i \mathbf{e}_j$. For a surface that passes through any point of the body, let $\mathbf{n}$ be the unit vector normal to the surface at $\mathbf{X}$. This normal has components n_i which are the direction cosines relative to the basis vectors $\mathbf{e}_i$; i.e. $\mathbf{n} = n_i \mathbf{e}_i$. At any point the traction $\mathbf{t(n)}$ (i.e. force per unit area) is a vector that depends on the stress $\boldsymbol{\sigma}$ and the orientation $\mathbf{n}$ of the surface passing through the point

$$\mathbf{t} = \boldsymbol{\sigma}\cdot\mathbf{n} = \mathbf{t}_j n_j$$

For a slender beam or a prismatic bar, we aim to find deflections of the centroidal axis — this is a representative point in the cross-section. The expressions for these deflections in terms of the applied loads and cross-sectional properties are obtained from equations of motion for a differential element that lies between two cross-sections as shown in Fig. 2.1; i.e. an element with length $\mathrm{d}S$ along the member. If the element is oriented such that the normal $\mathbf{n}$ has a direction $\mathbf{n} = \mathbf{e}_1$, then $\mathrm{d}S = \mathrm{d}X$ and the internal stresses give surface tractions $\mathbf{t}_1 = \sigma_{i1}\mathbf{e}_i$ acting on the cross-section. Hence a cross-section has stress resultants known as force $\mathbf{N} = N_j\mathbf{e}_j$ and a couple $\mathbf{M} = M_j\mathbf{e}_j$ that are obtained by integration over the cross-sectional area A of tractions and the first moment of tractions about the centroid,

$$N_i = \int \sigma_{ij}\,\mathbf{e}_j\cdot\mathbf{e}_i\,\mathrm{d}A\,, \qquad M_i = \int \mathbf{X}\times\mathbf{t}_1\,\mathrm{d}A = \int e_{ijk}\,\mathbf{e}_i X_j\,\sigma_{k1}\,\mathrm{d}A$$

These stress resultants have components parallel and transverse to the centroidal axis. Noting that for in-plane deflections some transverse components vanish ($N_y = M_z = 0$) while the nonzero stress resultants are given by

$$N = N_x = \int \sigma_{xx}\,\mathrm{d}A\,, \qquad Q = N_z = \int \sigma_{zx}\,\mathrm{d}A$$

$$T = M_x = \int \left(Y\sigma_{zx} - Z\sigma_{yx}\right)\mathrm{d}A\,, \qquad M = -M_y = -\int Z\sigma_{xx}\,\mathrm{d}A \tag{2.4}$$

These stress resultants are the components of the reduced generalized stress vector $Q_\alpha = \{N, Q, T, M\}$ where the axial force $Q_1 = N$, the shear force $Q_2 = Q$, the twisting moment $Q_3 = T$ and the bending moment $Q_4 = M$.

2.2.2 Equations of Motion

Forces acting on any differential beam element with cross-sectional area A and volume $\mathrm{d}V = A\,\mathrm{d}X$ are of two types — long range and short range. The long range forces, named *body forces*, are proportional to the mass of the element; typical examples are gravitational or electromagnetic forces between bodies that are some distance apart. A body force $\mathbf{g}$ has dimensions of force per unit volume. The other type of force is due to direct contact between bodies. *Contact forces* or tractions $\mathbf{t}(\mathbf{n})$ act at a surface in contact with a neighboring body; the surface has outward normal $\mathbf{n}$. Tractions have dimensions of force per unit area. In either case the force acting on one body is equal in magnitude but opposite in direction from the force acting on the second body of an interacting pair.

Unbalanced forces on any element result in acceleration $\mathrm{d}^2\mathbf{W}/\mathrm{d}t^2$ where $\mathbf{W}(\mathbf{X}, t)$ is a time dependent displacement field with components W_i. Hence for a differential element with volume $\mathrm{d}V$, an equation of motion is obtained from the rate of change of translational momentum

$$\frac{\partial \sigma_{ij}}{\partial X_j} + g_i = \frac{\partial}{\partial t}\left(\rho_v \frac{\partial W_i}{\partial t}\right), \qquad \sigma_{ij} = \sigma_{ji} \tag{2.5}$$

Assuming deformations in the cross-sectional plane are negligible, the acceleration of the axis through the centroids is representative of translational and flexural acceleration of a beam or a slender bar. For a beam with centroidal axis parallel to $\mathbf{e}_1$ and applied (body and contact) forces per unit length $\mathbf{g}$, the stress resultants acting on cross-sections of area A with normal direction $\mathbf{e}_1$ give

$$\frac{\partial}{\partial X}\int \mathbf{t}_1\,\mathrm{d}A + \mathbf{g} = \frac{\partial}{\partial t}\left(\rho\frac{\partial \mathbf{W}}{\partial t}\right)$$

where $\mathbf{W}$ is the displacement of the centroidal axis. Likewise, equating the differential element to the rate of change of moment of momentum about the centroid gives

$$\frac{\partial}{\partial X}\int \overline{\mathbf{X}}\times\mathbf{t}_1\,\mathrm{d}A = \frac{\partial}{\partial t}\int \rho_v\overline{\mathbf{X}}\times\left(\frac{\partial\boldsymbol{\theta}}{\partial t}\times\overline{\mathbf{X}}\right)\mathrm{d}A, \qquad \mathbf{0} = \int\left(\overline{\mathbf{X}}\times\mathbf{g}\right)\mathrm{d}A$$

where $\overline{\mathbf{X}}$ is the position vector from the centroid in the cross-sectional plane. Thus the *equations of motion* for infinitesimal in-plane deflections of a beam or bar can be expressed in terms of the stress resultants Q_α and corresponding inertia forces that depend on the distribution of momentum in the cross-section,

$$\frac{\partial N}{\partial X} + g_x = \rho\frac{\partial^2 W_x}{\partial t^2} \tag{2.6a}$$

$$\frac{\partial Q}{\partial X} + g_z = \rho\frac{\partial^2 W_z}{\partial t^2} \tag{2.6b}$$

$$\frac{\partial T}{\partial X} = \rho_v I_x\frac{\partial^2 \theta_t}{\partial t^2} \tag{2.6c}$$

$$\frac{\partial M}{\partial X} = \rho_v I_y\frac{\partial^2 \theta_m}{\partial t^2} \tag{2.6d}$$

Here $W_x(X)$ and $W_z(X)$ are axial and transverse displacements for the centroid of each section, respectively. The generalized rates of strain in these expressions can be related to stress resultants Q_α by a constitutive relation of the type developed in Chap. 1. If the stresses satisfy the equations of motion throughout the body and surface tractions on that part of the boundary where tractions are specified, the stress field is called *dynamically admissible*. A dynamically admissible set of forces in balance with accelerations $\ddot{W}_i^d$ is denoted as $\left(g_i^d, \rho\ddot{W}_i^d, Q_\alpha^d\right)$. If a dynamically admissible stress field also satisfies the yield condition, it is a *complete solution* only if the velocity field $\dot{W}_i^d$ is (1) kinematically admissible, (2) satisfies initial conditions and (3) gives deformations related to the stresses by the constitutive equations.

For quasistatic or slowly varying loads the accelerations are negligible so that Eq. (2.5) provides *equations of equilibrium*

$$\frac{\partial \sigma_{ij}}{\partial X_j} + g_i = 0, \qquad \sigma_{ij} = \sigma_{ji} \tag{2.7}$$

For infinitesimal in-plane deflections of a beam or bar, the equilibrium equations for stress resultants on cross-sections are as follows:

$$\frac{\partial N}{\partial X} + g_x = 0, \qquad \frac{\partial Q}{\partial X} + g_z = 0 \tag{2.8}$$

$$\frac{\partial T}{\partial X} = 0, \qquad \frac{\partial M}{\partial X} = -Q$$

A stress field $(g_i^s, \sigma_{ij}^s, Q_\alpha^s)$ that satisfies the equilibrium equations (2.7) in the interior of the body and surface tractions on any part of the boundary where contact forces are specified is called *statically admissible.* If a statically admissible stress field also satisfies the yield condition, the associated stress resultants cannot exceed the fully plastic stress resultants Q_α^p for a perfectly plastic solid, $Q_\alpha^s \le Q_\alpha^p$. This inequality is the cornerstone of a lower bound theorem for forces that induce quasistatic plastic deformation.

2.3 Principle of Virtual Velocity

A principle of virtual velocity (virtual power) is the dynamic counterpart of the principle of virtual work. For dynamic plasticity, the principle of virtual velocity and the requirement of positive dissipation by plastic deformation yield useful extremal principles for identifying dynamic modes and bounding dynamic solutions.

Here the principle of virtual velocity is established for a homogeneous body with volume V contained within a surface Γ. The body is loaded by prescribed surface tractions F_j and a body force g_i per unit volume. On the part of the surface Γ_s where surface tractions F_j are specified, the boundary condition is

$$F_j = n_i \sigma_{ij} \qquad \text{on} \quad \Gamma_s$$

where n_i are direction cosines of the unit outward normal vector $\mathbf{n}$. In the interior of the body the equation of motion (2.5) for a differential element relates the acceleration components $\partial^2 W_i / \partial t^2 \equiv \ddot{W}_i$ to dynamically admissible stresses σ_{ij}^d,

$$-\partial \sigma_{ij}^d / \partial X_j = -\rho_v \ddot{W}_i^d + g_i^d \tag{2.9}$$

A *virtual velocity* $\dot{W}_i^c$ is any kinematically admissible velocity field. By integrating the scalar product of $\dot{W}_i^c$ and Eq. (2.9) over the entire body V we obtain

$$-\int_V \frac{\partial \sigma_{ij}^d}{\partial X_j} \dot{W}_i^c \, dV = \int_V \left[-\rho_v \ddot{W}_i^d \dot{W}_i^c + g_i^d \dot{W}_i^c \right] dV \tag{2.10}$$

Notice that the virtual work rate by external tractions F_j is

$$\int_\Gamma F_j \dot{W}_j^c \, d\Gamma = \int_\Gamma n_i \sigma_{ij}^d \dot{W}_j^c \, d\Gamma = \int_V \frac{\partial(\sigma_{ij}^d \dot{W}_j^c)}{\partial X_i} dV \tag{2.11}$$

Then, since the stress tensor is symmetric

$$\sigma_{ij}^d \partial \dot{W}_i^c / \partial X_j = \sigma_{ij}^d \dot{\varepsilon}_{ij}^c$$

where the strain-rate tensor $\dot{\varepsilon}_{ij}^c = (\partial \dot{W}_i^c / \partial X_j + \partial \dot{W}_j^c / \partial X_i)/2$ is obtained from a kinematically admissible velocity field. Hence from Eqs (2.10) and (2.11) the *principle of virtual velocities* can be expressed as

$$\int_V \sigma_{ij}^d \dot{\varepsilon}_{ij}^c \, dV = \int_0^L Q_\alpha^d \dot{q}_\alpha^c \, dX = \int_V \left[-\rho_v \ddot{W}_i^d \dot{W}_i^c + g_i \dot{W}_i^c \right] dV + \int_\Gamma F_i \dot{W}_i^c \, d\Gamma \quad (2.12)$$

This principle relates any kinematically admissible velocity field $\dot{W}_i^c$ with any dynamically admissible stress field that is in balance with the applied forces. It is helpful to note that the principle does not require a complete solution. Equation (2.12) is most important for obtaining bounding inequalities for variables measuring system response to dynamic loads.

2.3.1 Rate of Change for Kinetic Energy of System

For any dynamically admissible stress field we can define a rate of work done on the system by external forces $\dot{E}_{in}^d$ and a *rate of dissipation of energy* by internal forces $\dot{D}^d$. These elements are defined as

$$\dot{E}_{in}^d \equiv \int_\Gamma F_i \dot{W}_i^d \, d\Gamma + \int_V g_i^d \dot{W}_i^d \, dV \quad \text{and} \quad \dot{D}^d \equiv \int_V \sigma_{ij}^d \dot{\varepsilon}_{ij}^d \, dV = \int_0^L Q_\alpha^d \dot{q}_\alpha^d \, dX$$

The dynamically admissible stress field is in balance with the rate of change of velocity $\dot{W}_i^d$ and this velocity field has kinetic energy K^d where

$$K^d \equiv \frac{1}{2} \int_V \rho_v \dot{W}_i^d \dot{W}_i^d \, dV$$

The rate of change of kinetic energy can be related to the rate of work done by external and internal forces if the dynamically admissible velocity field $\dot{W}_i^d$ is also kinematically admissible. (This is always true for the complete solution.) Then with the substitution $\dot{W}_i^d = \dot{W}_i^c$, the principle of virtual velocity (2.12) gives

$$\int_V \sigma_{ij}^d \dot{\varepsilon}_{ij}^d \, dV = \int_V \left[-\frac{\rho}{2} \frac{d\left(\dot{W}_i^d \dot{W}_i^d\right)}{dt} + g_i \dot{W}_i^d \right] dV + \int_\Gamma F_i \dot{W}_i^d \, d\Gamma \quad (2.13)$$

Finally, by substituting the deformations above into this expression of virtual velocity, we find that the dynamic solution satisfies a rate equation,

$$\dot{K}^d = \dot{E}_{in}^d - \dot{D}^d \quad (2.14)$$

The kinetic energy K^d increases if $\dot{E}_{in}^d > \dot{D}^d$ and decreases if $\dot{E}_{in}^d < \dot{D}^d$.

2.3.2 Rate of Change for Kinetic Energy of Kinematically Admissible Velocity Field $\dot{W}_i^c$

In regions that are plastically deforming the stresses are on the yield surface; elsewhere the state of stress is less than yield. For any kinematically admissible velocity field (KAVF), Drucker's postulate (1.31) gives $\sigma_{ij}^c \dot{\varepsilon}_{ij}^c \geq \sigma_{ij} \dot{\varepsilon}_{ij}^c$ where σ_{ij} is a *safe state of stress* that does not violate the yield criterion. Thus at any time the stresses at every point satisfy this inequality for the dissipation power or rate of dissipation. Noting that in plastically deforming regions the stresses σ_{ij}^c, Q_α^c are on the yield surface, we obtain an inequality,

$$\dot{D}^c = \int_V \sigma_{ij}^c \dot{\varepsilon}_{ij}^c \, dV \geq \int_V \sigma_{ij} \dot{\varepsilon}_{ij}^c \, dV \quad \text{or} \quad \dot{D}^c = \int_0^L Q_\alpha^c \dot{q}_\alpha^c \, dX \geq \int_0^L Q_\alpha \dot{q}_\alpha^c \, dX \quad (2.15)$$

Thus for any KAVF $\dot{W}_i^c$, the rate of plastic dissipation is greater than or equal to the rate of dissipation by a stress field that satisfies the yield condition, $\dot{D}^c \geq \dot{D}$. It is useful to recall that a statically admissible stress field σ_{ij} is safe.

The KAVF has kinetic energy K^c and a work rate by external forces $\dot{E}_{in}^c$, where

$$K^c = \frac{1}{2}\int_V \rho_v \dot{W}_i^c \dot{W}_i^c \, dV , \qquad \dot{E}_{in}^c = \int_\Gamma F_i \dot{W}_i^c \, d\Gamma + \int_V g_i \dot{W}_i^c \, dV \qquad (2.16)$$

Using a development similar to that for the principle of virtual velocity, the rate of work by external forces and 'safe' stresses acting on any KAVF can be defined as

$$\bar{J} = \int_V \sigma_{ij} \dot{\varepsilon}_{ij}^c \, dV - \int_\Gamma F_i \dot{W}_i^c \, d\Gamma - \int_V g_i \dot{W}_i^c \, dV \qquad (2.17)$$

Then, together with definitions (2.16), Drucker's postulate provides an inequality for rate of change of kinetic energy for any KAVF, $\dot{K}^c$. If a rigid–perfectly plastic body deforms while under the action of statically admissible applied loads, for any current velocity field $\dot{W}_i(X_j, t)$, the plastic dissipation exceeds the work rate of the active forces $\dot{D}^c > \dot{E}_{in}^c$ and so the kinetic energy K^c continually decreases. Equations (2.15) to (2.17) give the inequality

$$\bar{J} \leq -\dot{K}^c = \dot{D}^c - \dot{E}_{in}^c \qquad (2.18)$$

The complete solution is kinematically admissible but it has a rate of change of kinetic energy that is less than or equal to that of any other KAVF with the same kinetic energy K,

$$-\dot{K} \leq -\dot{K}^c \qquad (2.19)$$

Hence *as a rigid–perfectly plastic body decelerates, at every instant the velocity and deformation fields are evolving such that the rate of dissipation is a minimum consistent with satisfaction of the yield condition.*

This principle is sometimes used to calculate the location or extent of a deforming region in a rigid–plastic structure. The location of a hinge in a beam or the orientation of a hinge line in a plate is introduced as a variable; then an extremum of the global rate of dissipation gives the value of this variable that minimizes the rate of dissipation in the current configuration. This procedure should be used with caution, however, unless the state of stress is similar in a neighborhood of each deforming zone. Calladine [1985] has pointed to several examples of problems where this procedure does not provide the correct solution; his examples involve deformation fields with stretching in one part of the structure and bending in another part. In these problems, the stress approximations used in different regions are not equally accurate so the approximation 'error' can prejudice the analytical result.

2.3.3 Extremal Principles for Complete Solution

The principle of virtual velocity is the basis of two theorems for velocities in dynamically loaded rigid–perfectly plastic bodies. The theorem of Tamuzh [1962] is a kinematic principle while that of Martin [1972] is a dynamic principle. Each defines a functional that is an extremum (stationary value) for the actual response at any stage of motion.

***Tamuzh's Principle*:** For any KAVF $\dot{W}_i^c$ the corresponding accelerations $\ddot{W}_i^c$ and rates of change of strain-rate $\ddot{\varepsilon}_{ij}^c$ can be calculated. Also in those parts of the body that are deforming the KAVF can be obtained from the flow rule for a perfectly plastic material. Let T_i be any tractions that act on the free surface of the body during motion. Among all KAVFs $\ddot{W}_i^c$ (and corresponding stresses σ_{ij}^c), the actual acceleration and corresponding stresses result in an absolute minimum for the functional

$$J\left(\ddot{W}_i^c\right) = \frac{1}{2}\int_V \rho_v \ddot{W}_i^c \ddot{W}_i^c \, dV - \int_\Gamma T_i \ddot{W}_i^c \, d\Gamma + \int_V \sigma_{ij}^c \ddot{\varepsilon}_{ij}^c \, dV \tag{2.20}$$

If the acceleration field $\ddot{W}_i^c$ contains discontinuities, there are additional terms in the functional due to work done by stresses acting on the discontinuous surface (see Tamuzh [1962]). Tamuzh's principle is in fact an application of Gauss's principle of least constraint to a rigid–perfectly plastic body. This principle is discussed by Pars [1965, p 42].

***Martin's Principle*:** This theorem applies to any dynamically admissible acceleration field $\ddot{W}_i^d$ that is related through the laws of motion to a 'safe' stress field σ_{ij}^d (i.e. the stresses satisfy the yield criterion $\Psi(\sigma_{ij}^d) \le 0$). For any current velocity $\dot{W}_i$, the actual acceleration $\ddot{W}_i$ results in a minimum for the functional

$$J\left(\ddot{W}_i^d\right) = \int_V \rho_v \ddot{W}_i^d \dot{W}_i \, dV \tag{2.21}$$

In rigid–plastic structural dynamics, applications of this principle begin by finding dynamically admissible accelerations for some KAVF; i.e. the laws of motion are used to find accelerations $\ddot{W}_i^d$ for 'safe' stresses in plastically deforming regions. For this KAVF, these stresses can be obtained from the flow rule.

2.4 Bounds for Rigid–Perfectly Plastic Solids and Structures

For either static or dynamic loading of more complex elastoplastic structures, a complete solution may be beyond our grasp. In this case upper and lower bounds for response variables can be useful. These bounds are available for rigid–perfectly plastic materials since in this case all energy dissipation occurs in plastically deforming regions but in these regions there is no direct relationship between stress and deformation. The bounds are especially valuable if the upper and lower bounds are not far apart.

2.4.1 Upper and Lower Bounds on Static Collapse Force

For any particular distribution of applied force, the *plastic collapse force* is the smallest magnitude of load that results in a plastic collapse mechanism; i.e. the smallest force magnitude F_i in equilibrium with stresses Q_α that are conjugate to an increment of true plastic strain δq_α^p. A lower bound for the plastic collapse force of a perfectly plastic body is obtained from the principle of virtual work. This principle of statics is just the principle of virtual velocity (2.12) with negligible acceleration. Thus for an increment of deformation δq_α^p in a collapse mechanism, the difference between the work done by the actual and any statically admissible force gives

$$\int_\Gamma \left(F_i - F_i^s\right) \delta W_i \, d\Gamma = \int_S \left(Q_\alpha - Q_\alpha^s\right) \delta q_\alpha^p \, dS = \int_V \left(\sigma_{ij} - \sigma_{ij}^s\right) \delta \varepsilon_{ij}^p \, dV \geq 0 \qquad (2.22)$$

where the inequality comes from the condition of positive plastic dissipation or Drucker's postulate (1.31). Hence

$$\int_\Gamma F_i \delta W_i \, d\Gamma \geq \int_\Gamma F_i^s \delta W_i \, d\Gamma \qquad (2.23)$$

During any increment of displacement in a plastic collapse mechanism, the work done by forces in equilibrium with a statically admissible stress field is a *lower bound* for the work done by the collapse force F_i during this same increment of displacement. Recall that statically admissible stresses satisfy equilibrium, yield and boundary conditions, and notice that applications of this principle require statically admissible stresses only in those regions or on surfaces that are plastically deforming.

The complementary minimum principle considers work done by active forces during any kinematically admissible increment of displacement δW_i^c. Any kinematically admissible displacement field gives stresses Q_α^c and consequently, an applied force F_i^c that is in equilibrium with these stresses. The difference between work done by force F_i^c and that done by the actual force F_i in any kinematically admissible increment of displacement is given by the *principle of virtual work*,

$$\int_\Gamma \left(F_i^c - F_i\right) \delta W_i^c \, d\Gamma = \int_S \left(Q_\alpha^c - Q_\alpha\right) \delta q_\alpha^c \, dS = \int_V \left(\sigma_{ij}^c - \sigma_{ij}\right) \delta \varepsilon_{ij}^c \, dV \geq 0 \qquad (2.24)$$

where the inequality comes from Drucker's postulate. Therefore,

$$\int_\Gamma F_i^c \delta W_i^c \, d\Gamma \geq \int_\Gamma F_i \delta W_i^c \, d\Gamma \qquad (2.25)$$

During any increment of displacement in a KAVF, the work done by forces F_i^c in balance with dynamically admissible stresses Q_α^c is an *upper bound* for the work done by the collapse force F_i during this same increment of displacement. The dynamically admissible stresses Q_α^c are obtained from the flow law and thus satisfy yield in plastically deforming sections. Forces F_i^c are in balance with these stresses for negligible acceleration but the stresses need not satisfy equations of equilibrium in the body.

Equations (2.23) and (2.25) are a lower and an upper bound, respectively, for the plastic collapse force. Additional details on bounds for strain hardening materials can be found in Martin [1975].

2.4.2 Lower Bound on Dynamic Response Period

For a KAVF $\dot{W}_i^c\left(X_j, t\right)$, let the rate of dissipation by internal forces be $\dot{D}^c$ where

$$\dot{D}^c = \int_S Q_\alpha^c \dot{q}_\alpha^c \, dS = \int_V \sigma_{ij}^c \dot{\varepsilon}_{ij}^c \, dV \geq \int_V \sigma_{ij} \dot{\varepsilon}_{ij}^c \, dV$$

Suppose the body is subjected to both an impulsive velocity field $\dot{W}_{oi} \equiv \dot{W}_i(X_j, 0)$, that is suddenly applied at $t = 0$ and also to statically admissible active forces $F_i^s(X_j, t)$. These statically admissible forces prolong the response period but they are sufficiently small so there is no increase in the kinetic energy of the body. Thus the total dissipation occurring during the response period t_f satisfies an inequality that comes from the principle of virtual velocity (2.12) and Drucker's postulate

$$\int_0^{t_f} \dot{D}^c \, dt \geq \int_\Gamma \int_0^{t_f} F_i \dot{W}_i \, dt \, d\Gamma - \int_V \int_0^{t_f} \rho_v \ddot{W}_i \dot{W}_i^c \, dt \, dV \tag{2.26}$$

Martin [1964, 1965] obtained a *lower bound for the period of deformation* t_f^- by considering a kinematically admissible field that is independent of time $\dot{W}_i^c(X_j, t) = \dot{W}_i^c(X_j)$. It follows that the actual response period t_f is greater than the lower bound t_f^-,

$$t_f^- \equiv \frac{1}{\dot{D}^c} \left\{ \int_\Gamma \dot{W}_i^c \int_0^{t_f} F_i \, dt \, d\Gamma + \int_V \rho_v \dot{W}_{oi} \dot{W}_i^c \, dV \right\} \tag{2.27}$$

This bound depends on the distribution of the assumed velocity field $\dot{W}_i^c(X_j)$ but is independent of the velocity magnitude. The interaction requires that the direction cosines of the actual velocity field remain constant during the deformation period. The lower bound (2.27) does not require the traction $F_i(X_j, t)$ to be a separable function of space and time (see Stronge [1983] and Kaliszky [1970]).

2.4.3 Upper Bound on Dynamic Response Period

The upper bound on duration of motion follows from properties of the *principal or fundamental mode* for a structure. A mode is a velocity distribution that is an extremal for the rate of change of kinetic energy of the structure. Among all KAVFs that have the same kinetic energy K, the principal mode provides a minimum for the rate of change of kinetic energy $-dK/dt$. Thus the principal mode is the compatible velocity distribution that gives the longest duration of response. An example of the method of obtaining the principal mode is given in Sect. 2.5.5. Let the velocity in the principal mode be denoted by $V^*(t)\,\phi_i(X_j)$ where $V^*(t)$ is the speed of some characteristic point in the structure. Then for deformation in the principal mode of a rigid–perfectly plastic structure, the rate of dissipation is a linear function of velocity and the rate of change of kinetic energy is a constant. Consequently, in a mode form solution the velocity decreases linearly with time,

$$V^*(t) = V^*(0)\left[1 - t/t_f^+\right] \tag{2.28}$$

where t_f^+ is the duration of motion for the modal solution. For tractions F_i that are independent of time, deformation in the principal mode from any initial kinetic energy $K_0^* = K_0$ takes place during the period t_f^+

$$K_0^* = \int_0^{t_f^+} \left\{ \int_0^L Q_\alpha^* \dot{q}_\alpha^*(X, t) \, dX - \int_V F_i V^*(t)\, \phi_i \, dV \right\} dt \tag{2.29a}$$

Equation (2.29a) together with (2.28) give

$$t_f^+ = 2K_o / \dot{K}_o^* \tag{2.29b}$$

where the negative of the *initial* rate of change of kinetic energy in the mode is

$$\dot{K}_o^* \equiv \int_0^L \overset{*}{Q}_\alpha \overset{*}{\dot{q}}_\alpha (X,0)\,\mathrm{d}X - \int_\Gamma F_i V^*(0)\,\phi_i(X_j)\,\mathrm{d}\Gamma$$

The mode response period t_f^+ is an *upper bound on response duration* t_f since at any kinetic energy K, the complete solution has a rate of change of kinetic energy $|\dot{K}_o| \geq |\dot{K}_o^*|$. The bound on the response period has proved useful in a displacement bound for rate–dependent materials (see Symonds and Chon [1975]). This bound together with the lower bound (2.27) bracket the response period; i.e. $t_f^- \leq t_f \leq t_f^+$.

2.4.4 Lower Bound on Final Displacement

Since the true velocity and stress fields $\dot{W}_i$, Q_α are dynamically admissible, they can be used in the principle of virtual velocities (2.12) to give

$$\int_0^L Q_\alpha \dot{q}_\alpha^c \,\mathrm{d}X = -\int_V \rho_v \ddot{W}_i \dot{W}_i^c \,\mathrm{d}V + \int_\Gamma F_i \dot{W}_i^c \,\mathrm{d}\Gamma \tag{2.30}$$

where body forces g_i are considered to be negligible. Let the true initial velocity distribution be $\dot{W}_{oi} \equiv \dot{W}_i(X_j,0)$. In order to obtain a lower bound for components of final displacement $\Delta_{fi} \equiv \max\left[W_i(X_j,t_f)\right]$, consider a kinematically admissible velocity distribution $\dot{W}_{oi}^c \equiv \dot{W}_i^c(X_j,0)$; i.e.

$$\dot{W}_i^c = \begin{cases} \dot{W}_{oi}^c(1-t/t_f^-), & t<t_f^- \\ 0, & t>t_f^- \end{cases} \tag{2.31}$$

where t_f^- is a lower bound on the response time (2.27). The KAVF $\dot{W}_i^c$ gives generalized strain-rates $\dot{q}_\alpha^c(X_j,t)$ and initial strain-rates $\dot{q}_{o\alpha}^c \equiv \dot{q}_\alpha^c(X_j,0)$. With this assumed velocity field, Eq. (2.30) can be integrated over the estimated response period t_f^-. After integration by parts and noting that $\dot{W}_i^c(X_j,t_f)=0$ for $t>t_f^-$ we obtain

$$\frac{1}{t_f^-}\int_V \rho_v \dot{W}_{oi}^c W_i(X_j,t_f)\,\mathrm{d}V = \int_\Gamma\int_0^{t_f^-} F_i \dot{W}_{oi}^c(1-t/t_f^-)\,\mathrm{d}t\,\mathrm{d}\Gamma + \int_V \rho_v \dot{W}_{oi} \dot{W}_{oi}^c \,\mathrm{d}V$$

$$- \int_0^L\int_0^{t_f^-} Q_\alpha \dot{q}_{o\alpha}^c (1-t/t_f^-)\,\mathrm{d}t\,\mathrm{d}X \tag{2.32}$$

The lower bound results from two types of approximations:

1. In plastically deforming regions of the assumed deformation field where $\dot{q}_\alpha^c > 0$, any difference between the true stress Q_α and stress Q_α^c on the yield surface introduces an 'error' that separates the bound from the true solution. This stress difference is more significant in the early phase of deformation because of the time–dependence of the assumed field, Eq. (2.31).
2. The mean–value theorem for integrals is used to replace the final displacement field by the maximum displacement at the instant when motion ceases. This is

assumed to be the largest final displacement since elastic unloading effects are neglected. In general the error associated with this approximation increases with the number of spatial coordinates required to specify the deformation. This is called the geometric error.

With these approximations Eq. (2.32) gives a *lower bound for the largest magnitude of each component of final displacement* Δ_{fi}

$$\Delta_{fi} \int_V \rho_v \dot{W}^c_{oi}\, dV \geq \int_V \rho_v W_i \dot{W}^c_{oi}\, dV \geq t_f^- \left\{ \int_\Gamma \int_0^{t_f^-} F_i \dot{W}^c_{oi} (1 - t/t_f^-)\, dt\, d\Gamma + \int_V \rho_v \dot{W}_{oi} \dot{W}^c_{oi}\, dV - \frac{t_f^- D^c}{2} \right\} \quad (2.33)$$

This bound was obtained by Morales [1972], Morales and Neville [1970] and clarified by Wierzbicki [1972] who commented on its usefulness with an arbitrary distribution of initial velocity. In order to obtain a bound for any component of displacement W_i, that component must be included in the assumed initial velocity field $\dot{W}^c_{oi}$. The largest lower bound is obtained for an assumed velocity and coordinate directions that result in a maximum for

$$\int_V \rho_v \dot{W}^c_{oi} W_i(X_j, t_f)\, dV$$

By using the lower bound on response time t_f^-, Stronge [1983] obtained a corresponding lower bound for the maximum displacement,

$$\Delta_{fi} \geq \Delta_{fi}^- \equiv \frac{t_f^-}{2\int_V \rho_v \dot{W}^c_{oi}\, dV} \left\{ \int_\Gamma \int_0^{t_f^-} F_i \dot{W}^c_{oi} \left(1 - \frac{2t}{t_f^-}\right) dt\, d\Gamma + \int_V \rho_v \dot{W}_{oi} \dot{W}^c_{oi}\, dV \right\} \quad (2.34)$$

Accuracy of this bound depends on the difference between the yield surface and the true stresses in those parts of the assumed velocity field that are plastically deforming. Accuracy of the bound also depends on the dimensionality of the deformation field. This geometric error generally increases with the number of coordinates needed to represent the deformation field (Stronge [1985]).

2.4.5 Upper Bound on Final Displacement

An upper bound for the final displacement at a point in section $\bar{X}$ is obtained from a statically admissible stress field. Suppose the static collapse force F_i^s acting at section $\bar{X}$ of the structure is in equilibrium with the statically admissible stress distribution $Q_\alpha^s(X_j)$. Then for the structure deforming in the true velocity field $\dot{W}_i(X_j, t)$, at any time the difference between the rates of work done by statically admissible and actual applied forces $F_i(X_j, t)$ results in

$$\int_0^t F_i^s \dot{W}_i(\bar{X}, t)\, dt = \int_\Gamma \int_0^t F_i \dot{W}_i\, dt\, d\Gamma - \int_V \frac{\rho_v}{2} (\dot{W}_i \dot{W}_i - \dot{W}_{oi} \dot{W}_{oi})\, dV - \int_0^L \int_0^t (Q_\alpha - Q_\alpha^s) \dot{q}_\alpha\, dt\, dX$$

In this expression the last term depends on the difference between the dynamic generalized stresses Q_α and the statically admissible stress field. Note that this term is positive for a rigid–perfectly plastic material. This stress difference in the actual plastically deforming regions $\dot{q}_\alpha > 0$ influences the accuracy of this bound. By neglecting this difference we obtain an *upper bound Δ_{fi}^{+} for the component of final displacement* at the point where F_i^s is applied. Finally, at time t_f when the velocity vanishes, $\dot{W}_i(X_j, t_f) = 0$, the final displacement at the point where F_i^s is applied $\Delta_{fi} = W_i(X, t_f)$ satisfies

$$\Delta_{fi} \leq \Delta_{fi}^{+} \equiv \frac{1}{F_i^s(X)} \left\{ \int_\Gamma \int_0^{t_f} F_i \dot{W}_i \, dt \, d\Gamma + \int_V \frac{\rho_V}{2} \dot{W}_{0i} \dot{W}_{0i} \, dV \right\} \tag{2.35}$$

The upper bound Δ_{fi}^{+} was first obtained by Martin [1964] for impulsive loading where $\dot{W}_{0i} > 0$ and $F_i(X, t) = 0$. For this case the bound states that during the dynamic response period t_f, the work done by the static collapse force F_i^s is less than the initial kinetic energy. Ploch and Wierzbicki [1975] extended the bound to incorporate large deformations. The complementary upper bound for $\dot{W}_{0i} = 0$ and $F_i(X, t) > 0$ has had limited application since it requires the complete solution for the period of nonzero surface tractions. Robinson [1970] has obtained an upper bound on final deflection of elastic–perfectly plastic structures but again, this requires the complete solution if the load is not impulsive.

As mentioned, the accuracy of this bound on final displacement depends on the difference between the assumed statically admissible stress distribution and the actual state of stress in the deforming regions. This difference is largest when significant energy is dissipated in a transient or 'moving hinge' phase of motion; on the other hand, if the loading is impulsive and most of the initial kinetic energy is dissipated in a stationary (modal) configuration, the upper bound on the largest displacement is fairly accurate.

***Example*: Bounds for Impulsively Loaded Cantilever** For impact or blast loading the duration of loading (surface tractions) is often brief in comparison with the time required for deformation waves to redistribute energy in the structure. This loading can be represented as an initial velocity field imposed over part of the structure; subsequently the surface tractions are negligible. This is termed *impulsive loading*. In this case the energy imparted to the structure by the loading is equal to the initial kinetic energy $K_0 = \frac{1}{2}\int \rho_V \dot{W}_{0i} \dot{W}_{0i} \, dV$.

Consider a rigid–perfectly plastic cantilever of length L, with a cross-section that has mass per unit length ρ and fully plastic moment M_p. There is a concentrated mass $G = \gamma \rho L$ that is attached at the tip of the cantilever; the mass G may represent a colliding body during the contact phase of motion. If the particle at the tip is given an initial transverse velocity V_0, the kinetic energy is initially $K_0 = G V_0^2 / 2$.

A reasonable KAVF for estimating the dynamic response can be simple in-plane rotation about a hinge at the root of the cantilever. For coordinates indicated in Fig. 2.2, this gives

$$\dot{U}^c = -(1 - X/L) L \dot{\theta}_B^c \qquad \text{and} \qquad \dot{D}^c = -M_p \dot{\theta}_B^c$$

where $\dot{\theta}_B(t) = V(t)/L$ denotes the rate of rotation at the root. It will be shown that this velocity distribution also happens to be the primary mode of plastic deformation, $\phi_i^c = (1 - X/L)$. If the initial kinetic energy of this KAVF is equal to the imposed kinetic energy $G V_0^2 / 2$, the initial angular speed of the KAVF is

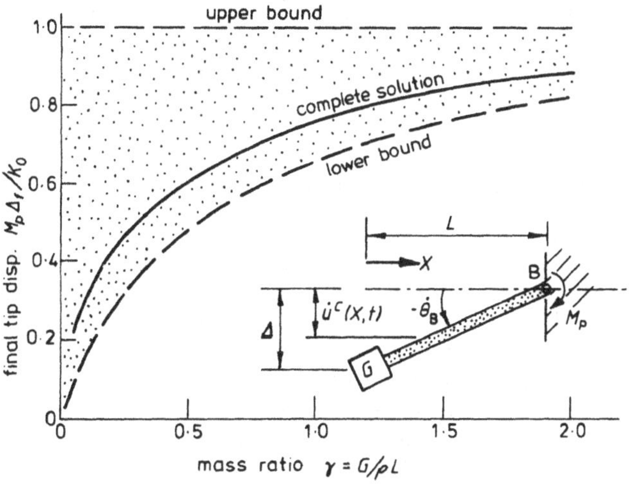

Fig. 2.2 Complete solution and bounds for final tip displacement of impulsively loaded cantilever as functions of the ratio of mass of striker at tip to mass of beam $\gamma = G/\rho L$.

$$\dot{\theta}_0^c = -\left[\frac{3GV_0^2}{\rho L^3(1+3\gamma)}\right]^{1/2}$$

Consequently, the response period is within the range

$$\frac{GLV_0}{M_p} \le t_f \le \frac{GLV_0}{M_p}\left[1+\frac{1}{3\gamma}\right]^{1/2}$$

The lower bound on final deflection at the tip Δ_f^- is obtained from the limit on response time t_f^- and the mode of deformation ϕ_i^c:

$$\Delta_f^- = \frac{t_f^- V_0}{2+\gamma^{-1}} = \frac{GLV_0^2\gamma}{M_p(1+2\gamma)}$$

The upper bound on final deflection at the tip is obtained by assuming that all the initial kinetic energy is absorbed by deformation in the static collapse mechanism of rotation about the root,

$$\Delta_f^+ = \frac{GLV_0^2}{2M_p}$$

The actual response of a rigid–plastic cantilever to impact at the tip involves a transient phase of motion followed by deformation in the primary mode. The part of the total kinetic energy dissipated in each of these phases depends on the mass ratio $\gamma = G/\rho L$. The analytical solution for final tip deflection Δ_f can be expressed as (see Sect. 4.5.4)

$$\Delta_f = \frac{GLV_0^2}{6M_p}\left[\frac{\gamma}{1+2\gamma} + 2\gamma\ln\left(\frac{1+2\gamma}{2\gamma}\right)\right]$$

This solution lies between the bounds, $\Delta_f^- \le \Delta_f \le \Delta_f^+$. If the cantilever has negligible mass in comparison with G, almost all of the energy is dissipated in modal deformation and both bounds converge to the analytical solution as shown in Fig. 2.2. For small mass ratios, however, the upper bound is not close to the actual deflection.

2.5 Dynamic Modes of Deformation

Modal solutions are of fundamental importance for explaining and estimating the deformation of structures that develop in response to impact or blast loading. For either rigid–perfectly plastic or rigid–strain hardening structures that are given a heavy blow over a segment of the structure, the deformation that develops divides into an initial transient phase where the pattern or location of deformation is continually changing and a modal phase where the pattern is constant. During the transient phase the pattern of deformation steadily evolves from the initial velocity distribution imposed at impact to a *mode configuration.* After attaining the velocity distribution of a stable mode configuration, the pattern of deformation remains constant for some period of time. In most cases a substantial part of the impact energy is dissipated in a mode configuration during the second phase of deformation.

2.5.1 Modal Solutions

Here we consider a rigid–perfectly plastic structure subjected to an initial velocity field $\dot{W}_{0i}(X_j)$ and time–independent surface tractions $F_i(X_j)$. Let $\dot{W}_i(X_j, t)$ be the velocity field that develops from $\dot{W}_{0i}$ and F_i . This velocity satisfies the field equations; i.e. the laws of motion, compatibility and constitutive relations, in addition to the initial condition $\dot{W}_i(X_j, 0) = \dot{W}_{0i}(X_j)$. In general, $\dot{W}_i$ is not a separable function of time and spatial coordinates.

A *modal solution* is a velocity field with separated functions for spatial and temporal variables

$$\dot{W}_i^*(X_j, t) = V^*(t)\, \phi_i^*(X_j) \tag{2.36}$$

which satisfies the field equations but not necessarily the initial condition. The function ϕ_i^* is called the *mode* or *mode shape*; the scalar velocity V^* of a characteristic point is specified such that $-1 \le \phi_i^* \le 1$. Notice that the modal velocity field $\dot{W}_i^*$ and the associated acceleration field $\ddot{W}_i^*$ have the same shape; i.e.

$$\ddot{W}_i^* = \dot{V}^* \phi_i^* \tag{2.37}$$

If no external forces act in the structure, mode shapes can be regarded as natural modes of dynamic plastic deformation since they are a property of the structure and independent of the initial velocity field. Modes exist also for structures subjected to statically admissible time–independent surface tractions F_i.

2.5.2 Properties of Modes

Impulsive loading on a rigid–plastic structure results in dynamic deformation that continually evolves towards a modal solution.

Proof for this principle of convergence to a mode was given by Martin and Symonds [1966]. Before convergence, the part of the initial kinetic energy dissipated in the transient phase increases with the difference between the initial momentum distribution and the mode shape. That is, convergence to the modal solution occurs at an early stage of dynamic response if the initial momentum distribution is not very different from a mode.

In order to find particular modes for a structure we employ the functional $J^*\left(\dot{W}_i^c\right)$ for the rate of change of kinetic energy of a kinematically admissible velocity field $\dot{W}_i^c$; i.e. the dissipation of power J^* for the kinetic energy $K^c = \int(\rho_v \dot{W}_i^c \dot{W}_i^c /2)\, \mathrm{d}V$ is given by,

$$J^*\left(\dot{W}_i^c\right) = \left\{\int_0^L Q_\alpha^c \dot{q}_\alpha^c \,\mathrm{d}X - \int_\Gamma F_i \dot{U}_i^c \,\mathrm{d}\Gamma\right\}\left(K^c\right)^{-1/2} \tag{2.38}$$

This is known as *Lee's functional.* It is used to identify dynamic modes for a rigid–plastic structure by means of the principle of minimum dissipation of specific power:[3]

> *Among all kinematically admissible velocity fields with the same kinetic energy, modal solutions will be extrema for the functional J^*.*

These extrema can be stationary maxima, minima or saddle points of J^*. Nonstationary extrema, however, also occur in rigid–plastic structures; these extrema occur at points where J^* is nonanalytic so the extrema do not require $\delta J^* = 0$ (see Symonds and Wierzbicki [1975] or Palomby and Stronge [1988]). Usually nonanalytic extrema correspond to nodes (i.e. plastic hinges) located at supports or other points of displacement constraint where the bending moment is large but shear force $Q \neq 0$.

There are multiple modes (in most cases an infinite number) for any structure with continuously distributed mass. They can be numbered in the same manner as elastic modes; for a rigid–plastic structure this number increases with the number of plastic hinges. *Stable modes* are local minima for J^* while maxima and saddle points are unstable. If the momentum distribution is very close to that of an unstable mode, initially the complete solution for the velocity field can converge to the nearby unstable mode. Nevertheless, the final stage of deformation is always in a stable mode unless the initial velocity distribution is identical to an unstable mode (see Martin [1983] and Martin and Lloyd [1983]). Among the stable modes there is one that gives the minimum specific dissipation of power J^*; this basic mechanism of rigid–plastic deformation is called the *primary* or fundamental *mode.*

To sum up, dynamic modes for rigid–plastic structures can be characterized as follows:

1. Mode shapes are properties of both the structure and any statically admissible steady load; thus modal velocity distributions ϕ_i^* are kinematically admissible and independent of time.
2. The velocity and acceleration fields of a modal solution are geometrically similar.

---

[3]The kinematic extremal principle for identifying modes was presented by Lee and Martin [1970] and Lee [1972]. Later Martin [1981] described a systematic method for finding modes of any rigid–plastic structure.

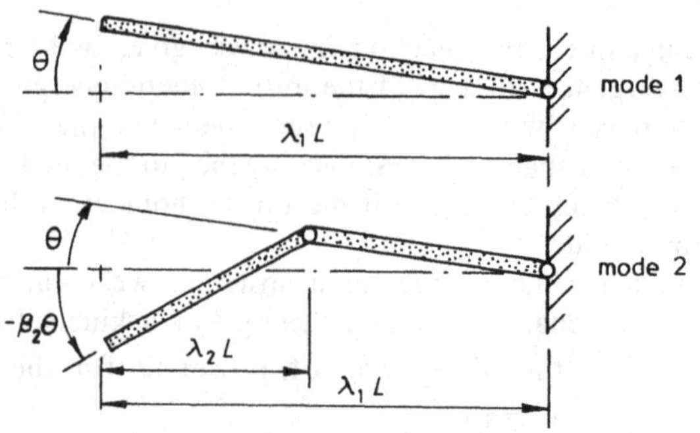

Fig. 2.3 First and second dynamic modes of deformation for rigid–perfectly plastic cantilever. The mode number increases with the number of plastic hinges. For a mode, the location of these hinges λ_i and their relative rotation β_i give an extremum for Lee's functional $J^*(\lambda_i, \beta_i)$.

3. A modal solution has a rate of energy dissipation that increases linearly with velocity; hence, during any period wherein time dependent forces are negligible, the velocities decrease linearly with time.

***Example*: Dynamic Modes for Rigid-Plastic Cantilever** The modes for a rigid–plastic structure are found by considering a mechanism of deformation with arbitrarily located plastic hinges; the number of degrees of freedom for the mechanism are determined by the number of hinges. The location and relative rates of rotation at these hinges are obtained from Lee's functional (2.38) evaluated with the velocity field of the mechanism. For the uniform cantilever shown in Fig. 2.3, the first mode is a mechanism with a single hinge at some distance $\lambda_1 L$, from the tip. The KAVF corresponding to a single hinge mechanism gives a minimum value for the specific dissipation power J^* if the hinge is at the root of the cantilever. This result is a nonanalytic extremal (i.e. $\partial J^*/\partial\lambda_1 \neq 0$) since the solution is at the boundary of the admissible region, $0 < \lambda_1 \leq 1$.

The second mode has two hinges. In order to identify the second mode, the shape parameters λ_1, λ_2 and β for the mechanism illustrated in Fig. 2.3 are found that provide an extremal for J^*. The mechanism has hinges located at $X = \lambda_1 L$ and $X = \lambda_2 L$ with rotation rates $\dot\theta$ and $\beta\dot\theta$, respectively. This mechanism has a transverse velocity distribution $\dot W_c$ where

$$\dot W_c = \begin{cases} 0 & \lambda_1 L < X \\ (\lambda_1 L - X)\dot\theta & \lambda_2 L < X < \lambda_1 L \\ (\lambda_1 L - \lambda_2 L)\dot\theta + (\lambda_2 L - X)\beta\dot\theta & X < \lambda_2 L \end{cases}$$

Consequently, for a uniform mass per unit length ρ, the kinetic energy K^c of the mechanism can be expressed as the sum of translational and rotational parts for each of the two segments; i.e.

$$K^c = \frac{\rho L^3 \dot\theta^2}{2}\left\{\frac{(\lambda_1 - \lambda_2)^3}{3} + \lambda_2\left(\lambda_1 - \lambda_2 + \frac{\beta\lambda_2}{2}\right)^2 + \frac{\beta^2\lambda_2^3}{12}\right\}$$

Likewise, for a rigid–plastic beam with yield moment M_p, the rate of dissipation for the two hinge mechanism can be expressed as

Table 2.1 Dynamic modes for uniform cantilever

Mode	Type of extremum	λ_1	λ_2	λ_3	β_2	β_3	$J^* \sqrt{\rho L^3} / M_p$
1	Nonanalytic min. (stable)	1.0	–	–	–	–	$\sqrt{6}$
2	Saddle point (unstable)	1.0	0.50	–	–2.00	–	$8\sqrt{6}$
3	Saddle point (unstable)	1.0	0.71	0.29	–1.41	2.0	$8\sqrt{6}(3\sqrt{2}-4)^{-1}$

$$\dot{D}^c = \begin{cases} \beta M_p \dot{\theta} & \beta > 1 \\ (2-\beta) M_p \dot{\theta} & \beta < 1 \end{cases}$$

These elements of the specific dissipation power $J^* = \dot{D}^c / \sqrt{K^c}$ give an extremum at a saddle point $\lambda_1 = 1$, $\lambda_2 = 0.5$ and $\beta = -2.0$. There are no other analytic extremes of J^* for two hinge mechanisms. Shapes of the first, second and third modes are listed in Table 2.1.[4] The first mode has the smallest specific dissipation power of any stable mode so it is the primary mode of deformation.

2.5.3 Mode Approximations for Structural Response to Impulsive Loading

The dynamic deformation of an impulsively loaded rigid–plastic structure generally involves an initial transient period of deformation where plastic 'hinges' travel away from points where the initial velocity field is discontinuous; these velocity discontinuities can be located at supports where displacement is constrained. During the transient period the hinges move such that the velocity distribution evolves towards a mode shape. After the velocity distribution converges to a mode, then deformation continues in the mode configuration. The energy dissipated in the modal phase of deformation can be a large part of the initial energy imparted to the structure; this is especially true if the structure is light but it contains small segments of heavy mass. In this case a modal solution can be a good approximation for the dynamic deformation of the structure.

'Best' Mode for Approximating Dynamic Deformation A continuous structure has a large number of dynamic modes of plastic deformation. The structure almost always finally deforms in the primary mode; therefore it is tempting to use this mode to approximate the structural response to impulsive loading. Lepik [1980] showed however, that there are initial velocity distributions that are close to the shape of higher modes (unstable extrema) which yield more accurate estimates for final displacements of rigid–plastic beams. The higher mode shape is a better choice for use in the modal approximation technique when most of the impact energy is dissipated in this mode before the velocity distribution finally switches to the primary mode. But what identifies the mode which best represents the transient velocity field resulting from impulsive loading?

[4]In obtaining hinge locations for a higher mode, it is helpful to recognize that no shear force is present at perfectly plastic hinges located away from displacement constraints. If a mode is calculated for no external force acting on the structure and no shear force is present at the hinges, then there is no acceleration of the center–of–mass of a rigid segment. In a modal velocity field, each interior segment simply rotates about it's own center–of–mass.

Symonds [1980] examined several possible criteria for identifying the mode that gives the best approximation for dynamic deformation of impulsively loaded rigid–plastic structures. He stated the following conjecture for selecting the mode:

Generally the most accurate modal approximation for final shape of the deformed structure is given by the mode that gives the largest lower bound for dynamic response period.

The bound on response period is expressed by Eq. (2.27); for impulsive loading it depends on the difference between the initial momentum distribution and the mode shape.

Slutsky et al. [1982] tested this conjecture for impulsively loaded beams and portal frames. They speculated that the largest lower bound on response time is also the criterion that gives the best mode for structures composed of rigid–viscoplastic or homogeneous viscoplastic materials.

Initial Velocity of Modal Solution A measure of the difference between the initial modal velocity field $V_0^* \phi_i^*(X_j)$ and the impulsively applied velocity $\dot{W}_{0i} \equiv \dot{W}_i(X_j, 0)$ is given by the functional $\tilde{\Delta}_0$,

$$\tilde{\Delta}_0 \equiv \frac{1}{2}\int \rho_v \left(\dot{W}_{0i} - V_0^* \phi_i^*\right)\left(\dot{W}_{0i} - V_0^* \phi_i^*\right) \mathrm{d}V$$

Since both the modal approximation and the actual velocity field are complete solutions (for different but 'equivalent' initial conditions), the difference $\tilde{\Delta}(t)$ between the modal approximation and the actual velocities is continually decreasing, $\mathrm{d}\tilde{\Delta}/\mathrm{d}t \le 0$. The initial difference $\tilde{\Delta}_0$ results from the mode shape not being geometrically similar to the initial velocity distribution. The initial speed of the modal solution V_0^* that minimizes this difference is given by $\mathrm{d}\tilde{\Delta}_0/\mathrm{d}V_0^* = 0$; thus

$$V_0^* = \frac{\int_V \rho_v \dot{W}_{0i} \phi_i^* \,\mathrm{d}V}{\int_V \rho_v \phi_i^* \phi_i^* \,\mathrm{d}V}$$

This initial speed for the modal velocity gives an initial kinetic energy for the modal solution that is less than that of the imposed velocity field, $K_0^* \le K_0$. Nevertheless, the approximate and actual velocities are often identical during the final phase of motion if the 'best' mode is the primary mode. This method of obtaining the initial modal velocity results in a best estimate rather than an upper or lower bound for the deformation. Alternatively, if the modal field is given the actual initial kinetic energy $K_0^* = K_0$, the approximation gives an upper bound on the final deflection Δ_f^+.

References

Calladine, C. [1985]. Analysis of large plastic deformation in shell structures. *Inelastic Behaviour of Plates and Shells* (ed. L. Gevilacqua, R. Feijoo et al.) IUTAM Sym., Rio de Janeiro. 69–101.

Kaliszky, S. [1970]. Approximate solutions for dynamically loaded inelastic structures and continua. *Int. J. Nonlinear Mech.* **5**, 143–158.

Kaliszky, S. [1985]. Dynamic plastic response of structures. *Plasticity Today* (ed. A. Sawczuk and G. Bianchi), Elsevier, London. 787–820.

Lee, L.S.S. [1972]. Mode responses of dynamically loaded structures. *J. Appl. Mech.* **39**, 904–910.

Lee, L.S.S. and Martin, J.B. [1970]. Approximate solutions of impulsively loaded structures of a rate sensitive material. *J. Appl. Math. Phys. (ZAMP)* **21**, 1011–1032.

Lepik, U. [1980]. On the dynamic response of rigid–plastic beams. *J. Struct. Mech.* **8**, 227–235.

Martin, J.B. [1964]. Impulsive loading theorems for rigid–plastic continua. *ASCE J. Engng. Mech. Div.* **90**, 27–42.

Martin, J.B. [1965]. Displacement bound principle for inelastic continua subjected to certain classes of dynamic loading. *J. Appl. Mech.* **31**, 1–6.

Martin, J.B. [1972]. Extremum principles for a class of dynamic rigid–plastic problems. *Int. J. Solids Structures* **8**, 1185–1204.

Martin, J.B. [1975]. *Plasticity: Fundamentals and General Problems.* M.I.T. Press, 386–403.

Martin, J.B. [1981]. The determination of mode shapes for dynamically loaded rigid–plastic structures. *Meccanica* **16**, 42–46.

Martin, J.B. [1983]. Convergence to mode form solutions in impulsively loaded piecewise linear rigid–plastic structures. *Int. J. Impact Engng.* **1**, 125–142.

Martin. J.B. and Lloyd, A.R. [1983]. Convergence to higher symmetric modes in impulsively loaded rigid plastic structures. *Int. J. Impact Engng.* **1**, 143–156.

Martin, J.B. and Symonds, P.S. [1966]. Mode approximations for impulsively loaded rigid–plastic structures. *ASCE J. Engng. Mech. Div.* (EM5), **92**, 43–46.

Morales, W.J. [1972]. Displacement bounds for blast loaded structures. *ASCE J. Engng. Mech. Div.* (EM4) **98**, 965–974.

Morales, W.J. and Neville, G.E. [1970]. Lower bounds on deformations of dynamically loaded rigid–plastic continua. *AIAA J.* **8**, 2043–2046.

Palomby, C. and Stronge, W.J. [1988]. Evolutionary modes for large deflections of dynamically loaded rigid–plastic structures. *Mech. Struct. Mach.* **16**(1), 53–80.

Pars, A. [1965]. *Treatise on Analytical Dynamics* Heinemann, London.

Ploch, J. and Wierzbicki, T. [1975]. On an extremum principle for mode form solutions of dynamically loaded continua and structures. *Int. J. Solids Structures* **17**, 630–640.

Robinson, D.N. [1970]. Displacement bound principle for elastic–plastic structures subjected to blast loading. *J. Mech. Phys. Solids* **18**, 65–80.

Slutsky, S., Chon, C.T. and Yeung, K.S. [1982]. On the 'best' mode form in the mode approximation technique using the finite element method. *J. Struct. Mech.* **10**, 117–131.

Stronge, W.J. [1983]. Lower bounds to large displacements of impulsively loaded rigid-plastic structures. *Int. J. Solids Structures* **19**, 1049–1063.

Stronge, W.J. [1985]. Accuracy of bounds on plastic deformation for dynamically loaded plates and shells. *Int. J. Solids Structures* **27**, 97–104.

Symonds, P.S. [1980]. The optimal mode in the mode approximation technique. *Mech. Res. Comm,* **7**, 1–6.

Symonds, P.S. [1984]. Twenty years of developments in inelastic dynamics. *ASCE Conference.*

Symonds, P.S. and Chon, C.T. [1975]. Bounds for finite deflections of impulsively loaded structures with time dependent plastic behaviour. *Int. J. Solids Structures* **11**, 403–423.

Symonds, P.S. and Wierzbicki, T. [1975]. On an extremum principle for mode form solutions in plastic structural dynamics. *J. Appl. Mech.* **42**, 630–635.

Tamuzh, V.P. [1962]. On a minimum principle in the dynamics on rigid–plastic bodies. *Prikladnaya Matematika Mekhanika* **26**, 715–722.

Van, The Vu and Sawczuk, A. [1983]. Lower bounds to large displacements of impulsively loaded plastically orthotropic structures. *Int. J. Solids Struct.* **19**, 189–205.

Wierzbicki, T. [1972]. Comment on lower bounds on deformations of dynamically loaded rigid–plastic continua. *A.I.A.A. J.* **10**, 364–365.

Chapter 3

Static Deflection

Flexural deformation in a slender structural element depends on the axial distribution of the bending moment and the flexural rigidity. Here a basic quasistatic loading on a uniform cantilever will be used to illustrate effects of nonlinear material behavior in plastically deforming structures. A cantilever with a transverse force at the tip has a bending moment that increases linearly with distance from the tip. The cantilever is wholly elastic and has linear behavior if the largest bending moment (at the root) is less than the yield moment. However, if the force is large enough there are sections in an elastic cantilever where the moment would be larger than the yield moment M_Y. At these sections the outer fibers of the cantilever are stretched or compressed beyond the yield strain ε_Y. In an elastic–perfectly plastic cantilever the stress in these fibers is limited to the yield stress Y; consequently, the flexural rigidity of the section decreases with increasing curvature. Plastic deformation in the outer fibers is irreversible, so sections that are partly plastic have been permanently deformed; if these sections are unloaded they do not return to their initial curvature.

This chapter focuses on the development of plasticity in a simple structure composed of elastic–plastic material when the structure is subjected to slowly increasing load; the effects of elastic and plastic deformation on deflections of the structure are described.

3.1 Small Elastic–Plastic Deflections

In this section, small static deflections of a cantilever are analyzed. Changes in inclination of sections are assumed to remain negligibly small; consequently, the distance of each section from the line of action of any transverse load remains constant. Also, the bending moment does not change due to rotation θ of the neutral axis. Mathematically this implies $\cos\theta \approx 1$ and $\sin\theta \approx \theta$. The simplest loading case is a transverse force F applied at the tip of a uniform cantilever of length L, as shown in Fig. 3.1. Deflections of sections of the cantilever can be determined as a function of an axial coordinate X measured from the tip. Transverse deflection $W(X)$ of the neutral axis depends on the force and properties of the

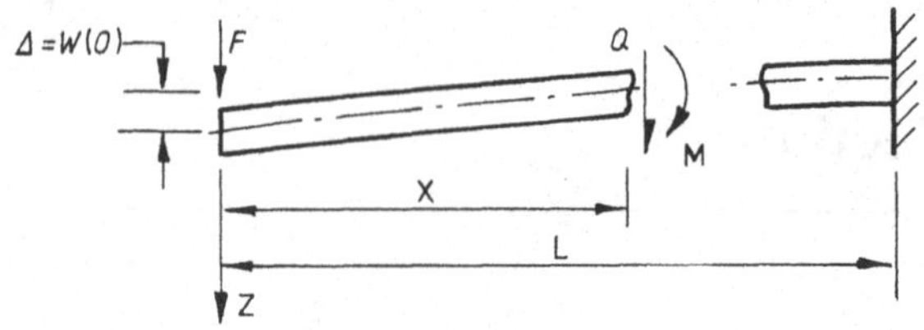

Fig. 3.1. Transverse force F at tip of cantilever.

cantilever. At any section the shear force $Q(X)$ and the bending moment $M(X)$ are given by

$$Q(X) = -F \tag{3.1}$$

$$M(X) = FX \tag{3.2}$$

The positive sense of these variables is indicated in Fig. 3.1. With this load the shear force is uniform and the bending moment increases linearly with distance from the tip so the largest moment occurs at the root where $X = L$.

In order to demonstrate the development of rate–independent plasticity with increasing load, it is useful to relate the applied force F to the largest force that can be supported by the structure if the material is perfectly plastic; this largest force is termed the *perfectly plastic collapse force* F_c. Plastic collapse of a structure occurs if there is a sufficient number of fully plastic hinges for the structure to become a mechanism; at each plastic hinge the magnitude of the bending moment equals the fully plastic bending moment M_p. For a cantilever, a single plastic hinge at any location results in a mechanism. With a transverse force applied at the tip of a uniform cantilever, the plastic collapse force F_c is the force magnitude that results in the moment M_p at the root; i.e.

$$F_c = \frac{M_p}{L} \tag{3.3}$$

Recollect from Table 1.1 that the ratios of fully plastic to yield moments for rectangular and circular cross-sections are 1.5 and 1.7, respectively. Since this transverse force results in only a single hinge at the root of the cantilever, the ratio of plastic collapse force to yield force is $F_c/F_Y = M_p/M_Y = \phi_m$, where ϕ_m is the shape factor of the cross-section for bending defined in Eq. (1.11). Hence, this ratio of forces is a structural characteristic that depends on the shape of cross-section.

3.1.1 Elastic Deflections

For small inclination of the neutral axis, the curvature is given by $\kappa = \kappa(X) \approx \mathrm{d}^2W/\mathrm{d}X^2$; if $|\kappa| \le \kappa_Y$ the outer fiber of the section remains elastic and

$$\frac{\mathrm{d}^2W}{\mathrm{d}X^2} \approx \kappa = \frac{M}{EI_0} \tag{3.4}$$

where EI_0 is the elastic flexural rigidity. By using the moment distribution specified by Eq. (3.2), the differential equation can be written as

$$\frac{d^2W}{dX^2} = \frac{FX}{EI_o} \tag{3.5}$$

and the boundary conditions of zero displacement and inclination at the root provide

$$W(L) = \frac{dW(L)}{dX} = 0 \tag{3.6}$$

At this point it is convenient to cast the problem in terms of a set of nondimensional variables and structural parameters which are defined as follows:

$$x = \frac{X}{L}, \qquad w = w(x) = \frac{W}{L}, \qquad f = \frac{F}{F_c} \quad \text{and} \quad \mu = \frac{F_c L^2}{EI_o} \tag{3.7}$$

In general the flexibility parameter μ is given by $\mu = (M_p / M_Y)L\kappa_Y = \phi_m L\kappa_Y$; for a rectangular cross-section with depth h, $\mu = 3L\kappa_Y/2 = 3(Y/E)(L/h)$. For forces that are similar in magnitude to F_c the limitation of this analysis to small inclinations implies that the flexibility parameter is in the range $0 < \mu < 1$; this parameter approaches unity only if the cantilever is extremely slender.

For compactness, denote differentiation with respect to the spatial variable by $(\,)' \equiv d(\,)/dx$. Thus, using nondimensional variables the differential equation (3.5) can be written as

$$w'' = f\mu x \tag{3.8}$$

If the cantilever is entirely elastic the elastic region extends to the root at $x/L = 1$ where displacement and rotation vanish, $w(1) = w'(1) = 0$. Hence we obtain

$$w' = \frac{f\mu}{2}(x^2 - 1), \qquad 0 \le x \le 1 \tag{3.9a}$$

$$w = \frac{f\mu}{6}(x^3 - 3x + 2), \qquad 0 \le x \le 1 \tag{3.9b}$$

In general, however, the elastic region will extend through only a segment near the tip of the cantilever. Suppose the cantilever is elastic in a region $0 \le x \le \chi$ and that at the limit of this region the neutral axis has inclination $w'(\chi)$ and deflection $w(\chi)$. Then the inclination and deflection in the elastic region can be expressed as

$$w'(x) = \frac{f\mu}{2}(x^2 - \chi^2) + w'(\chi), \qquad 0 \le x \le \chi \tag{3.10a}$$

$$w(x) = \frac{f\mu}{6}(x^3 - 3x\chi^2 + 2\chi^3) + (x - \chi)w'(\chi) + w(\chi), \qquad 0 \le x \le \chi \tag{3.10b}$$

The length χ of the elastic region depends on the tip force f and the flexibility of the section; thus the value of χ is determined by the curvature that causes yield

$$\chi = \frac{M_Y}{fM_p} = \frac{1}{f\phi_m} \le 1 \tag{3.11}$$

When $f = 1$ (i.e. $F = F_c$), the length of the elastic region reaches its maximum value, $\chi_{max} = 1/\phi_m$, which depends on the cross-section only. For instance, $\chi_{max} = 2/3$ and 0.589 for beams of rectangular and circular cross-sections, respectively.

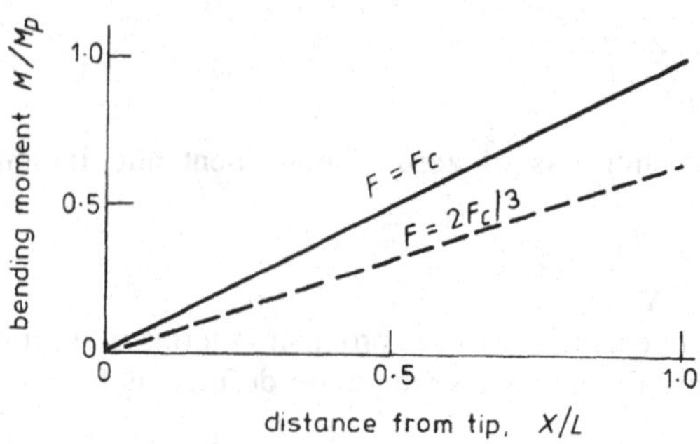

Fig. 3.2. Bending moment distribution in cantilever with rectangular cross-section for initial yielding – – – –, and fully plastic collapse ———.

3.1.2 Deflection of Elastic–Perfectly Plastic Cantilever

If the curvature of the loaded cantilever exceeds the yield curvature in a segment near the root, the moment–curvature relation (3.4) is not valid. In this case, the appropriate elastic–plastic moment–curvature relation depends on the cross-section of the cantilever. The analysis in this section is based on a rectangular cross-section. Given a rectangular cross-section, the curvature is related to the bending moment by the relation

$$\frac{\kappa}{\kappa_Y} = \left(3 - 2\frac{M}{M_Y}\right)^{-1/2} \tag{3.12}$$

With a transverse tip force, the bending moment increases linearly with distance from the tip, $M = FX$, as shown in Fig. 3.2. For a cantilever of rectangular cross-section $F_c = 3M_Y/2L$ and Eq. (3.12) gives

$$\frac{d^2W}{dX^2} = \kappa_Y\left[3\left(1 - \frac{FX}{F_cL}\right)\right]^{-1/2}$$

Since $L\kappa_Y = 2\mu/3$, this equation for elastoplastic curvature has a nondimensional form

$$w'' = \frac{2\mu}{3\sqrt{3}}(1 - fx)^{-1/2}, \qquad \frac{2}{3f} \le x \le 1 \tag{3.13}$$

For boundary conditions of vanishing deflection and inclination at the fixed end, $w(1) = w'(1) = 0$, integration of (3.13) gives

$$w'(x) = \frac{4\mu}{3\sqrt{3}f}\left[(1 - f)^{1/2} - (1 - fx)^{1/2}\right] \tag{3.14a}$$

$$w(x) = \frac{4\mu}{9\sqrt{3}f^2}\left[(3fx - f - 2)(1 - f)^{1/2} + 2(1 - fx)^{3/2}\right] \tag{3.14b}$$

This solution applies to the elastic–plastic region at the root; the region terminates at $x = \chi$ where plasticity vanishes. For a rectangular cross-section Eq. (3.11) gives $\chi = 2/3f$.

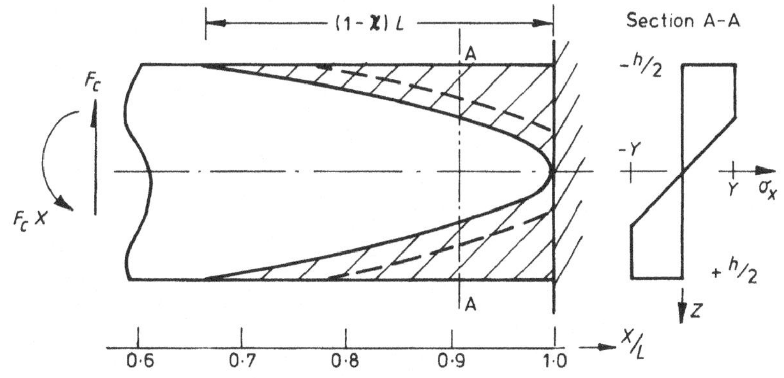

Fig. 3.3. Plastically deforming region at root of an elastic-perfectly plastic cantilever in case of F/F_c = 1.0 ——; and $F/F_c = 0.96$ - - - -.

The inclination and deflection of the neutral axis at the interface between elastic and elastic–plastic regions are obtained from Eqs (3.14a) and (3.14b); i.e.

$$w'(\chi) = \frac{4\mu}{3\sqrt{3}f}\left[(1-f)^{1/2} - \frac{1}{\sqrt{3}}\right] \tag{3.15a}$$

$$w(\chi) = \frac{4\mu}{9\sqrt{3}f^2}\left[\frac{2}{3\sqrt{3}} - f(1-f)^{1/2}\right] \tag{3.15b}$$

For tip forces $F \le 2F_c/3$, the bending moment $M(X)$ results in curvatures that are elastic at every section and deflections that are given by Eq. (3.9b). For values of force in the range $2F_c/3 < F < F_c$, the curvature near the root exceeds the yield curvature κ_Y throughout a segment of length $(1-\chi)L$. The stress distribution in this elastic–plastic segment is illustrated in Fig. 3.3. In this segment the outer fibers of the beam are strained beyond yield while the central core remains elastic. Figure 3.4a shows the curvature distributions for the yield force f = 2/3, and several larger loads that cause elastic–plastic behavior in a segment near the root. For the largest force displayed in Fig. 3.4, $F = 0.96\,F_c$, the plastic region has spread through 65% of the thickness at the root where the moment $M(L) = 0.96\,M_p$; only the central core of the cross-section remains elastic. The increasing curvature in the elastic–plastic region near the root is not especially large until the tip force is almost equal to the plastic collapse force F_c; consequently after the elastic–plastic segment develops, changes in deflection with increasing force are only slightly more rapid; this is shown in Fig. 3.4b.

3.1.3 Deflection of Elastic–Linear Strain Hardening Cantilever

Most materials that yield at a specific state of stress require an increase in stress to increase plastic strain after yield commences; this material behavior is termed *plastic strain hardening*. The stress strain relation for an elastic–linearly strain hardening material is illustrated in Fig 1.12. Heavily loaded structures composed of strain hardening materials develop sections where the bending moment exceeds the yield moment M_Y; in these sections the strains are partly elastic and partly

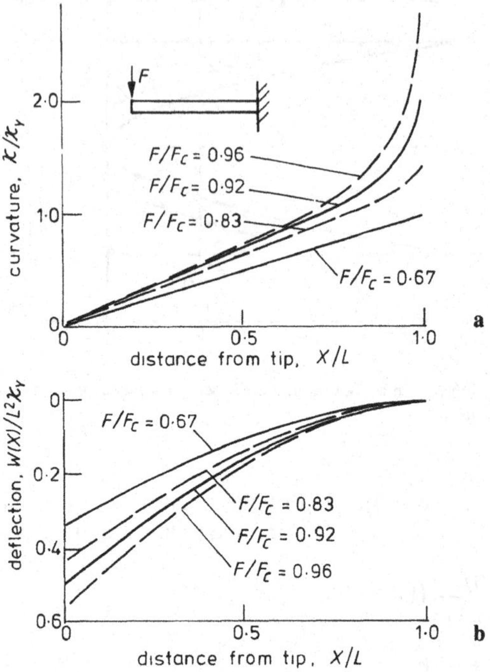

Fig. 3.4. Elastic–perfectly plastic cantilever: (a) distribution of curvature; (b) transverse deflection for forces $F_Y \le F \le F_C$.

plastic. If the material strain hardens, the elastic–plastic sections are somewhat stiffer than the elastic–perfectly plastic representation in Sect. 3.1.2 would suggest. So what are the conditions wherein the simpler elastic–perfectly plastic analysis is sufficiently accurate? In this section we compare curvature and deflection resulting from this perfectly plastic approximation with results obtained for a linearly strain hardening material and find that accuracy of the approximation depends on both the bending moment distribution and the load magnitude in comparison with the characteristic plastic load F_c. (Note that a linearly hardening material has no ultimate moment M_p for a section; nevertheless it is convenient to define a *characteristic load* $F_c = \phi_m M_Y / L$. For a rectangular cross-section $F_c = 3M_Y/2L$.)

In a structure composed of a material with elastic modulus E, yield strain $\varepsilon_Y = Y/E$ and plastic strain hardening modulus $E_t = \alpha E$, there is a relationship between bending moment M and section curvature κ given as

$$\frac{M}{M_Y} = \frac{(1-\alpha)}{2}\left[3-\left(\frac{\kappa_Y}{\kappa}\right)^2\right]+\alpha\frac{\kappa}{\kappa_Y} \tag{3.16}$$

Here the curvature at yield $\kappa_Y = 2\varepsilon_Y/h$ results in a yield moment $M_Y = EI_0\kappa_Y$. Equation (3.16) can be transformed to a nondimensional form by letting $\overline{m} \equiv M/M_Y$ and $\overline{k} = \kappa/\kappa_Y$. For a transverse force at the tip of a cantilever, $\overline{m} = (FX\phi_m)/(F_cL) = \phi_m fx$. Hence for a rectangular cross-section where $\phi_m = 3/2$, Eq. (3.16) can be rearranged as

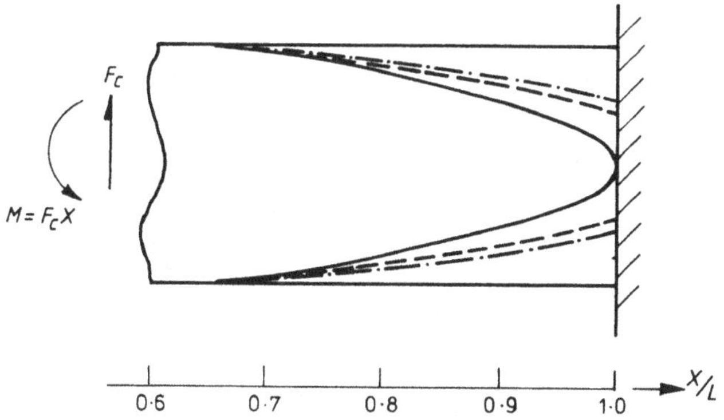

Fig. 3.5. Extent of plastically deforming region at root of elastic–linearly strain hardening cantilever for tip force $F = F_c$ and strain hardening coefficient $\alpha = 0$ ——— , $\alpha = 0.1$ - - - - and $\alpha = 0.3$ —·—·.

$$2\alpha\bar{k}^3 + 3(1-\alpha-fx)\bar{k}^2 - (1-\alpha) = 0 \qquad (3.17)$$

The cubic polynomial (3.17) has one real root for curvature $\bar{k}$ if the section is within the elastic–plastic segment $\chi < x < 1$ and the force $f \approx 1$. This root is found by letting

$$c_0 = \frac{(\alpha-1)}{2\alpha}, \qquad c_1 = -\left(c_0 + \frac{fx}{2\alpha}\right)^2, \qquad c_2 = \frac{c_0}{2} - \left(c_0 + \frac{fx}{2\alpha}\right)^3$$

Then the nondimensional curvature $\bar{k}$ is given by

$$\bar{k} = \left(-c_2 + \sqrt{c_2^2 + c_1^3}\right)^{1/3} + \left(-c_2 - \sqrt{c_2^2 + c_1^3}\right)^{1/3} + c_0 + \frac{fx}{2\alpha} \qquad (3.18)$$

The bending moment $\bar{m}$ increases linearly with distance from the tip where force F is applied. For $F > 2F_c/3$ there is a segment at the root where strain in the outer fibers exceeds the yield strain; the length of this elastic–plastic segment $(1 - \chi)L$ is independent of the hardening coefficient α. On sections within the elastic–plastic segment, however, the stress distribution does depend on the hardening coefficient α. Since the stiffness of the section increases with strain hardening, the height of the elastic core of any section increases with α ; this corresponds to less curvature in the elastic–plastic segment with increasing hardening coefficient α. These differences in the extent of the plastically deforming region are shown in Fig. 3.5.

For any load, a larger strain hardening coefficient α results in larger stiffness and consequently less curvature in the elastic–plastic region. Figure 3.6a shows the influence of hardening coefficient α on the curvature distribution in the cantilever for three force magnitudes $f = 2/3$, 1.0 and 1.2. The forces considered are limited because the present analysis is restricted to solutions where rotations of sections remain small. There is no ultimate force for the small deflection analysis of a structure composed of linearly strain hardening material.

Inclination and deflection of the neutral axis for the loaded cantilever can be obtained by integration of the curvature. Hence, for a rectangular section where $\chi = 2/3f$, the elastic–plastic segment of length $(1 - \chi)L$ has inclination $w' = \mathrm{d}w/\mathrm{d}x = \mathrm{d}W/\mathrm{d}X$ as follows

$$w'(x) = -L\kappa_Y \int_x^1 \bar{k}(\hat{x})\,\mathrm{d}\hat{x}, \qquad \chi \le x \le 1 \qquad (3.19)$$

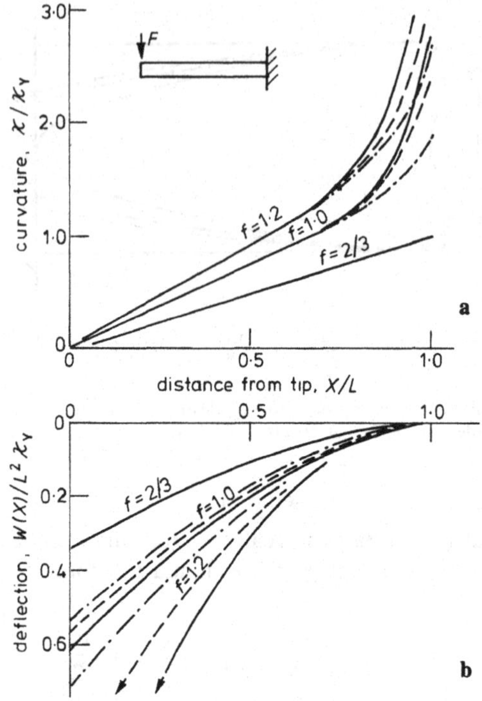

Fig. 3.6. Elastic–linearly strain hardening cantilever: (a) curvature and (b) transverse deflection for forces $f = F/F_c$ and hardening coefficient $\alpha = 0.05$ ———, $\alpha = 0.1$ - - - - and $\alpha = 0.3$ —·—.

$$w_f'(x) = -L\kappa_Y \int_x^1 \bar{k}_f(\hat{x})\mathrm{d}\hat{x}, \qquad \chi \le x \le 1 \tag{3.20}$$

The deflections shown in Fig. 3.6b for $f = 1.0$ and 1.2 were calculated by numerically integrating the curvature and inclination between the root ($x = 1$) and a section located at x. The integrals satisfy boundary conditions at the root, $w(1) = w'(1) = 0$. At the transition section with the elastic region, (3.19) and (3.20) provide the interface conditions $w'(\chi)$ and $w(\chi)$; these values are used in Eqs (3.10) to obtain the inclination and deflection for the elastic segment. It can be seen that the deflections shown in Fig. 3.6b are insensitive to the hardening coefficient α if $F \le F_c$.

The tip deflection $W(0)=W_0$ in Fig. 3.7 also illustrates this point. The tip deflection begins to vary from the elastic solution only if the applied force F is almost as large as F_c. This is a consequence of the linearly increasing bending moment distribution; with this moment distribution the large plastic deformations are localized to a short segment at the root of the cantilever.

3.1.4 Residual Deflection After Elastic Unloading

If load is slowly (quasistatically) applied to an unstressed structure and then later slowly removed, as force decreases during unloading the moment at every section also decreases. The moment–curvature relation for unloading is simply the linear

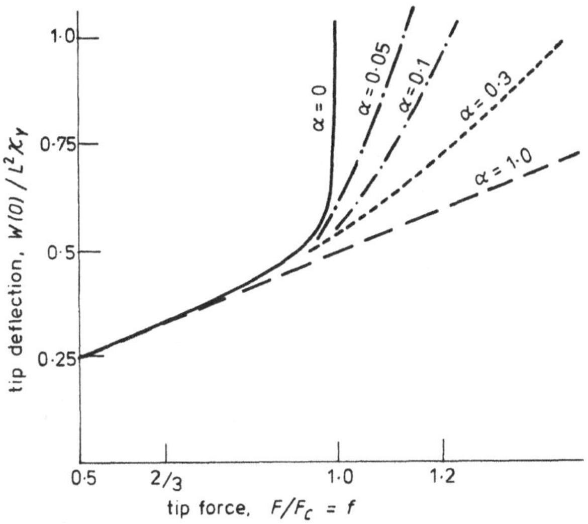

Fig. 3.7 Tip deflection W_0 of elastic–linearly strain hardening cantilever as a function of transverse force F.

elastic M–κ relation unless the change of the moment at a section is larger than $2M_Y$. If there is significant strain hardening and the moment is large, unloading can cause the outer fibers of the section to be strained into the range of *reverse plastic flow*. Reverse plastic flow during unloading can occur only if strain hardening is *kinematic*; i.e. in materials that exhibit the Bauschinger effect.

During unloading the bending moment at every section is reduced and finally vanishes. In elastic segments of the structure this unloading is easily accomplished by removing the elastic curvature that developed as a result of the load. For an initially straight cantilever, all sections that remain elastic are thereby restored to a straight configuration during unloading. In sections where the outer fibers are plastically deformed during loading however, unloading leaves behind a final or residual distribution of stress and a residual curvature $\kappa_f(X)$. This final curvature is determined by the curvature $\kappa_*(X)$ under largest load and the bending moment $M_*(X)$ that is in equilibrium with the largest magnitude of the applied load. In order to obtain the residual curvature κ_f, the elastic curvature due to M_* is subtracted from the curvature under load κ_*,

$$\kappa_f = \kappa_* - \frac{M_*}{EI_0} \tag{3.21}$$

In terms of the nondimensional curvature $\bar{k} = \kappa/\kappa_Y$ and bending moment $\bar{m} = M/M_Y$ this gives

$$\bar{k}_f = \bar{k}_* - \bar{m}_* \tag{3.22}$$

For an elastoplastic section where the moment–curvature relation is given by Eq. (3.16), the bending moments during loading and unloading can be equated to find the residual curvature; thus from Eqs (3.16) and (3.22),

$$\bar{k}_* - \bar{k}_f = \alpha\bar{k}_* + (1-\alpha)(3-\bar{k}_*^{-2})/2$$

Hence throughout the cantilever the residual curvature satisfies

$$\bar{k}_f = 0, \qquad \bar{k}_* \le 1 \tag{3.23a}$$

$$\bar{k}_f = (1-\alpha)(2\bar{k}_* - 3 + \bar{k}_*^{-2})/2, \qquad \bar{k}_* > 1 \tag{3.23b}$$

The nondimensional largest curvature $\bar{k}_*$ is obtained from Eq. (3.18). For an elastic–perfectly plastic cantilever, $\alpha = 0$ and Eq. (3.12) gives $\bar{k}_*^{-2} = 3 - 2\bar{m}_* = 3(1-fx)$; so the residual curvature distribution in the elastic–plastic region ($x > 2/3f$) is given by

$$\bar{k}_f = \frac{1}{\sqrt{3-2\bar{m}_*}} - \bar{m}_* \tag{3.24a}$$

or

$$\bar{k}_f = \left[3(1-fx)\right]^{-1/2} - 3fx/2, \qquad 2/3f < x \le 1 \tag{3.24b}$$

However, outside the elastic–plastic segment near the root, $x \le 2/3f$ and there is no residual curvature. Nevertheless, the elastic segment away from the root can have inclination, and its deflection increases linearly with distance from the root.

The residual inclination and deflection at any section can be obtained by integrating the residual curvature. For the cantilever with boundary conditions $w(1) = w'(1) = 0$, the inclination of a section located at $x \equiv X/L$ is

$$w_f'(x) = -L\kappa_Y \int_x^1 \bar{k}_f(\hat{x})\,d\hat{x} \tag{3.25}$$

This residual inclination w_f' only varies in sections of the cantilever where $x > 2/3f$; i.e. $w'(x) = w'(2/3f)$ for $x < 2/3f$.

In the plastically deformed segment near the root of an elastic–perfectly plastic cantilever, using Eqs (3.24b) and (3.25) it can be shown that

$$w_f' = L\kappa_Y\left[\frac{2}{\sqrt{3}f}\left(\sqrt{1-f} - \sqrt{1-fx}\right) + \frac{3}{4}f\left(1-x^2\right)\right] \tag{3.26}$$

At the transition section from elastic to elastic–plastic behavior the inclination or slope is

$$w_f'(2/3f) = L\kappa_Y\left(\frac{2}{\sqrt{3}f}\sqrt{1-f} + \frac{3f}{4} - \frac{1}{f}\right) \tag{3.27}$$

The nondimensional final deflection of the plastically deformed segment near the root is given by

$$w_f(x) = -\int_x^1 w_f'(\hat{x})\,d\hat{x} \tag{3.28}$$

After using (3.26) as the integrand, Eq. (3.28) gives

$$w_f(x) = -L\kappa_Y\left\{\frac{4}{3\sqrt{3}f^2}\left[(1-fx)^{3/2} - (1-f)^{3/2}\right] - \frac{2}{\sqrt{3}f}(1-x)\sqrt{1-f} - \frac{f}{4}\left(2-3x+x^3\right)\right\} \tag{3.29}$$

At the end of this segment ($x = 2/3f$) where curvature $\kappa = \kappa_Y$, the residual deflection is

$$w_f(2/3f) = L\kappa_Y\left[\frac{2}{27f^2}\left(1+6\sqrt{3}\right) - \frac{2(2+f)}{3\sqrt{3}f^2}\sqrt{1-f} - \frac{f}{2}(1-f)\right] \tag{3.30}$$

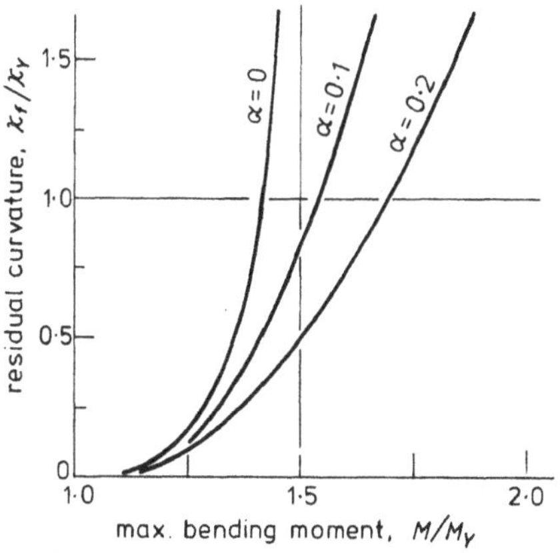

Fig. 3.8. Residual curvature of elastic–linearly strain hardening beams of rectangular cross-section depends on the largest bending moment $M(X)$ at the section.

So every section of the cantilever has flexure that increases as the largest moment $\overline{m}_*$ at that section increases; the flexure later springs back elastically as the moment is reduced during unloading. The residual curvature $\overline{k}_f$ for elastic–plastic, linearly strain hardening beams with rectangular cross-section is shown in Fig. 3.8. The curve shown for perfectly plastic material $\alpha = 0$ is given by expression (3.24). For materials with linear strain hardening ($\alpha > 0$), the results were obtained by Johnson and Yu [1981]. These authors also examined springback in elastic–plastic beams and plates with power law strain hardening.

3.1.5 Elastoplastic Beam–Columns

An oblique force acting on a slender member has both tangential and transverse components of applied force. The angle between the force and the normal to the neutral axis in the initial configuration is termed the *angle of obliquity*. Here it is considered that the direction of this force does not vary with deflection; the direction is fixed relative to the initial configuration. This combined loading results in a beam–column problem; if the angle of obliquity is large this problem enters the realm of buckling of eccentrically loaded columns.

Uniform Elastic Cantilever The deflection $W(X)$ of a cantilever due to both vertical force F_z and horizontal force F_x applied at the tip is illustrated in Fig. 3.9. The curvature $\mathrm{d}^2W/\mathrm{d}X^2$ at any elastic section is directly related to the bending moment $M(X)$

$$EI_0\frac{\mathrm{d}^2W}{\mathrm{d}X^2} = M = F_zX + F_x(\Delta - W) \tag{3.31}$$

where $\Delta \equiv W(0)$. Let $\omega^2 \equiv F_x/EI_0$ and the ratio r of the vertical to horizontal force is $r \equiv F_z/F_x$; thus

$$\frac{\mathrm{d}^2W}{\mathrm{d}X^2} + \omega^2W = \omega^2(\Delta + rX)$$

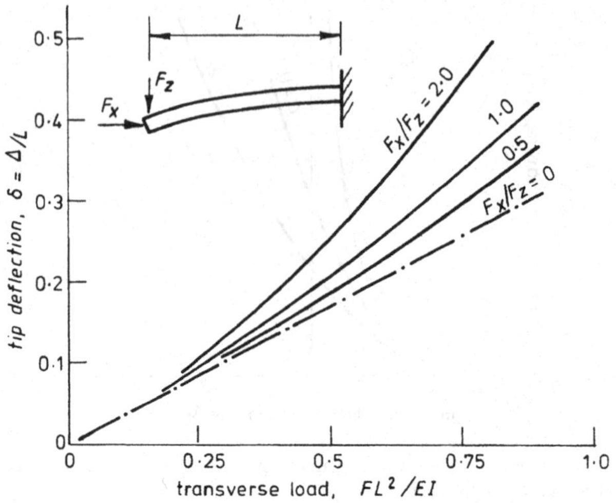

Fig. 3.9. Tip deflection of uniform elastic beam–column with horizontal force F_x and transverse (vertical) force F_z.

This ordinary differential equation has a solution

$$W = \Delta + rX + A\sin\omega X + B\cos\omega X$$

which satisfies the boundary conditions $W(L) = \mathrm{d}W(L)/\mathrm{d}X = 0$. Hence the deflection is

$$W = \Delta + rX - \frac{r\sin\omega X}{\omega\cos\omega L}, \qquad r > 0 \tag{3.32}$$

where

$$\Delta = \frac{r}{\omega}\left(\tan\omega L - \omega L\right), \qquad r = F_z / F_x$$

For small deflections, the deflection of the tip of an elastic cantilever increases linearly with vertical (transverse) force. A small additional horizontal force somewhat increases the deflection; however the increase is not large if $r > 1$ and the largest moment $M(L)$ is substantially less than M_Y. An upper bound on the range of applicability of this elastic theory is provided by the load that results in the limiting curvature κ_Y for yield at the root

$$\left[1 + \frac{r^2}{\omega L}\left(\tan\omega L - \omega L\right)\right]\frac{F_z L^2}{EI_0} < L\kappa_Y = \frac{2L\varepsilon_Y}{h} \tag{3.33}$$

For $\varepsilon_Y = 0.002$ we obtain $L\kappa_Y < 0.004L/h$. Another upper bound is provided by the Euler buckling load for an elastic column obtained with $F_z = 0$; i.e.

$$\frac{F_x L^2}{EI_0} < \frac{\pi^2}{4} \tag{3.34}$$

It may be noticed that although the above analysis has considered the additional bending moment which is produced by the axial (tangential) force and the tip deflection, it still employs the approximate expression $\kappa \approx \mathrm{d}^2W/\mathrm{d}X^2$ for the curvature, which is valid under the condition $(\mathrm{d}W/\mathrm{d}X)^2 \ll 1$. Hence, this analysis is applicable only when the changes in geometry of the beam–column are not too large. Large geometrical changes will be fully taken into account in the following section.

3.2 Large Elastic–Plastic Deflections

In deriving equilibrium equations used in the preceding sections of this chapter, the inclination of the neutral axis for bending has been assumed to remain small. In this section, effects of large deflections that involve large rotations from the initial configuration are considered for an elastic–plastic cantilever subjected to a transverse concentrated force F at the tip. Because the deflection is no longer required to be small in comparison with the length of the cantilever, the approximate expression $\kappa \approx \mathrm{d}^2W/\mathrm{d}X^2$ will be replaced by an exact expression for the curvature of the neutral axis of a deformed cantilever; that is,

$$\kappa = \frac{\mathrm{d}\theta}{\mathrm{d}S} = \frac{\mathrm{d}^2W/\mathrm{d}X^2}{\left[1+(\mathrm{d}W/\mathrm{d}X)^2\right]^{3/2}} \tag{3.35}$$

The problem of obtaining deformations will be reformulated using this expression for curvature.

3.2.1 Elastica: Large Elastic Deflection

When $M \le M_Y$, then $\kappa \le \kappa_Y$ is satisfied at every cross-section and the cantilever behaves elastically. With Eq. (3.35) in mind, the moment–curvature relation (3.5) can be expressed as

$$\frac{\mathrm{d}\theta}{\mathrm{d}S} = \frac{M}{EI_0} \tag{3.36}$$

Consider a coordinate system XOZ with the origin O located at the tip A of the *undeformed* cantilever, as shown in Fig. 3.10. The X axis is along the initial neutral axis of the cantilever and the Z axis is transverse to the neutral axis and in the plane of deflection. Let S denote the arc length of the deformed cantilever measured from the tip A and let $\theta(S) = \tan^{-1}(\mathrm{d}W/\mathrm{d}X)$ denote the inclination angle

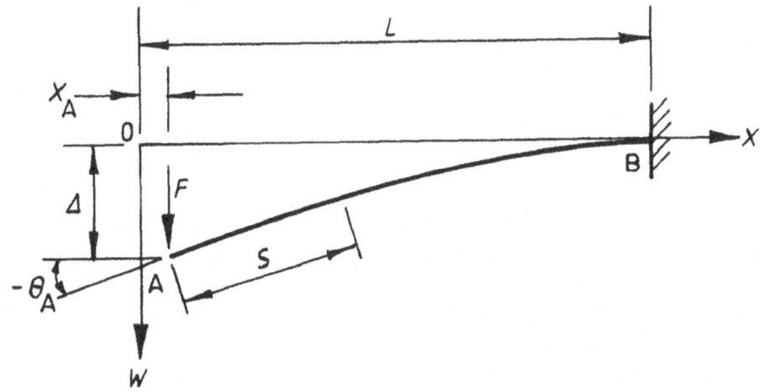

Fig. 3.10 Large elastic deflection of a uniform cantilever subjected to a concentrated force F at the tip.

measured clockwise from the X axis. Note that if the cantilever is loaded at the tip by a force F in the positive Z direction, θ is negative at the tip A and gradually increases to zero at root B.

Introduce nondimensional variables

$$x = \frac{X}{L}, \qquad w = \frac{W}{L}, \qquad s = \frac{S}{L}, \qquad \mu f = \frac{FL^2}{EI_0} \tag{3.37}$$

where f is nondimensional force and μ is the flexibility parameter defined in (3.7). Since the bending moment distribution is $M = F(X - X_A)$, the moment–curvature relation given in Eq. (3.36) results in

$$\frac{d\theta}{ds} = \mu f(x - x_A) \tag{3.38}$$

Kinematic relationships give

$$\frac{dx}{ds} = \cos\theta, \qquad \frac{dw}{ds} = \sin\theta \tag{3.39}$$

Equations (3.38) and (3.39) form a system of first-order ordinary differential equations for the unknown functions $\theta(s)$, $x(s)$ and $w(s)$. Equations of this type that describe large elastic deflections of flexible bars were first studied by Euler [1744] and termed *elastica* (e.g. see Frisch-Fay [1962] and Wang [1981]).

The differential equation for the elastica is obtained by differentiating (3.38) with respect to s and then using (3.39) to obtain

$$\frac{d^2\theta}{ds^2} = \mu f \cos\theta \tag{3.40}$$

Because $\frac{d}{ds}\left[\left(\frac{d\theta}{ds}\right)^2\right] = 2\frac{d\theta}{ds}\left(\frac{d^2\theta}{ds^2}\right)$, Eq. (3.40) leads to $d\left[\left(\frac{d\theta}{ds}\right)^2\right] = 2\mu f \cos\theta \, d\theta$.

It follows that

$$\frac{d\theta}{ds} = \sqrt{2\mu f}\,(\sin\theta - \sin\theta_A)^{1/2} \tag{3.41}$$

where θ_A denotes the inclination angle at A, ($\theta_A < 0$). This expression satisfies the boundary condition that requires the curvature to vanish at the tip.

Assuming that the total length of the neutral axis does not change during flexural deformation, and noting that $\theta = 0$ at root B where $s = 1$, we have

$$\int_{\theta_A}^{0}\left(\frac{d\theta}{ds}\right)^{-1} d\theta = \int_0^1 ds = 1 \tag{3.42}$$

Combining Eqs (3.41) and (3.42) results in

$$\sqrt{2\mu f} = \int_{\theta_A}^{0} \frac{1}{(\sin\theta - \sin\theta_A)^{1/2}}\, d\theta \tag{3.43}$$

In order to bring the right-hand side of Eq. (3.43) to the standard form of elliptic integrals, a new variable ϕ is introduced that satisfies the equation

$$1 - \sin\theta = (1 - \sin\theta_A)\sin^2\phi \tag{3.44}$$

with $0 \le \phi \le \pi/2$. Furthermore, define a parameter p such that

$$p^2 \equiv (1 - \sin\theta_A)/2 \tag{3.45}$$

In Eq. (3.43) we change from variable θ to ϕ, eventually to obtain

$$\sqrt{\mu f} = \int_{\phi_B}^{\pi/2} \frac{d\phi}{(1-p^2\sin^2\phi)^{1/2}} = K(p) - F(p,\phi_B) \tag{3.46}$$

where the value of ϕ at root B (i.e. where $\theta = 0$) is $\phi_B = \sin^{-1}(1/\sqrt{2}p)$. The terms on the right-hand side of Eq. (3.46) are an incomplete elliptic integral of first kind, $F(p,\phi) \equiv \int_0^\phi (1-p^2\sin^2\phi)^{-1/2}\,d\phi$ and the complete elliptic integral of first kind, $K(p) \equiv F(p,\pi/2)$. Thus Eq. (3.46) has one unknown variable — the modulus p; this can be found by trial and error with help of the tables of elliptic integrals, provided the value of $\mu f = FL^2/EI_0$ is specified.

In order to obtain the deflection of the cantilever for arbitrarily large inclination, we recall Eqs (3.39b) and (3.41) which give

$$dw = \sin\theta\, ds = \frac{\sin\theta\, d\theta}{\sqrt{2\mu f}(\sin\theta - \sin\theta_A)^{1/2}}$$

By changing variable from θ to ϕ, the tip deflection can be expressed as

$$\delta \equiv \frac{W_A}{L} = \frac{1}{\sqrt{\mu f}}\left[K(p) - F(p,\phi_B) - 2E(p) + 2E(p,\phi_B)\right] \tag{3.47}$$

where $E(p,\phi) \equiv \int_0^\phi (1-p^2\sin^2\phi)^{1/2}\,d\phi$ is the incomplete elliptic integral of the second kind and $E(p) \equiv E(p,\pi/2)$ is the complete elliptic integral of the second kind.

The horizontal displacement at the tip X_A, shown in Fig. 3.10, is expressed as

$$x_A \equiv \frac{X_A}{L} = 1 - \left[\frac{2(2p^2-1)}{\mu f}\right]^{1/2} \tag{3.48}$$

Details of the derivation can be found in the book by Frisch-Fay [1962].

Based on numerical results obtained by Bishop and Drucker [1945] and Frisch-Fay [1962], Fig. 3.11 shows vertical deflection δ and horizontal displacement of the tip X_A as functions of the external force F. Figure 3.12 sketches the shape of the cantilever and the inclination angle at the tip θ_A for various force magnitudes.

The applicability of this elastica solution is restricted by the requirement that the deformations remain elastic in all cross-sections of the beam; i.e. $F(L - X_A) \le M_Y$, or

$$f \equiv \frac{F}{F_c} \le \frac{M_Y}{F_c L(1-x_A)} = \frac{1}{\phi_m(1-x_A)}$$

Using Eqs (3.48) and (3.45), we find

$$f \le \frac{\mu}{2\phi_m^2|\sin\theta_A|}$$

or

$$\mu f \le \frac{\mu^2}{2\phi_m^2|\sin\theta_A|} = \frac{(\kappa_Y L)^2}{2|\sin\theta_A|}$$

Consequently, it is necessary that

$$\kappa_Y L \ge \sqrt{2\mu f|\sin\theta_A|} \tag{3.49}$$

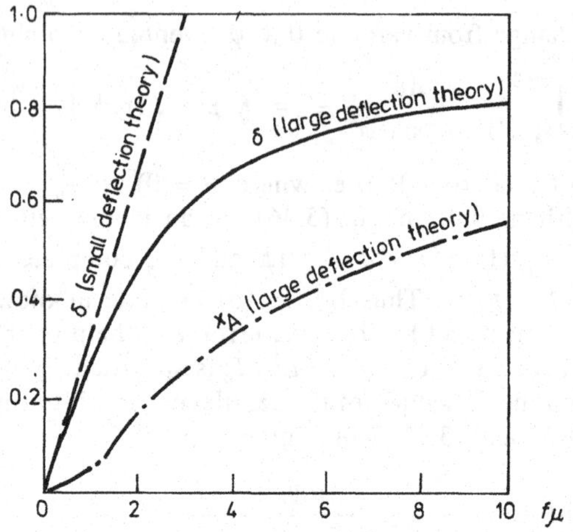

Fig. 3.11. Vertical deflection δ and horizontal displacement x_A of an elastic cantilever, as functions of force F at the tip.

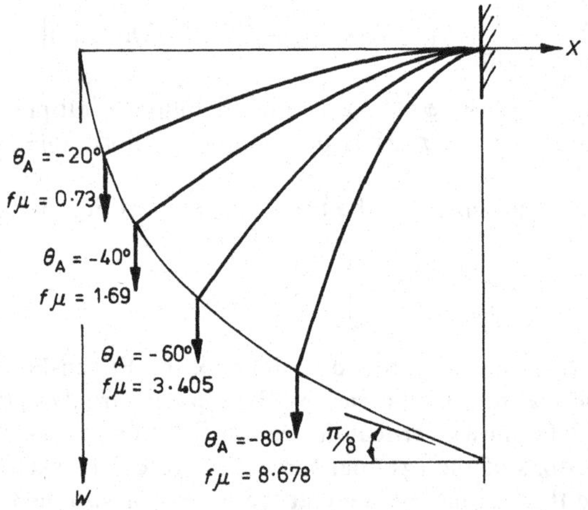

Fig. 3.12. Deformed shape and tip inclination of an elastic cantilever due to vertical force at the tip.

Particularly, for cantilevers of rectangular cross-section, $\kappa_Y L = M_Y L / EI_0 = 2YL/Eh$; hence, the elastica solution is valid only if the force is in the range

$$f|\sin\theta_A| \le \frac{2}{\mu}\left(\frac{YL}{Eh}\right)^2 \tag{3.50}$$

or the slenderness of the cantilever is in the range

$$\frac{L}{h} \ge \frac{E}{Y}\sqrt{\frac{1}{2}\mu f|\sin\theta_A|} \tag{3.51}$$

Utilizing results shown in Figs 3.11 and 3.12, we summarize the limitations for the elastica solution of a tip-loaded cantilever in Table 3.1.

This shows that the elastica theory is applicable only for long slender bars. Since the ratio of E to Y is of the order of 500 for most metals, the last line in

Table 3.1 Validity of elastica solution of tip–loaded cantilever

$-\theta_A$	20°	40°	60°	80°	Comment
$\delta \equiv \Delta/L$	0.21	0.42	0.61	0.79	
μf	0.73	1.69	3.405	8.678	See Fig. 3.12
$\kappa_Y L \geq$	0.707	1.474	2.428	4.134	
$\mu \geq$	1.06	2.21	3.64	6.20	Rectangular cross-section
$L/h \geq$	177	368	607	1034	E/Y =500 is assumed

Table 3.1 states that the elastica solution is applicable for tip deflections less than or equal to $\Delta = 0.21L$ only if the slenderness ratio $L/h \geq 177$. Application of the elastica solution up to $\Delta = 0.61L$ requires $L/h \geq 607$; this is a slenderness ratio far beyond the practical range of structural members. Therefore, applicability of the elastica solution is greatly limited by the fact that in engineering applications most beams exceed their elastic range before the deflections become large.

Consequently, in the range of large deflections, the effects of nonlinear geometric relations for beams and struts have been examined in conjunction with nonlinear elastic constitutive relations. Oden and Childs [1972] performed a finite deflection analysis of an Euler strut or elastica with moment–curvature relationship given by a hyperbolic tangent function. Prathap and Varadan [1976] studied the case of a strut which has a Ramberg–Osgood type constitutive relation. Lo and Das Gupta [1978] investigated the case in which the stress–strain relationship takes the form of a logarithmic function when the stress is larger than the elastic limit.

3.2.2 Plastica: Large Plastic Deflection

The discussion in Section 3.2.1 about applicability of the elastica theory is illustrated by Fig. 3.13. The elastica solution is valid only in the region below the curve; if the tip deflection δ of a cantilever exceeds the value on the curve, elastic–plastic deformation occurs in the beam. Therefore, a large deflection analysis is required for elastic–plastic beams.

In the following analysis, we assume that the material is elastic–perfectly plastic and the cantilever has a rectangular cross-section. The bending moment in the cantilever increases as the external force F increases. When F exceeds a value such that $M_{max} = M_B = F(L - X_A) \geq M_Y$, an elastoplastic region first appears in the neighborhood of root B; in this region the core of the section is elastic while the top and bottom layers are perfectly plastic, similar to that shown in Fig. 3.3. Suppose that for a certain force F, the plastic deformation develops within segment CB ($S_C \leq S \leq L$) as shown in Fig. 3.14; then a combination of the $M - \kappa$ relation (3.12) in the plastic region with the general definition of curvature (3.35) gives

$$\frac{d\theta}{dS} = \kappa_Y \left[3 - 2\frac{F(X - X_A)}{M_Y} \right]^{-1/2}, \qquad S_C \leq S \leq L$$

or in nondimensional form,

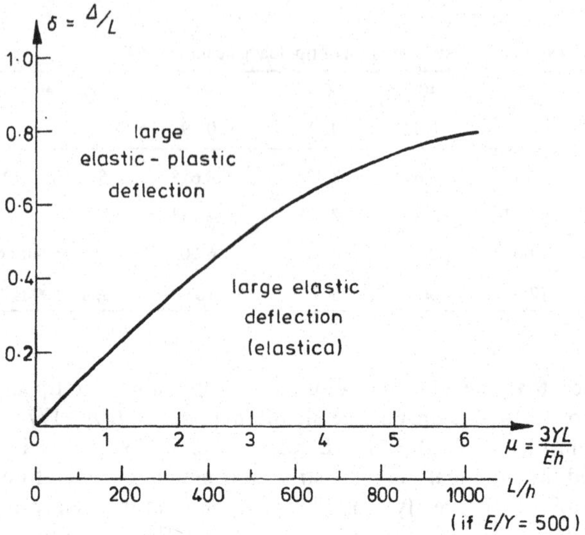

Fig. 3.13. Validity of the elastica solution of a cantilever subjected to a concentrated force at the tip.

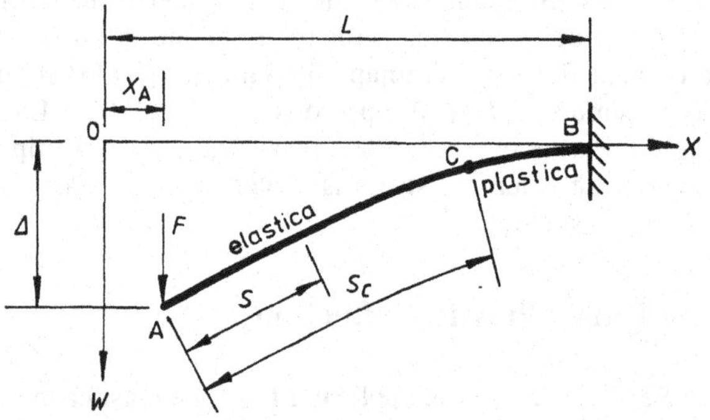

Fig. 3.14. Large elastic–plastic deflection of a uniform cantilever subjected to a concentrated force F at the tip; point C is the interface between the elastic and plastic deforming segments.

$$\frac{d\theta}{ds} = \frac{2\mu}{3\sqrt{3}}\left[1 - f(x - x_A)\right]^{-1/2}, \qquad s_C \le s \le 1 \tag{3.52}$$

where $s_C \equiv S_C / L$; while definitions of x, f and μ are given in Eq. (3.7). Combining (3.52) with the geometric relations (3.39)

$$\frac{dx}{ds} = \cos\theta, \qquad \frac{dw}{ds} = \sin\theta$$

we obtain a system of first-order ordinary differential equations for the unknown functions $\theta(s)$, $x(s)$ and $w(s)$. These equations govern large deflections of elastic–perfectly plastic bars — a problem that has been termed *plastica* by Yu and Johnson [1982].

The boundary conditions for plastica equations of a cantilever are

$$w = \theta = 0, \qquad \text{at} \quad s = 1 \qquad \text{(i.e. at root B)} \tag{3.53}$$

$$\theta = \theta_C, \quad \frac{d\theta}{ds} = \frac{2}{3}\mu, \quad \text{at} \quad s = s_C \quad \text{(i.e. at interface C)} \tag{3.54}$$

Although the plastica equations appear to be more complex than those for the elastica, a closed-form analytical solution is obtainable. Starting from (3.52), the second derivative of s with respect to θ can be obtained as

$$\frac{d}{d\theta}\left(\frac{ds}{d\theta}\right) = \frac{3\sqrt{3}}{2\mu}\frac{d}{d\theta}\left[1 - f(x - x_A)\right]^{1/2} = \frac{3\sqrt{3}}{4\mu}\left[1 - f(x - x_A)\right]^{-1/2}\frac{dx}{d\theta}$$

Noting that $dx/d\theta = (dx/ds)(ds/d\theta) = (ds/d\theta)\cos\theta$, gives

$$\frac{d}{d\theta}\left(\frac{ds}{d\theta}\right) = -\frac{27f}{8\mu^2}\cos\theta$$

After integrating this equation and using the boundary condition (3.54) at point C, we obtain

$$\frac{ds}{d\theta} = \frac{3}{2\mu} + \frac{27f}{8\mu^2}(\sin\theta_C - \sin\theta) \tag{3.55}$$

The inverse of (3.55) is the nondimensional curvature. The largest nondimensional curvature is at the root B where

$$L\kappa_B = \left.\frac{d\theta}{ds}\right|_{s=1} = \frac{2\mu}{3}\frac{1}{1 + \dfrac{9}{4\mu}\sin\theta_C} > \frac{2\mu}{3} = L\kappa_C \tag{3.56}$$

Notice that $\sin\theta_C < 0$ due to $\theta_C < 0$. After integrating (3.55) and applying the boundary conditions of vanishing deflection and rotation at the root of the cantilever, Eq. (3.53), the intrinsic coordinate s where the bar has rotation θ is obtained,

$$s = \frac{3}{2\mu}\theta + \frac{27f}{8\mu^2}(\theta\sin\theta_C + \cos\theta - 1) + 1 \tag{3.57}$$

Using $dw = \cos\theta\, ds = \cos\theta(ds/d\theta)d\theta$ and Eq. (3.55), the vertical deflection of the plastic segment CB in the cantilever is found to be

$$w = \frac{3}{2\mu}\sin\theta + \frac{27f}{8\mu^2}\left(\sin\theta_C\sin\theta - \frac{1}{2}\sin^2\theta\right) \tag{3.58}$$

This satisfies the boundary condition $w = \theta = 0$ at root B. Accordingly, the deflection at the terminus of the elastoplastic region, point C, is

$$w_C = \frac{3}{2\mu}\sin\theta_C + \frac{27f}{16\mu^2}\sin^2\theta_C \tag{3.59}$$

At this stage of the analysis, all quantities in plastic segment CB have been expressed in terms of parameter θ_C. The value of θ_C can be determined by the continuity conditions for θ and $(d\theta/ds)$ at point C as follows. By repeating the elastica analysis presented in Sect. 3.2.1, an equation similar to (3.41) is obtained; this is a relationship between the arc length s_C of the elastic segment and the inclination θ_C at the elastic–elastoplastic interface C,

$$s_C = \frac{1}{\sqrt{f\mu}}\left[K(p) - F(p, \phi_C)\right] \tag{3.60}$$

with

$$\phi_C = \sin^{-1}\left(\frac{1}{\sqrt{2}p}\sqrt{1 - \sin\theta_C}\right) \tag{3.61}$$

Substituting $\theta = \theta_C$ in (3.57) and using Eq. (3.60) and $s = s_C$ leads to an equation for θ_C,

$$\frac{1}{\sqrt{f\mu}}\left\{K(p) - F\left[p, \sin^{-1}\left(\frac{1}{\sqrt{2}p}\sqrt{1-\sin\theta_C}\right)\right]\right\}$$
$$= \frac{3}{2\mu}\theta_C + \frac{27f}{8\mu^2}(\theta_C \sin\theta_C + \cos\theta_C - 1) + 1 \quad (3.62)$$

By referring to Eqs (3.41), (3.44), (3.45) and (3.53), the continuity of $(d\theta/ds)$ at point C results in the parameter p that is related to tip rotation,

$$p = \left(\frac{1-\sin\theta_C}{2} + \frac{\mu}{9f}\right)^{1/2} \quad (3.63)$$

Recall that for a rectangular cross-section the flexibility parameter $\mu \equiv 3L\kappa_Y/2$, the external force $f \equiv F/F_c = FL/M_p$ and the values of θ_C and p can be found for any specified force F by an iteration procedure based on Eqs (3.62) and (3.63). Then with (3.47) and (3.59), the tip deflection of an elastic–plastic cantilever is found to be

$$\delta \equiv \frac{W_A}{L} = \frac{3}{2\mu}\sin\theta_C + \frac{27f}{16\mu^2}\sin^2\theta_C + \frac{1}{\sqrt{f\mu}}\left[K(p) - F(p,\phi_C) - 2E(p) + 2E(p,\phi_C)\right] \quad (3.64)$$

with ϕ_C given by (3.61).

Numerical results supporting the plastica analysis are depicted in Figs 3.15–3.17. Figure 3.15 shows the tip deflection δ as a function of the force parameter f if the flexibility parameter is taken to be $\mu = 0.75$, 1.5 and 4.5. Recall that small values of the flexibility parameter ($\mu < 1$) imply sufficient stiffness so that $\theta_A \ll 1$. The curve $\mu = 0$ pertains to a rigid–perfectly plastic approximation for $E \to \infty$.

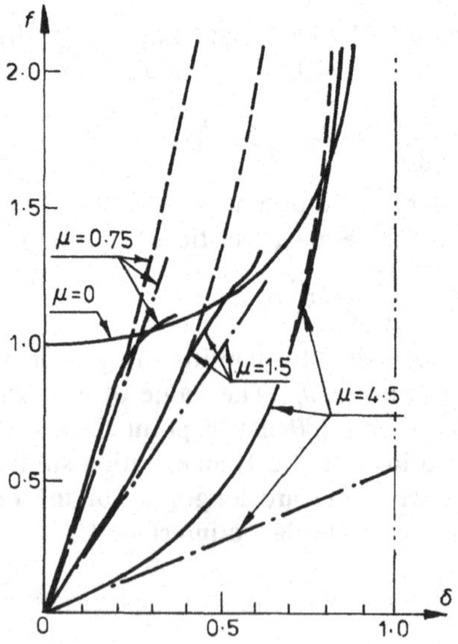

Fig. 3.15. Tip deflection as a function of the force applied at the tip according to plastica solution, ———; elastica solution, - - - - ; elastic small deflection solution, —·—.

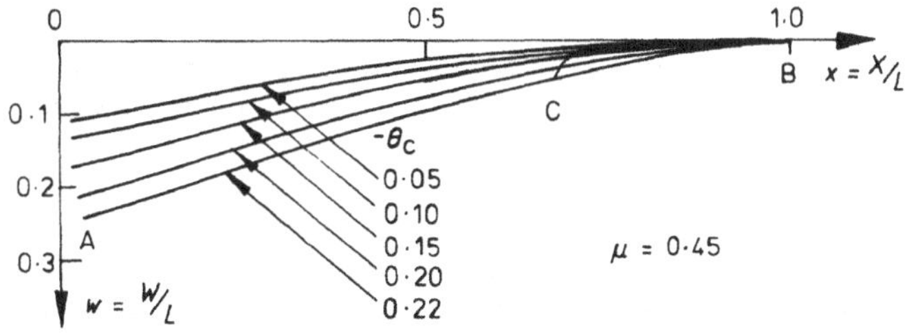

Fig. 3.16. Deformed shapes of an elastic–plastic cantilever with $\mu = 0.45$.

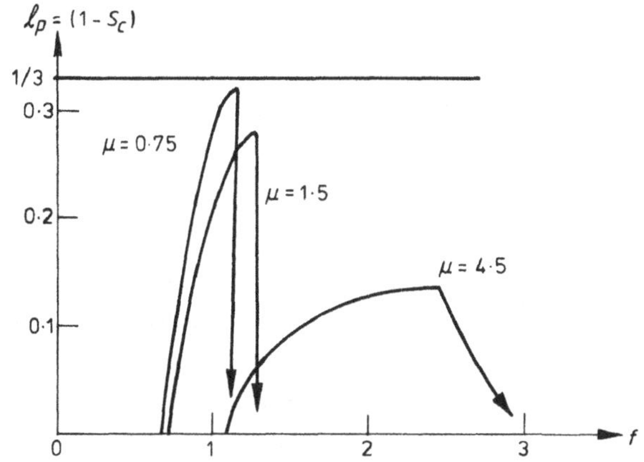

Fig. 3.17. Variation of the relative length of the plastic segment with force during large elastic–plastic deflections.

The deformed shape of a cantilever with $\mu = 0.45$ is shown in Fig. 3.16 where the evolution of the plastic segment with increasing θ_C is demonstrated by the locus of point C.

In studying the complete process of large elastic–plastic deflection of a cantilever, Wu and Yu [1986] found that with increasing force f the length of the plastic segment will reach a maximum and then decrease as the tip rotation becomes large. Hence a portion of the cantilever will undergo a loading–unloading process.

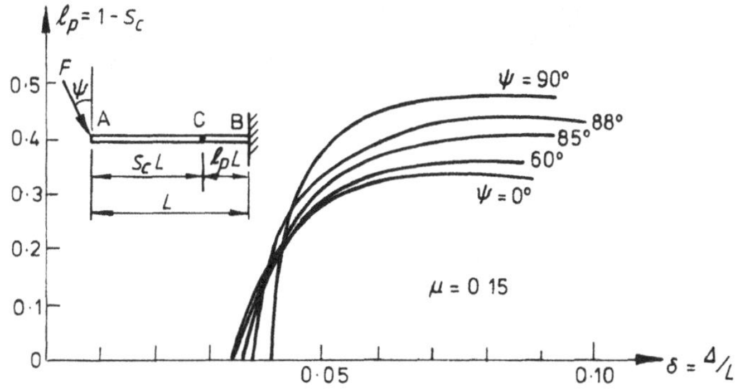

Fig. 3.18. Length of the plastic segment as a function of tip deflection; $\mu = 0.15$.

The variation of the length of the plastic segment, $l_p = (1 - s_c)$ is shown in Fig. 3.17 for the cases of $\mu = 0.75$, 1.5 and 4.5. The unloading phenomenon in elastic–plastic struts and bars that develops with large deflections was also studied by Monasa [1974, 1980].

The analysis of Wu and Yu [1986] was extended by Liu, Stronge and Yu [1989] by considering large deflections of an elastoplastic strain hardening cantilever. Luan and Yu [1991] also studied the effect of an inclined concentrated force. Figure 3.18 shows the variation in length of the plastic segment with tip deflection for a fairly large depth $\mu = 0.15$ and various angles of inclination ψ. All these analyses of flexure take both geometrical and material nonlinearities into account but they neglect the influence of the axial component of force on yielding.

References

Bishop, K.E. and Drucker, D.C. [1945]. Large deflections of cantilever beams. *Quart. Appl. Math.* **3**, 272–275.

Euler, L. [1744]. *Methodus Inveniendi Lineas Curvas*.

Frisch-Fay, R. [1962]. *Flexible Bars*. Butterworths, London.

Johnson, W. and Yu, T.X. [1981]. On springback after the pure bending of beams and plates of elastic work–hardening material — III. *Int. J. Mech. Sci.* **23**, 687–696.

Liu, J.H., Stronge, W.J. and Yu, T.X. [1989]. Large deflections of an elastoplastic strain hardening cantilever. *ASME J. Appl. Mech.* **56**, 737–743.

Lo, C.C. and Das Gupta, S. [1978]. Bending of a nonlinear rectangular beam in large deflection. *ASME J. Appl. Mech.* **45**, 213–215.

Luan, F. and Yu, T.X. [1991]. An analysis of the large deflection of an elastic–plastic cantilever subjected to an inclined concentrated force. *Appl. Math. Mech.* **12**, 547–555.

Monasa, F.E. [1974]. Deflections and stability behaviour of elasto–plastic flexible bars. *ASME J. Appl. Mech.* **41**, 537–538.

Monasa, F.E. [1980]. Deflections of postbuckled unloaded elasto–plastic thin vertical columns. *Int. J. Solids Struct.* **16**, 757–765.

Oden, J.T. and Childs, S.B. [1972]. Finite deflection of a nonlinearly elastic bar. *ASME J. Appl. Mech.* **37**, 48–52.

Prathap, G. and Varadan, T.K. [1976]. The inelastic large deformation of beams. *ASME J. Appl. Mech.* **43**, 689–690.

Wang, C.Y. [1981]. Folding of elastica: similarity solutions. *ASME J. Appl. Mech.* **48**, 199–200.

Wu, X. and Yu, T.X. [1986]. The complete process of large elastic–plastic deflection of a cantilever. *Acta Mechanica Sinica*, **2**, 333–347.

Yu, T.X. and Johnson, W. [1982]. The Plastica: the large elastic–plastic deflection of a strut. *Int. J. Non-Linear Mech.* **17**, 195–209.

Chapter 4

Dynamic Rigid–Plastic Response

Static deformation of elastoplastic structures applies only if the loading magnitude is less than the plastic collapse force F_c. With impact or explosive blast loading however, the structure can be subjected to an intense but short–duration pressure or force pulse that exceeds the plastic collapse force and initiates structural collapse. Here we begin with the simplest structural model — a rigid–perfectly plastic cantilever — and explain how characteristics of intense loads ($F > F_c$) influence the structural deformation. The analysis of this deformation uses dynamic equations of motion rather than the equilibrium equations used in previous chapters. First, the dynamic behavior of a rigid–perfectly plastic cantilever subjected to a load or a force pulse at the tip is examined. Then the slightly different response of a cantilever struck by a colliding body is studied in detail to reveal an additional transient phase of deformation. For both steady loads and impacts, an elementary model based on the rigid–perfectly plastic constitutive approximation is used to calculate the deformed shape as well as the energy dissipation during the process of deformation.

4.1 Step Loading

4.1.1 Static and Dynamic Loadings

For small deflections, steady forces on a rigid–plastic structure result in accelerations that remain constant. Since the distribution of acceleration does not vary, a structure that is initially at rest has a velocity distribution that is identical with the acceleration distribution. At every point the velocity simply increases linearly with time. The stationary pattern or distribution of acceleration that satisfies the equations of motion and the yield condition is in fact a dynamic mode for the structure with these applied forces.

Consider a uniform cantilever that is initially straight; the cantilever is subjected to a transverse force $F(t)$ at the tip, as shown in Fig. 4.1. In this section it is assumed that a constant force F is suddenly applied to a stationary cantilever; i.e. $F(t) = F$ for $t > 0$. This step loading is illustrated in Fig. 4.2.

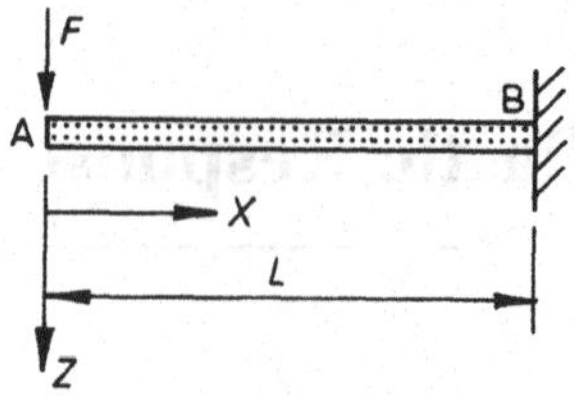

Fig. 4.1. Uniform cantilever subjected to a transverse force $F(t)$ at the tip.

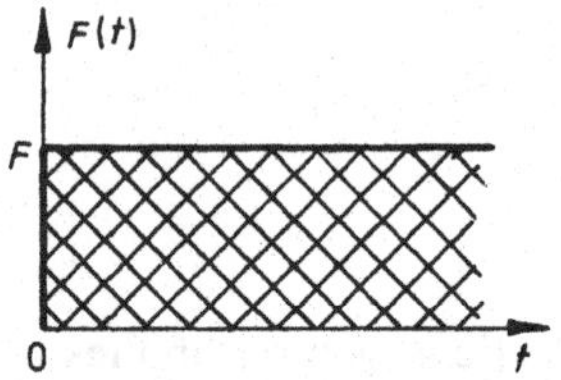

Fig. 4.2. Step force originating at $t = 0$.

The deflection of the cantilever is measured relative to a system of Cartesian coordinates X, Z with origin at the tip, as shown in Fig. 4.1. Static equilibrium of a beam element of differential length dX gives a relationship between distributed force per unit length of beam $g(X)$, shear force $Q(X)$ and bending moment $M(X)$

$$\frac{dQ}{dX} = -g(X) \tag{4.1}$$

$$\frac{dM}{dX} = -Q(X) \tag{4.2}$$

where the positive sense for g, Q and M is indicated in Fig. 4.3.

If a static transverse force F acts at the tip of a cantilever of length L, then the shear force and bending moment distributions resulting from this force are $Q(X) = -F$ and $M(X) = FX$ as discussed in Chap. 3. The plastic collapse load F_c for a transverse force at the tip is related to the structural parameters by

$$F_c = M_p / L \tag{4.3}$$

where M_p is the plastic bending moment of the beam. If the tip force is small, $F \leq F_c$ and there is no distributed load $g(X) = 0$, the above static solution for shear force Q and bending moment M does not violate the yield condition anywhere in the cantilever. Figure 4.4 illustrates the static shear and moment distributions from an applied force F equal to the collapse force F_c. If $F = F_c$, the root of the beam becomes a plastic hinge since the bending moment $M(L) = M_p$.

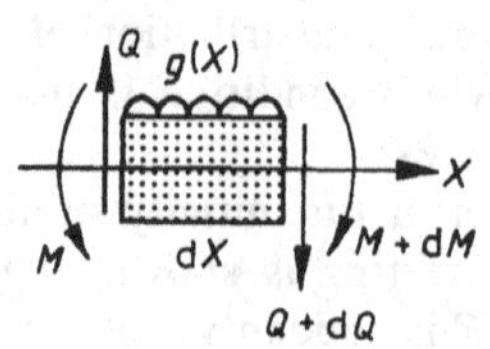

Fig. 4.3. Forces and moments on a beam element.

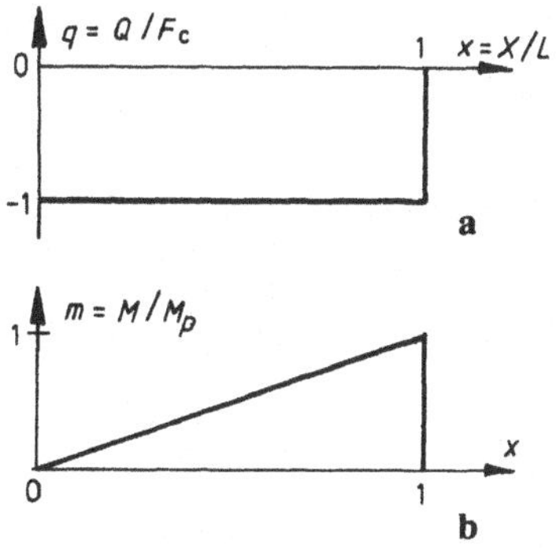

Fig. 4.4. Distribution of shear force and bending moment along a stationary cantilever when $F = F_c$: (a) shear force; (b) bending moment.

A perfectly plastic cantilever cannot support a static force F larger than the collapse load F_c. Forces that are larger than the fully plastic collapse force, $F > F_c$, cause dynamic deformation; in this case the extra load $(F - F_c)$ generates an accelerated motion of the beam.

4.1.2 Moderate Dynamic Load ($F_c < F \leq 3F_c$)

Let a transverse step force $F(t)$ be applied at the tip at time $t = 0$ and then held constant. If the magnitude of the force slightly exceeds the fully plastic collapse force F_c, there will be a plastic hinge located at the root of the cantilever (point B in Fig. 4.5) where the static bending moment takes its largest value; i.e. when a load only slightly exceeds the plastic collapse load, the spatial distribution of deformation is identical with that of static collapse. Thus if $F > F_c$ the cantilever accelerates as it rotates about this hinge at the root. Let ρ denote the mass per unit length of the cantilever and V denote the velocity at the tip; the translational equation of motion for the entire cantilever rotating about the root is given by

$$\frac{1}{2}\rho L \frac{\mathrm{d}V}{\mathrm{d}t} = F + Q_B \tag{4.4}$$

where the reaction shear force on the cantilever at the root, Q_B has a positive sense downwards (see Fig. 4.5). A second equation of motion for the rotating cantilever is obtained by taking moments about the plastic hinge at the root B:

$$\frac{1}{3}\rho L^2 \frac{\mathrm{d}V}{\mathrm{d}t} = FL - M_p \tag{4.5}$$

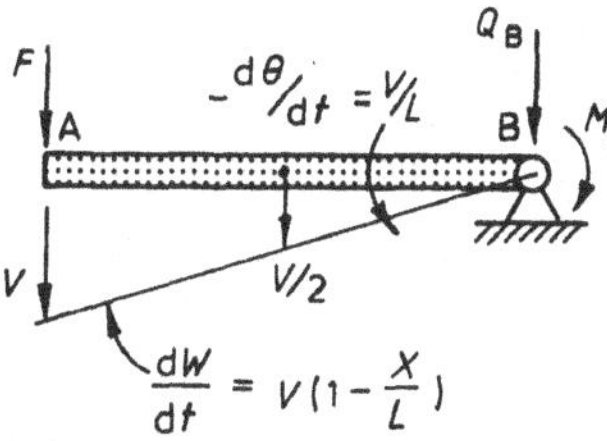

Fig. 4.5. Deformation mechanism of cantilever with a plastic hinge at the root.

This gives a constant rate of acceleration for the tip

$$\frac{dV}{dt} = \frac{3(FL - M_p)}{\rho L^2} = \frac{3(F - F_c)}{\rho L} \tag{4.6}$$

Alternatively, Eq. (4.6) can be derived by considering the kinetic energy K of the rotating cantilever. For a transverse displacement field $W(X, t)$ corresponding to rotation about a hinge at the root, the transverse velocity of any section is $\partial W/\partial t = V(1 - X/L)$; thus the kinetic energy is given by

$$K = \int_0^L \frac{1}{2}\rho\left(\frac{\partial W}{\partial t}\right)^2 dX = \frac{1}{6}\rho L V^2 \tag{4.7}$$

The rate of change of kinetic energy is related to the rate of work done by the external force and the rate of energy dissipation due to plastic deformation; that is

$$\frac{dK}{dt} = FV - \frac{M_p V}{L} = (F - F_c)V \tag{4.8}$$

where the angular velocity of a cantilever rotating about a plastic hinge at the root is $V/L = d\theta/dt$. Substituting (4.7) into (4.8) results in Eq. (4.6) as required.

At this point it is convenient to introduce a set of nondimensional variables and structural parameters to reformulate the present dynamic problem. Using the nondimensional variables previously introduced for static analysis,

$$x \equiv \frac{X}{L}, \qquad w \equiv \frac{W}{L}, \qquad f \equiv \frac{F}{F_c}, \qquad q \equiv \frac{Q}{F_c} \tag{4.9}$$

A nondimensional time is also defined as

$$\tau \equiv \frac{t}{T_0} \tag{4.10}$$

where T_0 is a *characteristic time* for the plastic cantilever:

$$T_0 \equiv L\sqrt{\frac{\rho}{F_c}} = L\sqrt{\frac{\rho L}{M_p}} \tag{4.11}$$

The characteristic time T_0 is related to the fundamental period T_1 for elastic vibration of the cantilever (e.g. see Meirovitch [1986], p 226), where

$$T_1 = 1.787 L^2 \sqrt{\frac{\rho}{EI}} \tag{4.12}$$

The ratio of the characteristic time to the fundamental period of vibration is a structural parameter,

$$\frac{T_0}{T_1} = 0.323\sqrt{\frac{Eh}{YL}} = 0.457\sqrt{\frac{1}{\kappa_Y L}} \tag{4.13}$$

where κ_Y is the elastic limit curvature of the beam. Consistent with this measure of time, a nondimensional velocity at the tip $v(\tau)$ can be defined as

$$v(\tau) \equiv \frac{\partial w(0,\tau)}{\partial \tau} \equiv \dot{w}(0,\tau) \tag{4.14}$$

where $\dot{(\ )} \equiv \partial(\ )/\partial\tau$. Equations (4.10) and (4.14) relate the nondimensional velocity to structural and load parameters, $v = V\sqrt{\rho/F_c}$.

To obtain the distribution of the shear force along the beam, note that the acceleration has the following distribution:

$$\ddot{w}(x) = \dot{v}(1-x) \tag{4.15}$$

Combining (4.6) with (4.4) and using nondimensional quantities, we find the reaction force at the root,

$$q_B = \frac{1}{2}(f-3) \tag{4.16}$$

The shear force $q(x)$ at any section can be obtained from the equation of motion for the tip segment together with the acceleration field (4.15); i.e.

$$q(x) = -f + \int_0^x \ddot{w}(\bar{x})\,d\bar{x} = -f + \dot{v}\left(x - \frac{x^2}{2}\right) \tag{4.17}$$

From the boundary condition $q(1) = q_B = (f-3)/2$ (see Eq. (4.16)), we obtain an expression for the tip acceleration,

$$\dot{v} = 3(f-1) \tag{4.18}$$

This is identical with the nondimensional expression for (4.6). Hence (4.17) gives the shear force distribution in the accelerating segment:

$$q(x) = \tfrac{1}{2}(f-3) - \tfrac{3}{2}(1-x)^2(f-1) \tag{4.19}$$

At the tip, $q_A = q(0) = -f$, which just balances the applied load.

The bending moment distribution is obtained by integrating the shear force (4.19) since $q(x) = -dm(x)/dx$; thus, for a plastic hinge at the root,

$$m(x) \equiv \frac{M}{M_p} = -\frac{1}{2}\left\{x(f-3) + \left[(1-x)^3 - 1\right](f-1)\right\} \tag{4.20}$$

This equation satisfies the boundary condition $m_B = m(1) = 1$. Note that in Chaps 4 and 5 where torque $\tilde{t}$ and axial force $\tilde{n}$ are not considered, the ratio of bending moment to fully plastic moment M/M_p is denoted by m instead of $\tilde{m}$ which is used in other chapters.

The distributions of shear force $q(x)$ and bending moment $m(x)$ that result from this applied force and acceleration field are shown in Fig. 4.6, for the case of $F_c < F \le 3F_c$. In this range of applied force, shear at the root is negative,

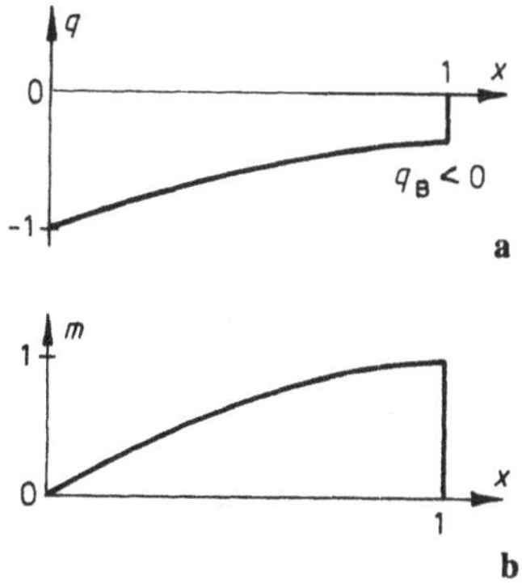

Fig. 4.6. Distribution of (a) shear force and (b) bending moment along the cantilever when the step force F is in range $F_c < F \le 3F_c$.

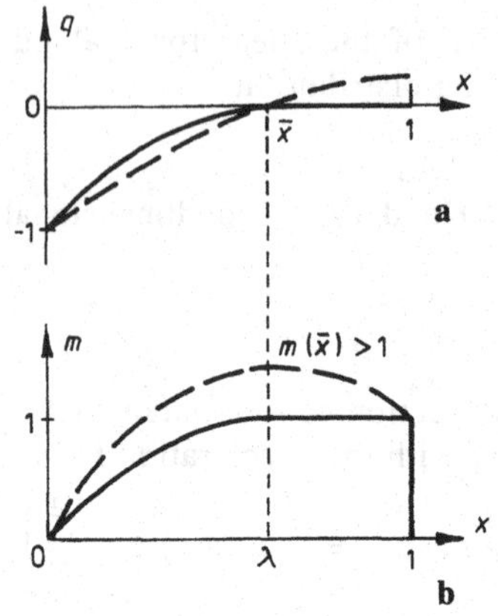

Fig. 4.7. Distribution of (a) shear force and (b) bending moment along the cantilever for a step force $F > 3F_c$. The broken lines are an inadmissible deformation mechanism with a plastic hinge at the root; the solid lines are the correct bending moment $m(x)$ and shear force $q(x)$, which result from a plastic hinge at an interior point $x = \lambda < 1$.

$q_B < 0$. If the shear force vanishes at any section, then the bending moment at that section is an extreme (maximum or minimum) since $q = -\mathrm{d}m/\mathrm{d}x = 0$. If both q_B and q_A are negative, the shear force does not change sign in the rotating segment $0 \le x \le 1$. Thus the bending moment steadily increases with increasing distance from the tip x and the largest magnitude of the moment occurs at the root where $M(L) = M_p$. Hence a suddenly applied force in the range $F_c < F \le 3F_c$ results in only a single plastic hinge located at the root where there is a reaction $Q_B \le 0$. When the force magnitude $F = 3F_c$, the shear force at the root vanishes.

4.1.3 Intense Dynamic Load ($F > 3F_c$)

Initially, suppose that the same velocity distribution applies for both moderate and intense forces; i.e. presume that there is a plastic hinge at the root. If $F > 3F_c$, then $Q_B > 0$. This assumption results in shear force $q(x)$ and bending moment $m(x)$ distributions that are shown as dashed lines in Fig. 4.7. In this case, shear force changes sign at an interior point located at

$$\bar{x} = 1 - \sqrt{\frac{f-3}{3(f-1)}} \tag{4.21}$$

and the bending moment has a maximum at this interior section where $q(\bar{x}) = -\mathrm{d}m(\bar{x})/\mathrm{d}x = 0$. Since the moment at the root $m(1) = 1$, this implies that the bending moment at $\bar{x}$ exceeds the plastic bending moment; i.e.

$$m(\bar{x}) > m(1) = 1 \tag{4.22}$$

This erroneous result is a consequence of assuming that the plastic hinge is located at the root irrespective of the magnitude of the applied force.

However, Eq. (4.22) and Fig. 4.7 indicate that for $F > 3F_c$ a plastic hinge must appear at an interior point of the cantilever instead of the root. Assume that the hinge forms at a point H located a distance Λ away from the tip, $\Lambda < L$. This

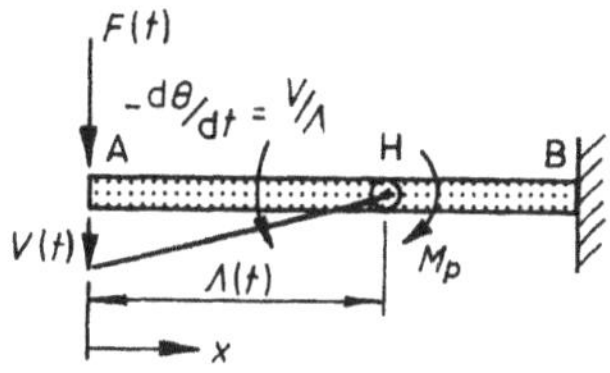

Fig. 4.8. Deformation mechanism of cantilever with plastic hinge at interior point H.

deformation mechanism is shown in Fig. 4.8. For this mechanism the angular acceleration of segment AH rotating about the hinge at H is $(dV/dt)/\Lambda$, while segment HB remains stationary. Since the bending moment distribution has a local maximum at the hinge, the hinge moment $M(\Lambda) = M_p$ and shear force $Q(\Lambda) = 0$. Hence by adopting $\lambda = \Lambda / L$ as the nondimensional measure of interior hinge position, the equations of motion for segment AH are found to be

$$\tfrac{1}{2}\lambda \dot{v} = f \qquad (4.23)$$

$$\tfrac{1}{3}\lambda^2 \dot{v} = f\lambda - 1 \qquad (4.24)$$

with unknowns $\dot{v}$ and λ. These equations replace Eqs (4.4) and (4.5). Starting from Eq. (4.23) the transverse acceleration of any section can be obtained as a function of hinge position λ,

$$\dot{v} = \frac{2f}{\lambda} \qquad (4.25)$$

$$\ddot{w} = \dot{v}\left(1 - \frac{x}{\lambda}\right) = \frac{2f}{\lambda}\left(1 - \frac{x}{\lambda}\right) \qquad (4.26)$$

Hence, noting that at the hinge the shear force $q_H = q(\lambda) = 0$ while the bending moment $m_H = m(\lambda) = 1$, the shear force and bending moment distributions are given by

$$q(x) = -f\left(1 - \frac{x}{\lambda}\right)^2 \qquad (4.27)$$

$$m(x) = 1 - \frac{f\lambda}{3}\left(1 - \frac{x}{\lambda}\right)^3 \qquad (4.28)$$

The distributions of shear force and bending moment, given by (4.27) and (4.28) are illustrated in Fig. 4.7 as solid lines. It is worthwhile mentioning that although $M(X) = M_p$ throughout segment HB near the root (i.e. $\Lambda \le X \le L$), this segment does not accelerate or deform since there is no transverse force (shear) in this segment.

The location of the plastic hinge can be identified from (4.28) with the boundary condition at the tip $m(0) = 0$; thus

$$\lambda = \frac{3}{f} \qquad (4.29)$$

The acceleration at the tip $\dot{v}$ can then be obtained using Eq. (4.25),

$$\dot{v} = \frac{2f^2}{3} \qquad (4.30)$$

If $f > 3$ (or $F > 3F_c$) the tip acceleration is proportional to F^2 rather than to $(F - F_c)$ as is the case for moderate loading; that is, if $f > 3$, $\dot{v}$ is no longer

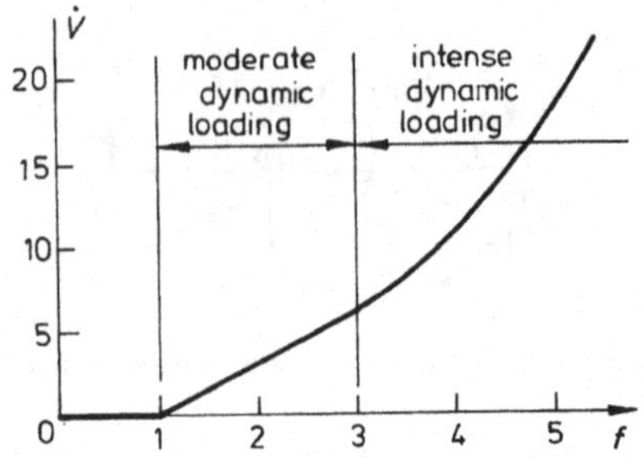

Fig. 4.9 The dependence of the tip acceleration upon the force magnitude of step loads.

proportional to $(f-1)$. For the full range of force magnitudes, this acceleration can be expressed as

$$\dot{v} = \begin{cases} 0, & 0 \le f \le 1 \\ 3(f-1), & 1 < f \le 3 \\ 2f^2/3, & 3 < f \end{cases} \tag{4.31}$$

Figure 4.9 shows the dependence of tip acceleration on the magnitude of the force. The rotation of the hinge as a function of time can be obtained directly by integration of (4.31).

Several useful remarks may be drawn from the analysis given in this section. A rigid–perfectly plastic cantilever with a suddenly applied load at the tip does not deform if $F < F_c$. Whenever $F > F_c$, there is a section where the bending moment equals the plastic bending moment; the segment outside this section begins to rotate. For a steady force the rate of angular acceleration and the location of the plastic hinge are both constant. The location of the plastic hinge is at the root for moderate loads $F_c < F \le 3F_c$, but if $F > 3F_c$ the plastic hinge is located between the ends at a distance $\Lambda = 3M_p/F$ from the tip.

4.2 Rectangular Pulse Loading

4.2.1 Three Phases in Response of Cantilever

Impact loads are frequently intense in comparison with the static collapse force; they are, however, of brief duration. Consequently, the structural deformations caused by these loads can depend on both the load magnitude and the load duration. Consider a transverse force at the tip of a uniform cantilever; the force is suddenly applied at time zero and subsequently suddenly removed at time t_d. This rectangular force pulse of magnitude F_o and duration t_d is shown in Fig. 4.10,

$$F(t) = \begin{cases} F_o, & 0 \le t \le t_d \\ 0, & t_d < t \end{cases} \tag{4.32}$$

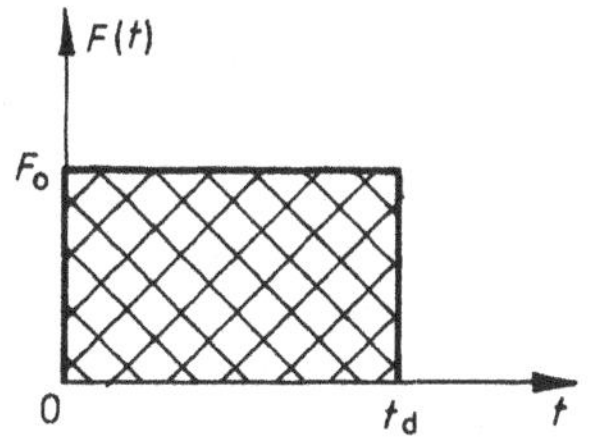

Fig. 4.10. A rectangular force pulse of magnitude F_0 and duration t_d.

During the loading period $0 \le t \le t_d$, the solution to this problem is identical with that of the step-loading case discussed in Sect. 4.1; that is, a stationary plastic hinge must form at the root if $F_c \le F_0 \le 3F_c$, or at an interior point $\Lambda_0 = 3M_p/F_0$ if $F_0 > 3F_c$. The response in this period can be termed *phase I*; during this phase the position of the hinge is denoted by H_0.

The force is suddenly removed at time $t = t_d$. Subsequently, the plastic hinge moves away from the section Λ_0, where it was located during phase I. In general therefore, we may assume a mechanism similar to that shown in Fig. 4.8, in which segment HB remains stationary and segment AH rotates about hinge H. Notice, however, that the hinge position $\Lambda = \Lambda(t)$ varies with time t if the force is not constant.

Recall that at a purely flexural plastic hinge in the interior of the member, there is no shear force. Hence, after the loading period, the translational momentum of the moving segment AH in the direction transverse to the longitudinal axis is constant,

$$\frac{1}{2}\rho\Lambda V = \int_0^t F(\bar{t})\,\mathrm{d}\bar{t} = \begin{cases} F_0 t, & 0 \le t \le t_d \\ F_0 t_d \equiv P_f, & t_d < t \end{cases} \tag{4.33}$$

where the total impulse imparted during loading $P_f \equiv \int_0^{t_d} F(t)\mathrm{d}t$. A second equation of motion is obtained by considering the change in the moment of momentum of segment AH about the initial tip position during the period before the hinge moves to a location Λ,

$$\int_0^{\Lambda} \rho \frac{\partial W}{\partial t} X \,\mathrm{d}X = M_p t \tag{4.34}$$

Substituting the velocity distribution $\mathrm{d}W/\mathrm{d}t = (1 - X/\Lambda)V$ into (4.34) results in

$$\frac{1}{6}\rho\Lambda^2 V = M_p t \tag{4.35}$$

These equations of motion can be obtained directly in terms of the momentum because the stresses are uniform and invariable in the rigid segment of the cantilever ahead of the *travelling hinge*; this is a consequence of the shear force vanishing at the hinge.

During the early phase of motion while the force is constant, $0 \le t \le t_d$, the two equations of motion (4.33) and (4.35) provide the hinge location,

$$\Lambda = \Lambda_0 = \frac{3M_p}{F_0} = \text{const} \qquad \text{if } F_0 > 3M_p \tag{4.36}$$

This stationary hinge for the loading phase has a nondimensional expression

$$\lambda_0 \equiv \frac{\Lambda_0}{L} = \frac{3}{f} \tag{4.37}$$

with $f = F_0/F_c$. Accordingly, during the loading phase the tip velocity v can be expressed as

$$v = \frac{2}{3} f^2 \tau \tag{4.38}$$

which is the same as (4.30). Phase I ends at time $\tau_1 = \tau_d \equiv t_d/T_0$, where the characteristic time T_0 is defined in Eq. (4.11) and subscript 1 pertains to phase I. The tip velocity at the end of phase I is

$$v_1 = v(\tau_d) = \frac{2}{3} f^2 \tau_d = \frac{2}{3} p_f f \tag{4.39}$$

where the nondimensional total impulse p_f is defined by

$$p_f \equiv \frac{P_f}{F_c T_0} = \frac{P_f}{L\sqrt{F_c \rho}} = f\tau_d \tag{4.40}$$

Consequently, the nondimensional deflection of the tip at time τ_1 is

$$\delta_1 \equiv \frac{\Delta(\tau_1)}{L} = \int_0^{\tau_1} v(\tau)\,\mathrm{d}\tau = \frac{p_f^2}{3} \tag{4.41}$$

Thus at the instant when the loading phase terminates, the tip deflection is proportional to the square of the applied impulse. At this instant, the rotation angle at the hinge location $\mathrm{H_0}$ is

$$\theta_1 = -\frac{1}{9} p_f^2 f \tag{4.42}$$

Note that the rotation angle is defined by $\theta \equiv \mathrm{d}w/\mathrm{d}x$, so here it has a negative sign.

When the load is removed the plastic hinge begins to travel away from the point λ_0 where it was located during phase I; it moves towards the root. The period of hinge travel immediately after time t_d is termed *phase II*; during this phase the hinge location is given by Eqs (4.33) and (4.35),

$$\lambda \equiv \frac{\Lambda}{L} = \frac{3\tau}{p_f} \tag{4.43}$$

The variation in hinge location with time is shown in Fig. 4.11; after the force is suddenly removed the hinge travels towards the root at a constant speed during phase II; i.e.

$$\dot{\lambda} \equiv \frac{\mathrm{d}\lambda}{\mathrm{d}\tau} = \frac{3}{p_f} = \text{const} \tag{4.44}$$

This hinge motion in a uniform member maintains a constant transverse momentum while the transverse velocity at the tip is slowed. The tip velocity during this travelling hinge phase is obtained by substituting Eq. (4.43) into (4.35)

$$v = \frac{6\tau}{\lambda^2} = \frac{2p_f^2}{3\tau} \tag{4.45}$$

while during this phase the tip deflection is given by

$$\delta(\tau) = \delta(\tau_d) + \int_{\tau_d}^{\tau} v(\tau)\,\mathrm{d}\tau = \frac{p_f^2}{3}\left[1 + 2\ln\left(\frac{\tau}{\tau_d}\right)\right] \tag{4.46}$$

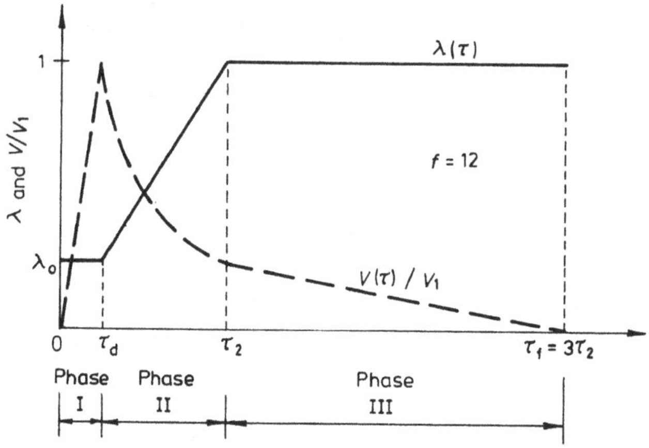

Fig. 4.11. The variation of the tip velocity and the hinge position with time for an intense rectangular pulse, $f = F_0 / F_c = 12$.

Phase II terminates when the travelling hinge reaches the root B ($\lambda = 1$) at time τ_2 where

$$\tau_2 = \frac{p_f}{3} \tag{4.47}$$

At this instant, the velocity and deflection at the tip are

$$v_2 \equiv v(\tau_2) = 2p_f \tag{4.48}$$

$$\delta_2 \equiv \delta(\tau_2) = \frac{p_f^2}{3}\left[1 + 2\ln\left(\frac{f}{3}\right)\right] \tag{4.49}$$

respectively. Note that both τ_2 and v_2 depend only upon the total impulse p_f, and not upon the magnitude or duration of the load.

After the travelling hinge reaches the root (i.e. for $\tau > \tau_2$) the pattern of deformation again changes and *phase III* takes place. In this phase of response, the plastic hinge is fixed at the root while the cantilever rotates about the root as a rigid body. Taking the change in the moment of momentum about the root for the period after τ_2,

$$v(\tau) = 3(p_f - \tau) \qquad \frac{p_f}{3} \le \tau \le p_f \tag{4.50}$$

Equations (4.38), (4.45) and (4.50) express the tip velocity during three successive phases. The variation of the tip velocity with time is shown in Fig. 4.11 as a broken line, for force $f = 12$. In this figure the tip velocity is illustrated in comparison with the velocity at the end of the loading phase, v_1.

Motion of the cantilever ceases when the velocity at the tip vanishes,

$$\tau_f = \tau_3 = p_f = 3\tau_2 \tag{4.51}$$

From Eq. (4.47) we see that the duration of the final phase with a stationary hinge at the root is twice the sum of the durations of phases I and II, i.e. $\tau_3 - \tau_2 = 2\tau_2$. Integration of (4.50) gives the increase in tip deflection during phase III. The final nondimensional deflection at the tip is found to be

$$\delta_f = \delta(\tau_f) = \delta_2 + \int_{\tau_2}^{\tau_f} v(\tau)\,\mathrm{d}\tau = \frac{p_f^2}{3}\left[1 + 2\ln\left(\frac{f}{3}\right) + 2\right]$$

where the three terms in the bracket represent the contributions from phases I, II and III, respectively. The part due to the final phase of rotation at the root is twice as large as that during the loading phase where the hinge is closer to the tip (if $f > 3$). The final tip deflection δ_f is

$$\delta_f = p_f^2\left[1+\frac{2}{3}\ln\left(\frac{f}{3}\right)\right], \qquad f > 3 \tag{4.52}$$

The final rotation angle at the root at termination of phase III is

$$\theta_3 = -\frac{2}{3}p_f^2 \tag{4.53}$$

4.2.2 Deformed Shape

To find the final deformed shape of the cantilever, first consider the plastic curvature $\kappa(X)$, caused by the travelling hinge. This curvature is generated during phase II. Across the travelling hinge there is a difference in angular velocity $\mathrm{d}\theta/\mathrm{d}t$; since there is a one-to-one correlation between hinge position and time during phase II, the angular velocity can be expressed as a function of hinge position rather than time,

$$\frac{\mathrm{d}\theta}{\mathrm{d}t} = \frac{\mathrm{d}\theta}{\mathrm{d}\Lambda}\frac{\mathrm{d}\Lambda}{\mathrm{d}t} = \kappa(\Lambda)\frac{\mathrm{d}\Lambda}{\mathrm{d}t} = \frac{V}{\Lambda} \tag{4.54}$$

By referring to Eqs (4.43) and (4.44), the nondimensional curvature $k(\lambda)$ that develops during phase II is found to be

$$k \equiv \kappa L = \frac{v}{\lambda\dot{\lambda}} = \frac{2p_f^2}{3\lambda^2} \tag{4.55}$$

In segment $\lambda_o \le x \le 1$, changes in curvature occur only as the plastic hinge transits any section. Thus, for a cross-section with coordinate x, the curvature is

$$k(x) = \frac{\partial^2 w(x)}{\partial x^2} = \frac{2p_f^2}{3x^2} \tag{4.56}$$

Although the hinge speed is constant in a uniform member, the curvature generated by this hinge decreases with increasing distance from the tip since the angular velocity $\mathrm{d}\theta/\mathrm{d}t$ is decreasing.

Integrating Eq. (4.56) twice with respect to x and using $w_2(1) = \partial w_2(1)/\partial x = 0$ at the root, we find the nondimensional deflection produced in phase II as

$$w_2(x) = \frac{2p_f^2}{3}\left(\ln\frac{1}{x}+x-1\right), \qquad \lambda_o \le x \le 1 \tag{4.57}$$

The rotational angles produced in phases I and III have already been given in (4.42) and (4.53), respectively. Summing deflections obtained from each of these terms, we obtain a final nondimensional deflection

$$w_f(x) = \begin{cases} p_f^2\left[1-\dfrac{x}{\lambda_o}+\dfrac{2}{3}\ln\left(\dfrac{1}{\lambda_o}\right)\right], & 0 \le x \le \lambda_o \\[2ex] \dfrac{2}{3}p_f^2\ln\left(\dfrac{1}{x}\right), & \lambda_o \le x \le 1 \end{cases} \tag{4.58}$$

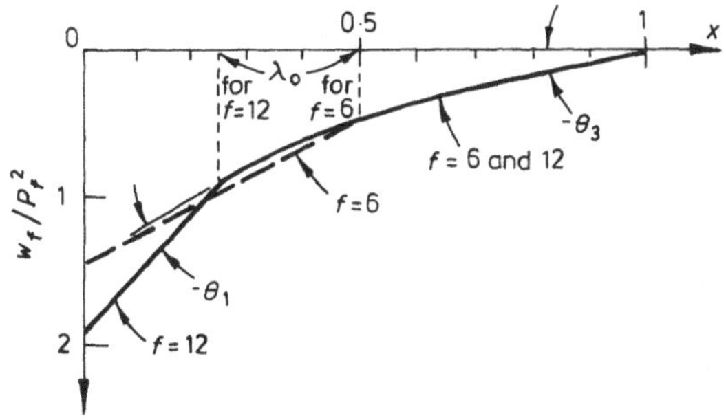

Fig. 4.12. Final deformed shapes of a cantilever subjected to rectangular pulses ($f = 6$ and $f = 12$).

with $\lambda_o = 3/f$. Figure 4.12 shows a typical final deformed shape of a cantilever for force ratios $f = F_o/F_c = 6$ and 12.

In the case of impulsive loading, $p_f = \text{const}$ although $F_o \to \infty$ while $\lambda_o \to 0$. For an impulsive load the final deflection of the cantilever approaches a logarithmic curve

$$w_f \to \frac{2}{3} p_f^2 \ln\left(\frac{1}{x}\right) \tag{4.59}$$

It is worth noting that for any force magnitude the final tip deflection Δ_f of a uniform cantilever with rectangular cross-section is

$$\Delta_f \propto \frac{P_f^2}{b^2 h^3} \tag{4.60}$$

where the cross-section has width b and depth h. Thus the final tip deflection is independent of the beam length L and the slenderness L/h.

4.2.3 Energy Dissipation

While a force acts on a plastically deforming structure, part of the work done on the structure by the force goes immediately into plastic work of deformation and part increases the kinetic energy of the structure. Ultimately all the input energy E_{in} is dissipated by plastic work.

The cantilever has energy dissipation focused at stationary hinges H_o and B; at each of these sections the dissipation is proportional to the rotational angle. The remainder of the input energy is dissipated by the plastic curvature (4.56) produced in phase II. Thus, the nondimensional energy dissipation in each phase d_i can be calculated as follows:

$$d_1 \equiv \frac{D_1}{M_p} = |\theta_1| = \frac{1}{9} p_f^2 f \tag{4.61}$$

$$d_2 \equiv \frac{D_2}{M_p} = \int_{\lambda_o}^{1} k(x)\,\mathrm{d}x = \frac{2}{9} p_f^2 (f-3) \tag{4.62}$$

$$d_3 \equiv \frac{D_3}{M_p} = |\theta_3| = \frac{2}{3} p_f^2 \tag{4.63}$$

The input energy, i.e. the work done by the pulse, is

$$E_{in} = F_0 \Delta(t_d) = \frac{P_f^2 F_0}{3\rho M_p} \tag{4.64}$$

or, in nondimensional form

$$e_{in} \equiv \frac{E_{in}}{M_p} = \frac{1}{3} p_f^2 f \tag{4.65}$$

so that the part of the input energy e_{in} that is dissipated during each phase of motion has the following proportionality:

$$\left(\frac{d_1}{e_{in}}\right) : \left(\frac{d_2}{e_{in}}\right) : \left(\frac{d_3}{e_{in}}\right) = \frac{1}{3} : \left(\frac{2}{3} - \frac{2}{f}\right) : \frac{2}{f} \tag{4.66}$$

It is observed, therefore, that (1) phase I always dissipates 1/3 of the total input energy, independent of the magnitudes of F_0 and t_d, provided $F_0 > 3F_c$ and (2) the energy dissipated in phases II and III is distributed between these phases according to the force ratio $f = F_0 / F_c$. Hence, as the force ratio f increases, the proportion of total energy dissipated in phase II increases at the expense of phase III. The distribution of energy dissipation along the cantilever is shown in Fig. 4.13; the dependence of energy dissipation on the magnitude of the force f is illustrated in Fig. 4.14.

If $1 \le f \le 3$ (i.e. $F_c \le F \le 3F_c$) then the hinge will remain at the root B throughout the response; in other words, in the case of moderate loading, phase II does not appear and all of the input energy is dissipated at the root. In this case, it can be shown that the final tip deflection is

$$\delta_f = \frac{3}{2} p_f^2 \left(1 - \frac{1}{f}\right), \qquad 1 < f < 3 \tag{4.67}$$

Together with Eq. (4.52), this provides the dependence of the final tip deflection on the magnitude of the applied force f; this is shown in Fig. 4.15. Since

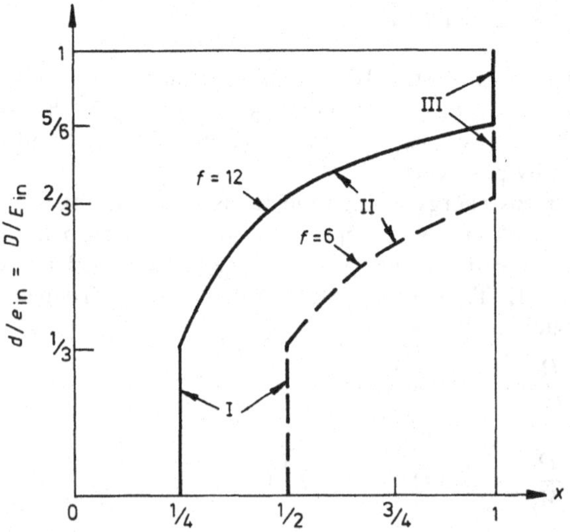

Fig. 4.13. Accumulation of energy dissipation along a cantilever subjected to rectangular pulses ($f = 6$ and $f = 12$).

$f\tau_d = p_f$, Eqs (4.67) and (4.52) also indicate how the final deflection varies with the pulse duration t_d.

4.2.4 Synopsis

1. If $f = F_0/F_c > 3$, the response of a cantilever to a rectangular pulse consists of three phases. Phase II is most significant if $f >> 3$. This intermediate phase is characterized by a travelling hinge triggered by unloading of the external force. The hinge travels away from the loaded segment at a constant speed.

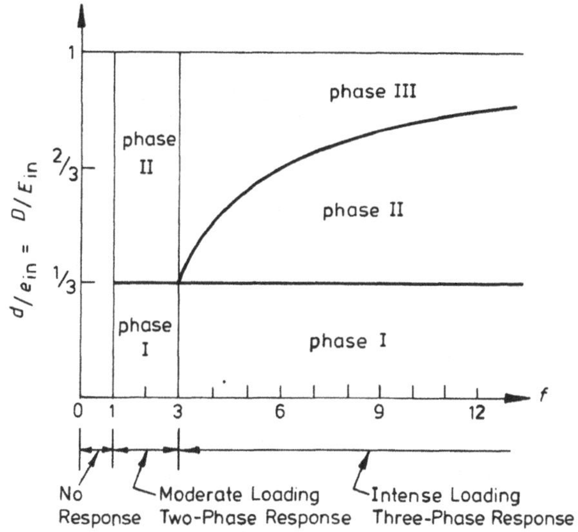

Fig. 4.14. Dependence of energy dissipation on the force magnitude f of rectangular pulse.

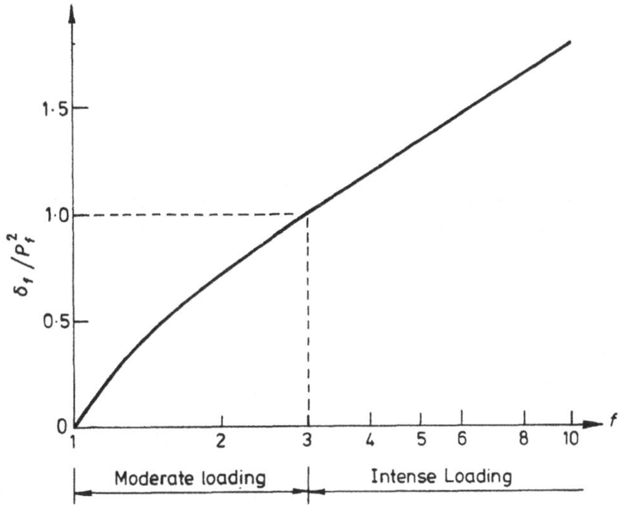

Fig. 4.15. Dependence of the final tip deflection on the force magnitude f of rectangular pulse.

2. Deformation associated with the three phases of response consists of (I) a rotation about the initial stationary hinge H_0; (II) a continuous change of curvature in segment H_0B; and (III) a rotation about the root B. As a result, the final deformed shape of the cantilever consists of kinks at H_0 and B and a plastically curved segment H_0B.
3. Although time t_2, at which the travelling hinge reaches the root, is independent of the force magnitude F_0, the energy dissipation in phase II depends on the force. As the force intensity increases, a larger part of the input energy is dissipated in the curved segment of the cantilever.

The main analytical results obtained in this section can be summarized in Table 4.1.

Table 4.1 Phase transition conditions for response of cantilever to rectangular pulse, $F_0 / F_c > 3$

Phase	Time	Hinge position	Tip velocity	Tip deflection
	$\tau = 0$	$\lambda_0 = 3/f$	$v = 0$	$\delta = 0$
Phase I				
	$\tau_1 = \tau_d$	$\lambda_0 = 3/f$	$v_1 = 2p_f f/3$	$\delta_1 = p_f^2/3$
Phase II				
	$\tau_2 = p_f/3$	$\lambda = 1$	$v_2 = 2p_f$	$\delta_2 = \frac{1}{3}p_f^2\left[1 + 2\ln\left(\frac{f}{3}\right)\right]$
Phase III				
	$\tau_f = p_f = 3\tau_2$	$\lambda = 1$	$v_f = 0$	$\delta_f = p_f^2\left[1 + \frac{2}{3}\ln\left(\frac{f}{3}\right)\right]$

4.3 Features of Travelling Hinges

In the last section, it was shown that a rigid–perfectly plastic cantilever has all deformation concentrated in a plastic hinge. The hinge is stationary if the applied force is constant or if the momentum distribution has converged to a mode form; if the force varies in its magnitude, however, there is a transient phase of motion wherein the hinge travels along the cantilever. The travelling hinge is an important concept in rigid–plastic structural dynamics; this concept was first introduced by Lee and Symonds [1952]. They analyzed flexural deformation in a rigid–plastic free–free beam subjected to a force pulse at the mid-point. In the present section, some general features of the travelling hinge will be discussed.

At a stationary plastic hinge the displacement is continuous but the inclination is discontinuous. Rotation of the hinge forms a kink in the neutral axis at the hinge location. In contrast, a travelling hinge leaves behind a residual curvature that is finite at every point. While both the displacement and inclination are continuous at a travelling hinge, the inclination $\theta = \mathrm{d}W/\mathrm{d}X = \mathrm{d}w/\mathrm{d}x$ may have weak discontinuities; that is the derivative with respect to time $\mathrm{d}\theta/\mathrm{d}t$ and that with respect to a spatial coordinate $\mathrm{d}\theta/\mathrm{d}x$ (the angular velocity and the curvature) may be discontinuous although θ is continuous.

In general, let $\Lambda(t)$ denote the location of a plastic hinge. Furthermore, for any function $\Phi(X)$ let Φ^+ and Φ^- denote the values of the function on either side of the plastic hinge position; thus,

$$\Phi^+ \equiv \Phi(\Lambda+0) \equiv \lim_{\varepsilon\to 0}\Phi(\Lambda+\varepsilon), \quad \Phi^- \equiv \Phi(\Lambda-0) \equiv \lim_{\varepsilon\to 0}\Phi(\Lambda-\varepsilon)$$

Accordingly, $[\Phi] \equiv \Phi^+ - \Phi^-$ denotes the jump or discontinuity in magnitude of $\Phi(X)$ that develops as the hinge transits the point $X = \Lambda(t)$.

At a travelling plastic hinge the transverse deflection $w(\lambda,\tau)$ is continuous so

$$[w] \equiv w^+ - w^- = 0 \tag{4.68}$$

Furthermore if shear deformation is negligible, then in order to develop continuous displacements, the velocity $\dot{w} = \partial w/\partial\tau$ of adjacent material particles must be continuous; i.e.

$$[\dot{w}] \equiv \dot{w}^+ - \dot{w}^- = 0 \tag{4.69}$$

Equations (4.68) and (4.69) apply to both stationary and travelling hinges.

At a hinge, continuity of displacements requires that a derivative with respect to time of (4.68) must also vanish:

$$\frac{\mathrm{d}}{\mathrm{d}\tau}[w(\lambda,\tau)] = 0 \tag{4.70}$$

where $\lambda = \lambda(\tau)$ is regarded as a function of nondimensional time τ. Thus, (4.68) gives

$$\begin{aligned} 0 &= \frac{\mathrm{d}}{\mathrm{d}\tau}[w] = \frac{\mathrm{d}}{\mathrm{d}\tau}\left[w^+(\lambda,\tau) - w^-(\lambda,\tau)\right] \\ &= \left(\frac{\partial w^+}{\partial\tau} - \frac{\partial w^-}{\partial\tau}\right) + \left(\frac{\partial w^+}{\partial\lambda}\frac{\partial\lambda}{\partial\tau} - \frac{\partial w^-}{\partial\lambda}\frac{\partial\lambda}{\partial\tau}\right) \\ &= \left(\frac{\partial w^+}{\partial\tau} - \frac{\partial w^-}{\partial\tau}\right) + \dot{\lambda}\left(\frac{\partial w^+}{\partial\lambda} - \frac{\partial w^-}{\partial\lambda}\right) \end{aligned}$$

That is,

$$[\dot{w}] + \dot{\lambda}[w'] = 0 \tag{4.72}$$

where velocity $\dot{w}$ and inclination $w' = \partial w/\partial x$ are derivatives for a material particle that is being transited by the hinge. Moreover, $\dot{\lambda}$ is the travelling speed of the plastic hinge. Hence at a hinge, we obtain from (4.69) and (4.72),

$$\dot{\lambda}[w'] = 0 \tag{4.73}$$

Equation (4.73) indicates that (1) for a stationary hinge $\dot{\lambda} = 0$, so $[w'] \neq 0$, i.e. the slope is discontinuous at the hinge; and (2) for a travelling hinge $\dot{\lambda} \neq 0$ so $[w'] = 0$, i.e. the slope remains continuous at a travelling hinge.

Similarly, if transverse velocity is continuous, differentiation of Eq. (4.69) yields

$$[\ddot{w}] + \dot{\lambda}[\dot{w}'] = 0 \tag{4.74}$$

Hence, (1) for a stationary hinge, $[\ddot{w}] = 0$ must hold, i.e. the acceleration remains continuous; and (2) for a travelling hinge, both acceleration $\ddot{w}$ and the angular velocity $\dot{w}' = \dot{\theta}$ can be discontinuous.

In the case of a travelling hinge, starting from $[w']=0$ and carrying out a similar differentiation, we obtain

$$[\dot{w}'] + \dot{\lambda}[w''] = 0 \tag{4.75}$$

Eliminating $[\dot{w}']$ from Eqs (4.74) and (4.75) gives

$$[\ddot{w}] = \dot{\lambda}^2[w''] \tag{4.76}$$

Note that (4.75) and (4.76) apply only to travelling hinges. Expressions (4.73)–(4.76) provide the features of discontinuities at plastic hinges. These are summarized in Table 4.2.

Table 4.2 Discontinuities at plastic hinges

	Stationary hinge $\dot{\lambda}=0$	Travelling hinge $\dot{\lambda}\neq 0$
Deflection w	C	C
Slope or inclination $w'=\theta$	D	C
Curvature $w''=k$	D[a]	D
Velocity $\dot{w}$	C	C
Angular velocity $\dot{w}'=\dot{\theta}$	D	D
Acceleration $\ddot{w}$	C	D

Note: C, continuous; D, discontinuous.
[a] $k(\lambda+0)=k(\lambda-0)=0,\ k(\lambda)\to\infty$.

A plastic hinge forms at a cross-section of a rigid–perfectly plastic cantilever, where the yield criterion $|M(\Lambda)| = M_p$ is satisfied. Since the energy dissipation rate at the plastic hinge is positive (i.e. $\dot{D} = M_p|\dot{\theta}| > 0$), the associated flow rule requires $\dot{\theta} \neq 0$. Indeed, Table 4.2 indicates that a discontinuity in angular velocity is a common and essential feature of plastic hinges, no matter whether they are stationary or travelling.

The expressions concerning discontinuities at travelling hinges can be useful in analysis. For instance, the configuration shown in Fig. 4.8 contains a static and undeformed segment HB; hence (4.75) leads to

$$\dot{w}'(\lambda-0) + \dot{\lambda}w''(\lambda-0) = 0 \tag{4.77}$$

By substituting $\dot{w}'(\lambda-0) = -v/\lambda$ and $w''(\lambda-0) = k(\lambda)$ into (4.77), one obtains Eq. (4.55) that was employed to find the final shape of the deformed cantilever.

4.4 General Pulse Loading

4.4.1 General Considerations

Now consider a force pulse $F(t)$ at the tip of a cantilever. Suppose the pulse has arbitrary shape but is always non-negative, i.e. $F(t) \geq 0$. Since any force that is smaller than the static collapse load F_c does not deform the initially static rigid–plastic cantilever, it is convenient to take t = 0 as the instant when $F(t)$ first reaches F_c, as shown in Fig. 4.16.

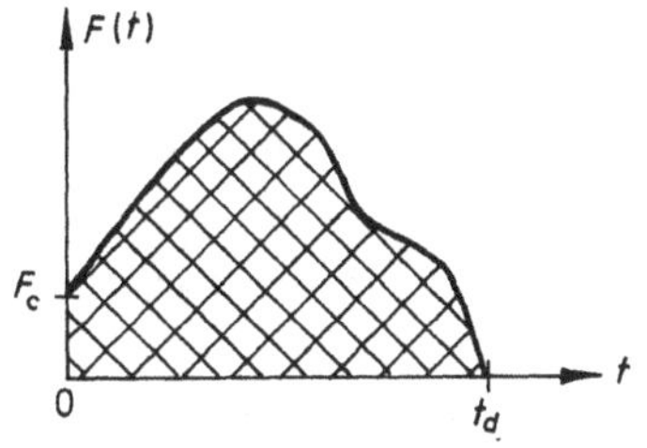

Fig. 4.16. General force pulse, with $t = 0$ taken as the instant when $F(t)$ first reaches F_c and $F(t) \geq 0$ for the entire loading duration $0 \leq t \leq t_d$

In order to determine how hinge position $\Lambda(t)$ and tip velocity $V(t)$ depend upon $F(t)$, we assume that at time t a single hinge H is located at spatial coordinate $X = \Lambda(t)$. Then similarly to the analysis in Sect. 4.1, the equations of motion for segment AH (refer to Fig. 4.8) can be written as

$$\frac{\mathrm{d}}{\mathrm{d}t}\left(\frac{1}{2}\rho\Lambda V\right) = F(t) \tag{4.78}$$

$$\frac{\mathrm{d}}{\mathrm{d}t}\left(\frac{1}{6}\rho\Lambda^2 V\right) = M_p \tag{4.79}$$

By employing the same nondimensional variables defined in Eqs (4.9), (4.10) and (4.14), the above expressions are recast as

$$\frac{\mathrm{d}}{\mathrm{d}\tau}\left(\frac{1}{2}\lambda v\right) = f(\tau) \tag{4.80}$$

$$\frac{\mathrm{d}}{\mathrm{d}\tau}\left(\frac{1}{6}\lambda^2 v\right) = 1 \tag{4.81}$$

These equations can be solved for hinge location $\lambda(\tau)$ and tip velocity $v(\tau)$ in terms of the *impulse* that has been applied at any time $p(\tau)$,

$$\lambda(\tau) = \frac{3\tau}{p(\tau)} \tag{4.82}$$

$$v(\tau) = \frac{2}{3\tau}(p(\tau))^2 \tag{4.83}$$

where

$$p(\tau) \equiv \int_0^\tau f(\bar{\tau})\mathrm{d}\bar{\tau} \tag{4.84}$$

The travelling speed of the hinge is found by differentiating Eq. (4.82),

$$\dot{\lambda}(\tau) \equiv \frac{\mathrm{d}\lambda}{\mathrm{d}\tau} = 3\frac{p(\tau) - \tau f(\tau)}{(p(\tau))^2} \tag{4.85}$$

while the direction of travel is determined by

$$\mathrm{sgn}(\dot{\lambda}) = \mathrm{sgn}[p(\tau) - \tau f(\tau)] \tag{4.86}$$

Hence, the travelling hinge moves towards the root or towards the tip depending on whether $[p(\tau) - \tau f(\tau)]$ is positive or negative.

If the force $f(\tau)$ is a continuous function of time τ, then the mean value theorem applied at any instant τ_1 gives

$$\frac{p(\tau_1)}{\tau_1} \equiv \frac{1}{\tau_1}\int_0^{\tau_1} f(\tau)\mathrm{d}\tau = f(\tau^*), \qquad 0 \le \tau^* \le \tau_1 \tag{4.87}$$

The following deductions can be made immediately:

1. For a force that monotonically increases with time, $\dot{f}(\tau) > 0$ and $f(\tau^*) < f(\tau_1)$; hence (4.86) and (4.87) indicate that $\dot{\lambda} < 0$ (i.e. the hinge moves towards the tip).
2. If $f(\tau)$ is a monotonically decreasing function of time τ, then $\dot{f} < 0$ and $f(\tau^*) > f(\tau_1)$; consequently, $\dot{\lambda} > 0$ (i.e. the hinge moves towards the root).
3. If $f(\tau) =$ const, then $\dot{\lambda} = 0$; i.e. while a constant force acts the hinge is stationary.
4. If the magnitude of a pulse first increases and later decreases, the travelling hinge may change direction. In this case the turning point can be determined as the time when the impulse satisfies the following condition,

$$\int_0^{\tau} f(\bar{\tau})\mathrm{d}\bar{\tau} = \tau f(\tau) \tag{4.88}$$

Figure 4.17a shows that (4.88) is the same as making the area of rectangle OD_1DD_2 equal to the area underneath portion OD of the pulse curve. Figure 4.17b is an example: for this triangular pulse the turning point of the travelling hinge is at $t_1 = t_d/\sqrt{2} = 0.707t_d$. The hinge moves from the root towards the tip in the period $0 < t < t_1$, and then back towards the root during $t > t_1$.

5. After the termination of any pulse $t > t_d$ the final impulse p_f has already been applied so $p(\tau) = p_f \equiv \int_0^{\tau_d} f(\tau)\,\mathrm{d}\tau > 0$ while $\tau f(\tau) = 0$. Thus during this period the hinge moves away from the loading point, $\dot{\lambda} > 0$ according to Eq. (4.86).

Hence, after a pulse of any shape has terminated, the distribution of translational momentum spreads and becomes more uniform. Equation (4.82) indicates that in a member with uniformly distributed properties, the hinge reaches the root at time $\tau = p_f/3$ when the tip velocity $v = 2p_f$; thereafter the hinge remains stationary at the root. It can be easily proved that the total rotation angle at the root always is equal to

$$\theta_{Bf} = -\frac{2}{3}p_f^2 \tag{4.89}$$

irrespective of the pulse shape if the total impulse is large enough, $p_f > 3\tau_d$.

6. Since the final rotation angle at the tip A equals the sum of the integral of the curvatures generated by the travelling hinge along the entire cantilever and the rotation angles generated at fixed hinges (e.g. at the root), it is concluded that

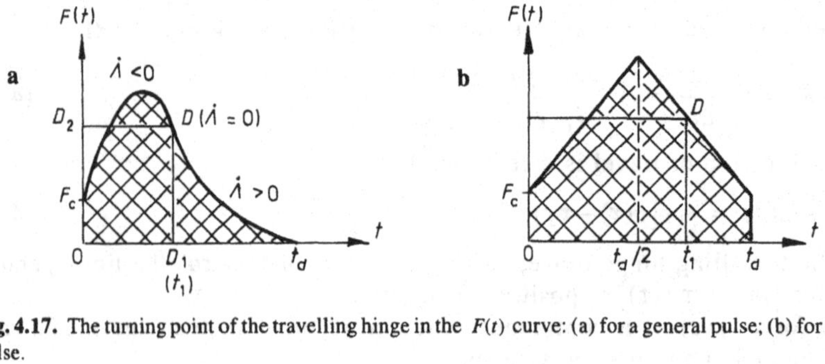

Fig. 4.17. The turning point of the travelling hinge in the $F(t)$ curve: (a) for a general pulse; (b) for a triangular pulse.

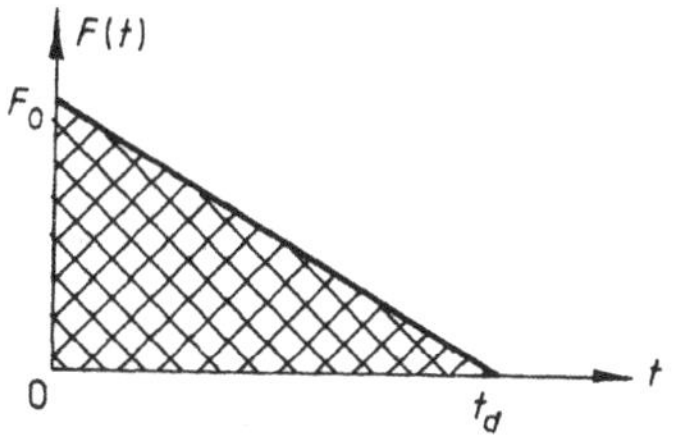

Fig. 4.18. A linearly decaying pulse with duration t_d.

$$\theta_{Af} = -\frac{E_{in}}{M_p} \equiv -e_{in} \tag{4.90}$$

where E_{in} and e_{in} are respectively the dimensional and nondimensional work done by an applied pulse of any shape.

In related experiments, the final hinge rotations θ_{Bf} and θ_{Af} can be easily measured on deformed specimens, so according to Eqs (4.89) and (4.90) these measured angles may provide checks on the final impulse P_f, and the total work E_{in} done by the pulse. This point is discussed further in Chap. 7.

4.4.2 Example: Linearly Decaying Pulse

Consider a cantilever that is subjected to a linearly decaying force pulse at the tip, as shown in Fig. 4.18. That is,

$$F(t) = \begin{cases} F_0\left(1 - \dfrac{t}{t_d}\right), & 0 \le t \le t_d \\ 0, & t \ge t_d \end{cases} \tag{4.91}$$

Taking $f_0 = F_0 / F_c$ and $p_f = f_0 t_d / 2$, a phase-by-phase analysis can be performed similar to that in Sect. 4.2. It is found that the response is characterized by the force magnitude f_0:

1. If $f_0 = F_0 / F_c < 1$, the cantilever remains static and undeformed.
2. If $1 < f_0 < 2$, a stationary plastic hinge appears at the root. The motion begins at $\tau = 0$ and then ceases before the pulse terminates; i.e. $\tau_f < \tau_d$.
3. If $2 < f_0 < 3$, a stationary plastic hinge appears at the root. The motion continues until $\tau_f > \tau_d$.
4. If $3 < f_0 < 6$, a plastic hinge appears and moves towards the root from an initial position $\lambda_0 = 3/f_0$. This travelling hinge reaches the root before the pulse terminates. Subsequently, the cantilever simply rotates about the hinge at the root.
5. If $f_0 > 6$, the early response is the same as that in case of $3 < f_0 < 6$, but the travelling hinge does not get to the root until after the pulse terminates; then there is a stationary hinge at the root until the motion finally ceases. These results are summarized in Table 4.3. Figure 4.19 shows the variation of the hinge position and the tip velocity with time for the case of $f_0 = 12$.

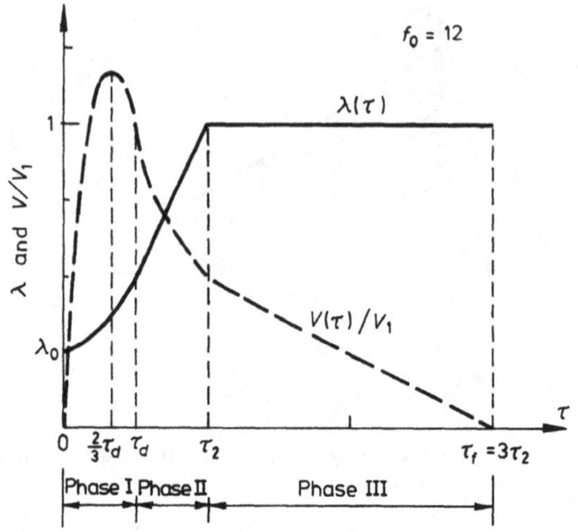

Fig. 4.19. Variation of tip velocity and hinge position with time for a linearly decaying pulse of magnitude $f_0 = F_0 / F_c = 12$.

Table 4.3 Phase transition conditions for response of cantilever to linearly decaying pulse with initial amplitude $f_0 > 6$

Phase	Time τ	Hinge position λ	Tip velocity v	Tip deflection δ
	$\tau = 0$	$\lambda_0 = 3/f_0$	$v = 0$	$\delta = 0$
Phase I	$\tau = \frac{2}{3}\tau_d$	$\lambda = \frac{9}{2f_0}$	$v_{max} = \frac{32}{81} p_f f_0$	$\delta = \frac{88}{243} p_f^2$
$(f > 0,\ \dot{\lambda} > 0)$				
	$\tau_1 = \tau_d$	$\lambda_1 = \frac{6}{f_0}$	$v_1 = \frac{1}{3} p_f f_0$	$\delta_1 = \frac{11}{18} p_f^2$
Phase II				
$(f = 0,\ \dot{\lambda} = 0)$				
	$\tau_2 = \frac{p_f}{3}$	$\lambda_2 = 1$	$v_2 = 2\,p_f$	$\delta_2 = p_f^2\left[\frac{11}{18} + \frac{2}{3}\ln\left(\frac{f_0}{6}\right)\right]$
Phase III				
$(f = 0,\ \dot{\lambda} = 0)$				
	$\tau_f = p_f = 3\tau_2$	$\lambda_f = 1$	$v_f = 0$	$\delta_f = p_f^2\left[\frac{23}{18} + \frac{2}{3}\ln\left(\frac{f_0}{6}\right)\right]$

4.4.3 Equivalent Replacement of Arbitrary Pulse

Although some general expressions have been given for a pulse of arbitrary shape (e.g. Eqs (4.82) and (4.83)), the analysis is very tedious and error prone. It is also impossible or impractical to determine or reproduce details of the actual loading pulse in many engineering problems. It is significant, therefore, to study the effect of pulse shape on the dynamic plastic deformation of structures. Based on the rigid–perfectly plastic idealization for dynamically loaded structures, Youngdahl [1970, 1971] proposed a set of equivalence parameters which effectively eliminate

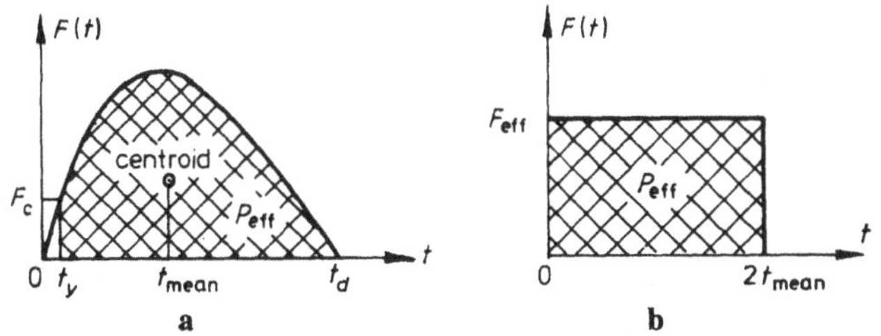

Fig. 4.20. Youngdahl's equivalence parameters for a general pulse: (a) definitions of P_{eff} and t_{mean}; (b) the equivalent rectangular pulse.

the effect of pulse shape for the final plastic deformation.

Youngdahl's equivalence parameters are described in Fig. 4.20. An effective total impulse is defined as

$$P_{\mathrm{eff}} = \int_{t_y}^{t_f} F(t)\,\mathrm{d}t \tag{4.92}$$

where t_y and t_f are the times when plastic deformation begins and ends. Then an effective load is defined by

$$F_{\mathrm{eff}} = \frac{P_{\mathrm{eff}}}{2t_{\mathrm{mean}}} \tag{4.93}$$

where t_{mean} is the centroid of the effective pulse. This is given by

$$t_{\mathrm{mean}} = \frac{1}{P_{\mathrm{eff}}}\int_{t_y}^{t_f} (t - t_y)F(t)\,\mathrm{d}t \tag{4.94}$$

In Eqs (4.92) and (4.94), t_y can be determined as the time when the force pulse first equals the plastic collapse force $F(t_y) = F_c$. The termination time t_f is not known *a priori*, however, Youngdahl [1970] proposed that t_f can be calculated from the following approximation:

$$P_{\mathrm{eff}} \approx F_c(t_f - t_y) \tag{4.95}$$

After examining solutions to four classical problems in dynamic plasticity, including the simply supported beam, circular plate and cylindrical shell, Youngdahl concluded that the effect of the pulse shape is virtually eliminated if the equivalent impulse P_{eff} and effective load F_{eff} are used to characterize the pulse. This suggests equivalence between pulses, so that an arbitrary pulse can be replaced by an 'equivalent' rectangular pulse, as shown in Fig. 4.20b. The equivalent pulses result in approximately the same final deflection for plastically deformed structures. This point has been discussed further by Krajcinovic [1972] and Zhu et al. [1986].

As an example, consider the linearly decaying pulse with initial amplitude $F > 6F_c$, which was analyzed in Sect. 4.4.2. It is easily shown that $t_y = 0$, $t_f = t_d$, $P_{\mathrm{eff}} = P_f$ and the mean time for the pulse $t_{\mathrm{mean}} = t_d/3$. Consequently, the magnitude of the effective rectangular pulse F_{eff} is

$$F_{\mathrm{eff}} = \frac{P_{\mathrm{eff}}}{2t_{\mathrm{mean}}} = \frac{F_o t_d/2}{2t_d/3} = \frac{3}{4}F_o \tag{4.96}$$

Substituting Eq. (4.92) into (4.52) and (4.67), the final tip deflection produced by this equivalent rectangular pulse is found to be

$$\delta_f(\text{equiv. rect.}) = \begin{cases} \frac{3}{2}p_f^2\left(1-\frac{1}{f_{\text{eff}}}\right) = \frac{3}{2}p_f^2\left(1-\frac{4}{3f_0}\right), & \frac{4}{3} < f_0 \le 4 \\ p_f^2\left[1+\frac{2}{3}\ln\left(\frac{f_{\text{eff}}}{3}\right)\right] = p_f^2\left[1+\frac{2}{3}\ln\left(\frac{f_0}{4}\right)\right], & 4 < f_0 \end{cases} \quad (4.97)$$

where f_0 is defined by the initial magnitude of the linearly decaying pulse.

Figure 4.21 compares the numerical calculation for final tip deflection obtained with the linearly decaying and the equivalent rectangular pressure pulses. It shows that (4.97) provides an excellent approximation for the final tip deflection developed in response to the linearly decreasing force described in Table 4.3. For $f_0 > 6$, Table 4.3 and Eq. (4.97) give

$$\delta_f(\text{lin. decay}) = p_f^2\left[\frac{23}{18}+\frac{2}{3}\ln\left(\frac{f_0}{6}\right)\right] = 0.08327+\frac{2}{3}\ln(f_0) \quad (4.98)$$

$$\delta_f(\text{equiv. rect.}) = p_f^2\left[1+\frac{2}{3}\ln\left(\frac{f_0}{4}\right)\right] = 0.07580+\frac{2}{3}\ln(f_0) \quad (4.99)$$

These equations represent parallel straight lines that are almost coincident as shown in Fig. 4.21. Table 4.4 compares the accuracy of final displacements obtained with the equivalent pulse for a range of different force magnitudes.

4.5 Impact on Cantilever

In previous sections, the load applied at the tip of a cantilever is assumed to be a prescribed force pulse; i.e. the load is a specified function of time. However, dynamic loads due to impact are not a prescribed function of time; the time dependence of the applied load in a collision depends on the compliance of the colliding bodies. Nevertheless, if a high speed missile strikes a structure or if two

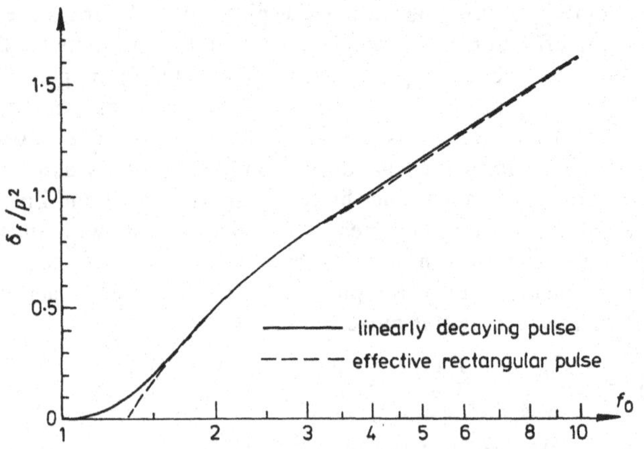

Fig. 4.21. Final tip deflection obtained with the linearly decaying and equivalent rectangular pulses.

Table 4.4 Comparison of tip deflections developed in response to linearly decaying and equivalent rectangular pulses

f_0	δ_f (linear decay)	δ_f (equiv. rect.)	Error (%)
1.0	0.0	0.0	0.0
1.5	0.1975	0.1667	-15.6
2.0	0.50	0.50	0.0
2.5	0.70	0.70	0.0
3.0	0.8333	0.8333	0.0
4.0	1.0139	1.0000	- 1.3
6.0	1.2777	1.2703	- 0.58
12.0	1.7399	1.7324	- 0.43
30.0	2.3507	2.3433	- 0.32

vehicles collide, the total impulse imparted to each body may be known from the change in relative velocities. The impulse is just the integral of the interaction force during the collision period. The impulse on each body is equal to the change in momentum of the body during collision. The impulse imparted to a structure by a high speed collision results in a sudden change in velocity for only the part of the body that is in contact with the colliding mass.

4.5.1 Problem and Assumptions

Suppose a particle with mass G strikes the tip of a uniform cantilever; the particle is initially travelling at velocity V_0 and it remains attached to the tip after impact, as shown in Fig. 4.22. This problem was first analyzed by Parkes [1955]. The analysis in the following sections assumes:

1. The constitutive model is rigid–perfectly plastic so effects of elasticity, strain hardening and strain-rate are all neglected (these effects are discussed in Chap. 5).
2. The collision imparts an initial velocity to a particle of mass G; the size, shape and deformability of the collision region are all neglected.
3. At impact the cantilever is at rest except for the colliding particle; this has velocity V_0 in the direction transverse to the undeformed axis of the cantilever.
4. After impact the colliding mass adheres to the tip of the cantilever.
5. Deflections remain small in comparison with the length of the member.
6. In a plastically deforming region, axial force has negligible effect on the bending moment.

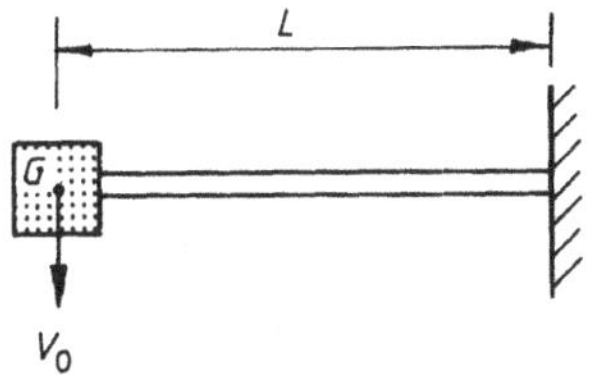

Fig. 4.22. Cantilever with mass G attached at the tip; this mass is given an initial velocity V_0.

4.5.2 Changing Pattern of Deformation

In order to study the dynamic deformation of the cantilever, let $V = V(t)$ denote the velocity of the tip (i.e. mass G) at any time $t > 0$ and $F(t)$ the contact force applied to the particle by the tip of the cantilever. $F(t)$ is just the shear force on the section adjacent to the colliding mass. This shear force is equal to the contact force on the particle; it can be obtained from Newton's second law,

$$F(t) = -G\frac{\mathrm{d}V}{\mathrm{d}t} \tag{4.100}$$

As in previous sections, it is convenient to introduce nondimensional variables,

$$T_0 \equiv L\sqrt{\frac{\rho}{F_c}} = L\sqrt{\frac{\rho L}{M_p}}, \quad \tau \equiv \frac{t}{T_0}, \quad \dot{(\ )} = \frac{\partial(\ )}{\partial \tau}$$

$$f(\tau) \equiv \frac{F}{F_c}, \quad v \equiv \frac{V}{L/T_0} = V\sqrt{\frac{\rho}{F_c}}, \quad v_0 \equiv \frac{V_0}{L/T_0} = V_0\sqrt{\frac{\rho}{F_c}}$$

where ρ is the mass per unit length of the cantilever, L is the length of the cantilever and $F_c = M_p / L$ is the static plastic collapse load for a transverse force at the tip of the cantilever. Let γ be the *mass ratio* between the concentrated mass at the tip and that of the remainder of the cantilever

$$\gamma \equiv \frac{G}{\rho L} \tag{4.101}$$

Then Eq. (4.100) can be recast in nondimensional form as

$$f(\tau) = -\gamma \dot{v} \tag{4.102}$$

Transient Phase of Motion (Phase I) A suddenly applied velocity over part of the structure is termed impulsive loading. This concept is useful for analyzing dynamic deformation due to impact. With impulsive loading, the velocity distribution is discontinuous at the instant of impact. After impact, the velocity distribution that maintains continuity of displacements consists of a disturbance that propagates away from points of velocity discontinuity; i.e. away from the moving particle at the tip. This disturbance is driven by the shear force $f(\tau)$ at the interface with the heavy particle. During the response period, the particle continually decelerates so the shear force decreases as a function of time. With no axial force, the uniform rigid–perfectly plastic cantilever has plastic deformation only at an isolated plastic hinge. This pattern of deformation is similar to that for a pulse load of continuously decreasing force (Sect. 4.4). Hence the equations for rates of change of transverse momentum and moment of momentum about the tip can be obtained from Eqs (4.80) and (4.81) together with (4.102):

$$\frac{\mathrm{d}}{\mathrm{d}\tau}\left(\frac{1}{2}\lambda v\right) = -\gamma \dot{v} \tag{4.103}$$

$$\frac{\mathrm{d}}{\mathrm{d}\tau}\left(\frac{1}{6}\lambda^2 v\right) = 1 \tag{4.104}$$

where the tip velocity $v = v(\tau)$ and the hinge location $\lambda = \lambda(\tau)$ are both unknown functions of time τ. Integrating (4.103) and (4.104) and using the initial conditions $v(0) = v_0$ and $\lambda(0) = 0$ gives

$$\lambda v = 2\gamma(v_o - v) \tag{4.105}$$

$$\lambda^2 v = 6\tau \tag{4.106}$$

Thus, the tip velocity and time when the hinge transits any location λ, are given by

$$v = v_o\left(1+\frac{\lambda}{2\gamma}\right)^{-1} \tag{4.107}$$

$$\tau = \frac{v_o}{6}\lambda^2\left(1+\frac{\lambda}{2\gamma}\right)^{-1} \tag{4.108}$$

Expressions (4.107) and (4.108) provide the relationships between v, λ and τ in the initial phase of deformation where a hinge travels away from the impact point at the tip. This stage of response is termed *phase I*. Differentiating (4.108) with respect to λ gives

$$\frac{d\tau}{d\lambda} = \frac{v_o}{6}\lambda\left(2+\frac{\lambda}{2\gamma}\right)\left(1+\frac{\lambda}{2\gamma}\right)^{-2} \tag{4.109}$$

so that in phase I the travelling speed of the hinge is found to be

$$\dot{\lambda} = \frac{d\lambda}{d\tau} = \frac{6}{v_o}\left(1+\frac{\lambda}{2\gamma}\right)^2\left[\lambda\left(2+\frac{\lambda}{2\gamma}\right)\right]^{-1} \tag{4.110}$$

Equations (4.108) and (4.110) indicate that at the initial instant $\tau = 0$, $\lambda = 0$ but $\dot{\lambda} = \infty$. With increasing time λ increases but $\dot{\lambda}$ decreases; that is, the hinge travels from the tip towards the root of the cantilever with decreasing speed.

Phase I terminates when the hinge reaches the root of the cantilever. Taking $\lambda = 1$ in (4.108) results in the time of arrival τ_1 and tip velocity v_1 at this instant,

$$\tau_1 = \frac{v_o}{3}\frac{\gamma}{1+2\gamma} \tag{4.111}$$

$$v_1 \equiv v(\tau_1) = \frac{2\gamma}{1+2\gamma}v_o \tag{4.112}$$

when the hinge is travelling at speed

$$\dot{\lambda}(\tau_1) = \frac{3}{v_o}\frac{(1+2\gamma)^2}{\gamma(1+4\gamma)} \tag{4.113}$$

The hinge starts moving away from the impact point at an indefinitely large speed.

Modal Phase (Phase II) For time greater than τ_1, the motion of the cantilever–mass system is a rigid body rotation about a hinge at the root B; this period is termed *phase II*. The stationary pattern of deformation during this phase is the same as that of the primary mode of the structure. During phase II the rate-of-change in the moment of momentum about the root is

$$\gamma\dot{v} + \frac{1}{3}\dot{v} = -1 \tag{4.114}$$

or

$$\dot{v} = -\frac{3}{1+3\gamma} = \text{const} \tag{4.115}$$

This implies a constant rate of deceleration for the particle at the tip. Since

$v_1/v_o = (1+1/2\gamma)^{-1}$, the duration of phase II is

$$\tau_2 - \tau_1 = \frac{2\gamma(1+3\gamma)}{3(1+2\gamma)} v_o \tag{4.116}$$

The angle of rotation at the root during phase II is

$$\theta_{Bf} = -\frac{1}{2} v_1 (\tau_2 - \tau_1) = -\frac{2}{3} v_o^2 \gamma^2 \frac{(1+3\gamma)}{(1+2\gamma)^2} \tag{4.117}$$

The response is completely terminated at

$$\tau_f = \tau_2 = \tau_1 + \frac{2\gamma(1+3\gamma)}{3(1+2\gamma)} v_o = \gamma v_o = p_f \tag{4.118}$$

It is interesting to find that the nondimensional response time τ_f in this problem is equal to the nondimensional total impulse $p_f = P_f / L\sqrt{\rho F_c}$; this is exactly the same as that for a cantilever without the tip mass which is subjected to either a rectangular or a linearly decreasing pulse at its tip (see Eq. (4.51) and Table 4.3). This direct correspondence between response period and applied impulse occurs because at the terminal location of the travelling hinge, the bending moment equals the fully plastic moment throughout the response period. After the applied force vanishes, the moment of momentum about this terminal location decreases linearly with time; thus, this part of the response period is directly proportional to total applied impulse. For these elementary pulse shapes the loading period is also directly proportional to the total impulse, so the total response period τ_f is proportional to the total applied impulse p_f. Note that in the analysis above, the nondimensional response variables are functions of only the *mass ratio* γ defined by Eq. (4.101).

The variation of the tip velocity and the position of the plastic hinge with time for mass ratios $\gamma = 0.2$, 1.0 and 5.0 are shown in Fig. 4.23. It is seen that the hinge speed decreases with increasing distance from the impact point since the moment of inertia of the moving segment about the hinge is continuously increasing.

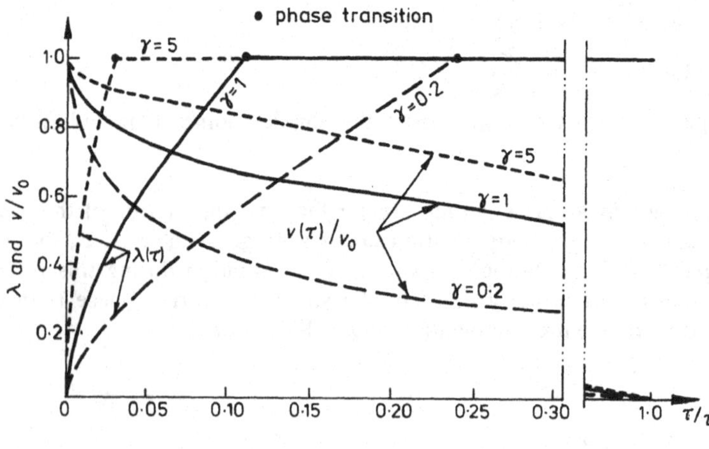

Fig. 4.23. Variation of tip velocity and hinge position with time for cantilevers struck by a rigid mass, for mass ratio $\gamma = G/\rho L$ equal to 0.2, 1.0 and 5.0.

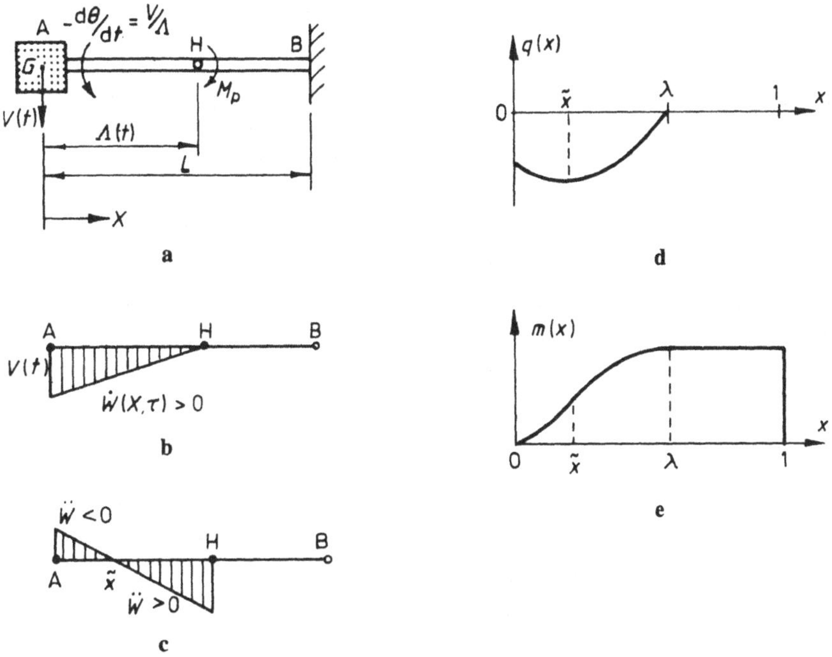

Fig. 4.24. Transient phase of dynamic response for cantilever struck by a rigid mass at the tip; (a) deformation mechanism with a travelling hinge H at $x = \lambda(\tau)$; (b) velocity distribution along the cantilever; (c) acceleration distribution; (d) shear force distribution; (e) bending moment distribution.

4.5.3 Acceleration, Force and Bending Moment

Phase I is the transient phase of response. In a rigid–perfectly plastic structure this phase has a rigid segment AH that rotates about hinge H, while segment HB remains static and undeformed (Fig. 4.24b). The velocity field is

$$\dot{w} = v\left(1 - \frac{x}{\lambda}\right), \qquad x \le \lambda \tag{4.119}$$

where $\lambda = \lambda(\tau)$, and the acceleration field is

$$\ddot{w} = \dot{v}\left(1 - \frac{x}{\lambda}\right) + \frac{vx}{\lambda^2}\dot{\lambda}, \qquad x \le \lambda \tag{4.120}$$

Note that the acceleration contains an additional term caused by the increasing length of the rotating segment, i.e $\dot{\lambda} > 0$. Differentiating Eq. (4.107) with respect to time τ gives the acceleration of the particle at the tip,

$$\dot{v} = -\frac{v_0 \dot{\lambda}}{2\gamma\left(1 + \frac{\lambda}{2\gamma}\right)^2}, \qquad x \le \lambda \tag{4.121}$$

Using Eq. (4.110), expressions (4.121) and (4.120) yield the acceleration at the tip as well as that at any section of the cantilever,

$$\dot{v} = -\frac{6}{\lambda(\lambda + 4\gamma)}, \qquad x \le \lambda \tag{4.122}$$

$$\ddot{w} = \frac{6}{\lambda(\lambda+4\gamma)}\left(-1+\frac{2(\lambda+\gamma)}{\lambda^2}x\right), \qquad x \le \lambda \tag{4.123}$$

Expression (4.122) indicates that the acceleration of the tip mass decreases with time from infinite at $\tau = 0$ to $\dot{v} = -6/(4\gamma+1)$ at the end of phase I when $\tau = \tau_1$. Equation (4.123) shows that the transverse acceleration at the tip and that at the travelling hinge are in opposite directions

$$\ddot{w}(0) = \dot{v} = -\frac{6}{\lambda(\lambda+4\gamma)} < 0, \qquad \ddot{w}(\lambda) = \frac{6(\lambda+2\gamma)}{\lambda^2(\lambda+4\gamma)} > 0 \tag{4.124}$$

That is, the material at any interior section is first jerked into motion as the hinge passes; sometime later the sign of acceleration changes from positive to negative. Hence, as shown in Fig. 4.24c, the acceleration changes sign at an interior point of segment AH. In fact, taking $\ddot{w}(\tilde{x}) = 0$ leads to

$$\tilde{x} = \frac{\lambda^2}{2(\lambda+\gamma)} \rightarrow \begin{cases} \lambda/2, & \gamma \rightarrow 0 \\ 0, & \gamma \rightarrow \infty \end{cases} \tag{4.125}$$

where the expressions on the right are asymptotic limits for very small or large mass ratios.

The distribution of shear force within the rotating segment can be found by integrating the acceleration field in the same manner as in Sect. 4.2; that is,

$$q(x) = -f + \int_0^x \ddot{w}(\bar{x})\,\mathrm{d}\bar{x} = -\frac{6\gamma}{\lambda(\lambda+4\gamma)}\left(1+\frac{x}{\gamma}-\frac{\lambda+\gamma}{\lambda^2\gamma}x^2\right), \quad x \le \lambda \tag{4.126}$$

with

$$q(0) = -\frac{6\gamma}{\lambda(\lambda+4\gamma)} = -f, \qquad q(\lambda) = 0$$

$$q_{\min} = q(\tilde{x}) = -\frac{3(\lambda+2\gamma)^2}{2\lambda(\lambda+\gamma)(\lambda+4\gamma)} < -f$$

where $\tilde{x}$ is given by Eq. (4.125) as the zero point of the acceleration field. The shear force distribution $q(x)$ is sketched in Fig. 4.24d.

Integration of the shear force with respect to x and satisfying the boundary conditions $m(0) = 0$ and $m(\lambda) = 1$ gives the distribution of bending moment within the rotating segment

$$m(x) = \frac{6\gamma x}{\lambda(\lambda+4\gamma)}\left(1+\frac{x}{2\gamma}-\frac{\lambda+\gamma}{3\lambda^2\gamma}x^2\right), \qquad x \le \lambda \tag{4.127}$$

This bending moment distribution is sketched in Fig. 4.24e. The dynamic analysis for a rigid–perfectly plastic cantilever gives $m = 1$ in segment HB. Although the segment ahead of the travelling hinge is at yield it does not deform because there is no shear force in this segment.

The shear reaction between the particle and the tip of the uniform cantilever can be obtained by combining Eqs (4.102), (4.115) and (4.122):

$$f(\tau) = \begin{cases} \dfrac{6\gamma}{\lambda(\lambda+4\gamma)}, & 0 \le \tau \le \tau_1 \\[2ex] \dfrac{3\gamma}{1+3\gamma}, & \tau_1 \le \tau \le \tau_f \end{cases} \tag{4.128}$$

where $\lambda = \lambda(\tau)$ is determined by (4.108). As shown in Fig. 4.25, this reaction

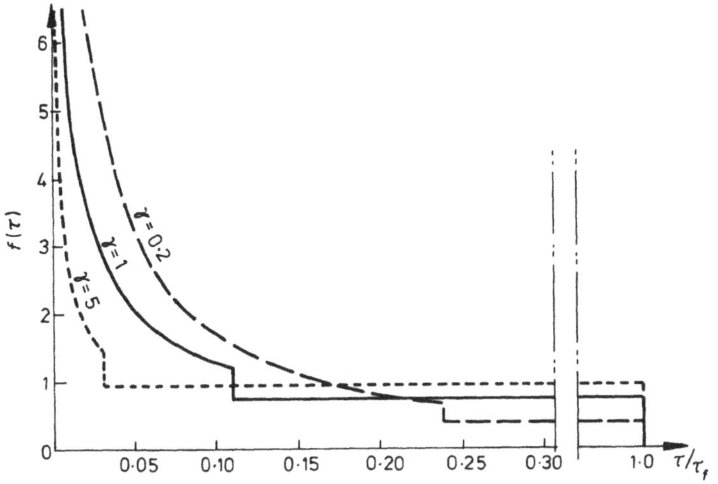

Fig. 4.25. Magnitude of shear force acting on mass at tip of cantilever.

force decreases monotonically with time, and it has a discontinuity at the time of phase transition. Note that for a mass at the tip, this shear force does not vanish until the entire response terminates; i.e. $\tau_d = \tau_f$. In this sense the response is different from the general pulse discussed in Sect. 4.4 for which $\tau_d < \tau_f$.

4.5.4 Deformed Shape

The curvature generated by the travelling hinge can be determined from the angular speed of the rigid segment near the tip and the speed of the hinge. At any time, the nondimensional curvature k along the cantilever is given by Eq. (4.55),

$$k \equiv \kappa L = \frac{v}{\lambda\dot{\lambda}} \tag{4.129}$$

where k is the curvature developed at the hinge. Substituting the tip velocity (4.107) and hinge speed (4.110) into (4.129) gives the final curvature

$$k_f(x) = w_f''(x) = \frac{2}{3}v_0^2\gamma^2\frac{(x+4\gamma)}{(x+2\gamma)^3} \tag{4.130}$$

The final curvature distribution for $\gamma = 0.2$, 1.0 and 5.0 is plotted in Fig. 4.26. If the colliding mass is small the final curvature is large throughout the cantilever, whereas if $\gamma >> 1$, almost all input energy is dissipated by rotation at the root during the modal phase of response. Integrating (4.130) and making use of the previously determined rotation angle at the root $w_f'(1) = \theta_B$ (see Eq. (4.117)), one obtains the variation of the inclination angle along the cantilever when the response terminates:

$$\theta_f(x) \approx w_f'(x) = -\frac{2}{3}v_0^2\gamma^2\frac{(x+3\gamma)}{(x+2\gamma)^2} \tag{4.131}$$

Hence, the final rotation at the tip θ_{Af} is found to be

$$\theta_{Af} = \theta_f(0) = -\frac{1}{2}v_0^2\gamma = -\frac{K_0}{M_p} \equiv -e_0 \tag{4.132}$$

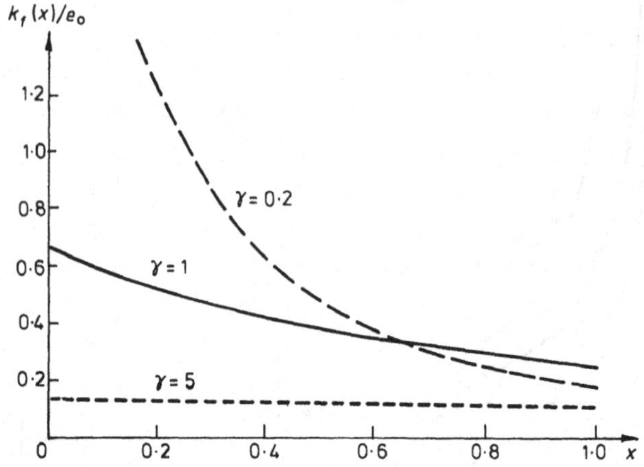

Fig. 4.26. Curvature distribution along cantilever following transit of travelling hinge.

where $K_0 = GV_0^2/2 = E_0$ is the initial kinetic energy and e_0 is the corresponding nondimensional energy parameter. Expression (4.132) shows that again the tip rotation is proportional to e_0 in agreement with the general statement (4.90).

The final deflection w_f can be obtained by integrating Eq. (4.131) with respect to x and using $w(1) = 0$ at the root. Hence the final deformed shape of the cantilever is

$$w_f(x) = \frac{1}{3}v_0^2\gamma^2\left[\frac{1}{1+2\gamma} - \frac{x}{x+2\gamma} + 2\ln\left(\frac{1+2\gamma}{x+2\gamma}\right)\right] \tag{4.133}$$

The final tip deflection, therefore, is

$$\delta_f \equiv w_f(0) = \frac{1}{3}v_0^2\gamma^2\left[\frac{1}{1+2\gamma} + 2\ln\left(1+\frac{1}{2\gamma}\right)\right] \tag{4.134}$$

Figure 4.27 shows the dependence of the final deflection at the tip δ_f on the mass ratio γ. This deflection increases in proportion to the input energy e_0. In limiting cases, expression (4.133) gives the final deformed shape of the cantilever:

1. If the colliding mass is much larger than the mass of the cantilever, $\gamma >> 1$ and

$$w_f(x) \approx \frac{1}{2}\gamma v_0^2(1-x) \tag{4.135}$$

Roughly speaking this implies that the cantilever merely rotates about the root as a rigid bar so the final tip deflection is approximately equal to the non-dimensional initial kinetic energy

$$\delta_f = w_f(0) \approx \frac{1}{2}\gamma v_0^2 = e_0 \tag{4.136}$$

2. If the colliding mass is much smaller than the mass of the cantilever, $\gamma << 1$ and

$$w_f(x) \approx \frac{2}{3}\gamma^2 v_0^2 \ln\left(\frac{1}{x}\right) \qquad \frac{x}{\gamma} >> 1 \tag{4.137}$$

In this case the cantilever is deformed into a logarithmic curve with a final tip deflection

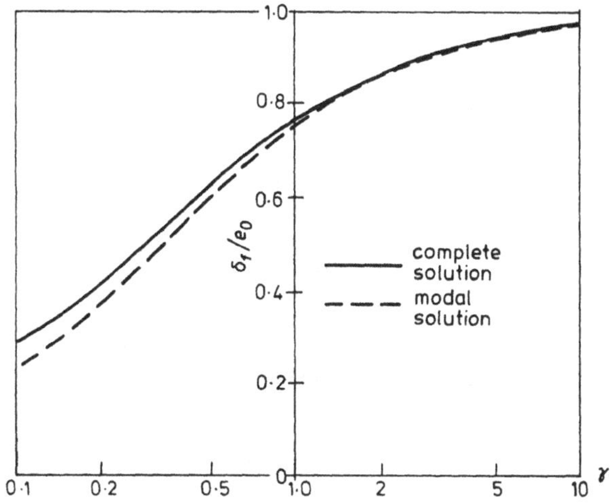

Fig. 4.27. Dependence of final tip deflection on the mass ratio γ:
—— complete solution, Eq. (4.134);
– – – modal solution, Eq. (4.154).

$$\delta_f = w_f(0) \approx \frac{2}{3}\gamma^2 v_0^2 \ln\left(\frac{1}{2\gamma}\right) \tag{4.138}$$

For a lightweight tip mass, the final tip deflection is proportional to the square of the initial impulse rather than the impact energy. The final shapes of cantilevers with mass ratios $\gamma = 0.2$, 1.0 and 5.0 are plotted in Fig. 4.28. As mass ratio γ increases, a larger part of the impact energy is dissipated in developing plastic curvature at the root during a modal phase of response.

4.5.5 Energy Dissipation

All the input energy for a rigid–plastic structure is ultimately dissipated by plastic work (Sect. 4.2.3). For impulsive loading, the input energy is the initial kinetic

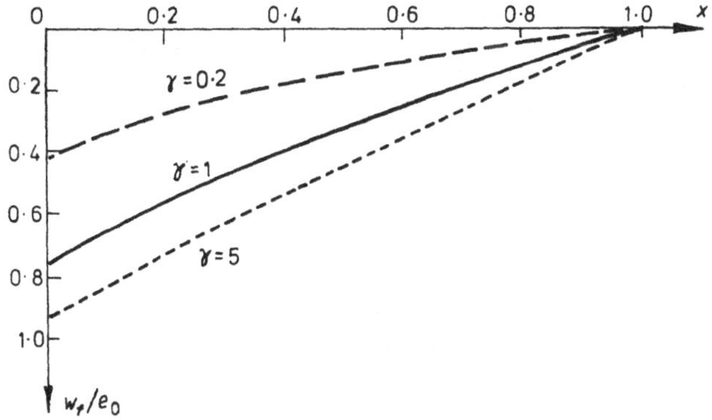

Fig. 4.28. Final deformed shapes of cantilevers with mass ratio γ = 0.2, 1.0 and 5.0.

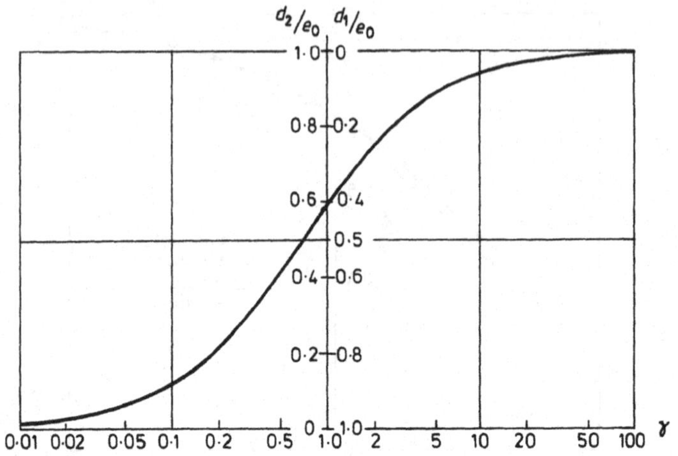

Fig. 4.29. Input energy dissipated in phases I and II as function of mass ratio γ.

energy of the tip mass; i.e. $E_o = K_o = GV_o^2/2$. Thus the nondimensional energies dissipated in phases I and II are

$$d_1 \equiv \frac{D_1}{M_p} = |\theta_{Af}| - |\theta_{Bf}| \tag{4.139}$$

$$d_2 \equiv \frac{D_2}{M_p} = |\theta_{Bf}| \tag{4.140}$$

where

$$D_i = \int_{t_{i-1}}^{t_i} M_p |\dot{\theta}(\lambda, t_f)| \, dt = \int_{\lambda_{i-1}}^{\lambda_i} M_p \kappa(x, t_f) \, dx, \qquad i = 1, 2$$

Hence, the part of the impact energy that is dissipated in phase II is

$$\frac{d_2}{e_o} = \frac{d_2}{d_1 + d_2} = \frac{|\theta_{Bf}|}{|\theta_{Af}|} = \frac{4\gamma(1+3\gamma)}{3(1+2\gamma)^2} \tag{4.141}$$

The total energy dissipation can be separated into work done in generating distributed curvature and the subsequent work done in modal deformation; the relative size of these parts depends merely on the mass ratio γ as shown in Fig. 4.29. The larger the mass ratio, the more energy is dissipated in modal deformation during phase II. Figure 4.30 shows the part of the input energy that has been dissipated before the travelling hinge transits a section located a distance x from the tip. The part of the energy remaining when the hinge reaches the root depends on the mass ratio γ.

4.5.6 Modal Approximation

The above analysis indicates that the complete response of a rigid–perfectly plastic cantilever that is hit at the tip by a heavy particle consists of two phases: phase I is a *transient* motion characterized by a travelling plastic hinge moving from the tip towards the root of the cantilever; and phase II is a rotation of the cantilever about the root as a rigid body. In phase I the velocity field changes its shape with

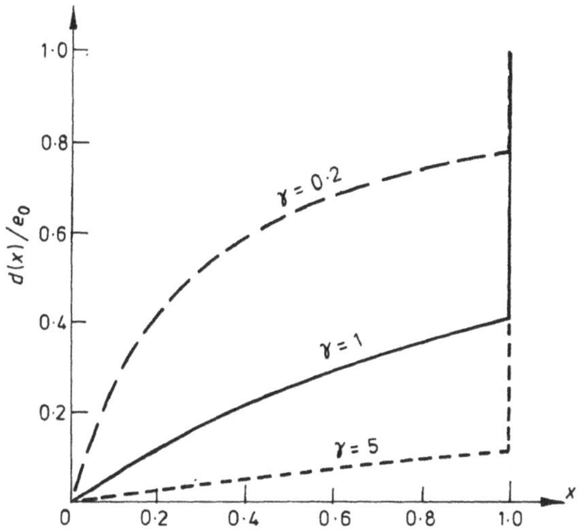

Fig. 4.30. Accumulation of energy dissipation along cantilever struck by a heavy particle for mass ratio $\gamma = 0.2$, 1.0 and 5.0.

time; during phase II, however, the deformation occurs in a *mode-form* or *modal* motion where the velocity field does not change in shape with time.

Equation (4.141) and Fig. 4.29 show that most of the input energy is dissipated either in phase I if the mass ratio $\gamma = G/\rho L << 1$, or in phase II if $\gamma >> 1$. Equations (4.111) and (4.118) imply that the relative duration of the transient phase is

$$\frac{\tau_1}{\tau_f} = \frac{1}{3(2\gamma + 1)} \tag{4.142}$$

Thus, when $\gamma >> 1$, phase I constitutes only a small fraction of the response period. Expression (4.135) confirms that when $\gamma >> 1$, the final deformed shape of the cantilever is virtually determined by phase II alone. This suggests that a modal solution could be a good approximation for the final deformed shape provided $\gamma >> 1$.

A modal approximation consists of a kinematically admissible velocity field that is a separable function of spatial and temporal independent variables,

$$\dot{w}^*(x,\tau) = v^*(\tau)(1-x) \equiv v^*(\tau)\phi^*(x) \tag{4.143}$$

where the superscript * pertains to the modal solution, and $\phi^*(x) = 1 - x$ is the shape function of the modal field for the cantilever. The modal acceleration field is obtained by differentiating Eq. (4.143) with respect to time τ:

$$\ddot{w}^*(x,\tau) = \dot{v}^*(\tau)(1-x) \equiv \dot{v}^*(\tau)\phi^*(x) \tag{4.144}$$

The acceleration and the velocity fields are shown in Fig. 4.31.

Referring to the modal velocity field given in (4.143) and finding the rate of change in the moment of momentum about the root, one obtains an equation similar to (4.114). It follows that

$$\dot{v}^*(\tau) = -\frac{3}{1+3\gamma} \tag{4.145}$$

and accordingly,

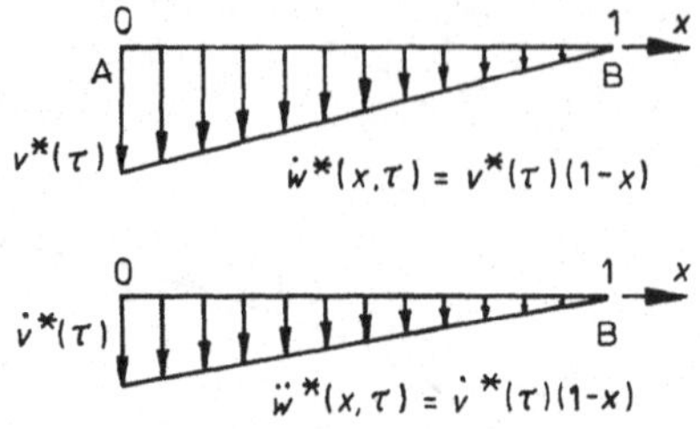

Fig. 4.31. Modal velocity and acceleration fields for cantilever loaded at tip.

$$\ddot{w}^*(x,\tau) = -\frac{3}{1+3\gamma}(1-x) \tag{4.146}$$

Integration of (4.146) with respect to time yields

$$\dot{w}^*(x,\tau) = v_0^* - \frac{3}{1+3\gamma}(1-x)\tau \tag{4.147}$$

$$w^*(x,\tau) = v_0^*\tau - \frac{3(1-x)}{1+3\gamma}\frac{\tau^2}{2} \tag{4.148}$$

for initial conditions $\dot{w}^*(x,\,0) = v_0^*$ and $w^*(x,\,0) = 0$.

In order to entirely determine the modal solution, the initial characteristic velocity for a mode v_0^* has to be specified. This is required because the modal velocity distribution is different from the initial velocity distribution. General considerations for the modal approximation are given in Chap. 2, where the so-called 'min Δ_0 technique' for determining the initial velocity of modal solutions is described. For the present problem the 'min Δ_0 technique' gives an approximate magnitude v_0^* for the initial tip velocity in a mode,

$$v_0^* = \frac{\int_0^1 \rho(x)v_0\phi^*(x)\,dx}{\int_0^1 \rho(x)\phi^*(x)\phi^*(x)\,dx} = \frac{\gamma v_0}{\gamma + \int_0^1 (1-x)^2\,dx} = \frac{3\gamma v_0}{1+3\gamma} \tag{4.149}$$

where $\rho(x)$ is the distribution of mass (including the distributed mass along the cantilever and the concentrated mass attached at the tip). It is worthwhile to point out that the initial modal velocity given in (4.149) can be obtained by equating the initial and the modal moment of momentum about the root B.

$$GV_0L = \left(G + \frac{1}{3}\rho L\right)V_0^*L \tag{4.150}$$

Here, conservation of moment of momentum about the plastic hinge has been used to obtain V_0^*; the equivalence between this method and the 'min Δ_0 technique' was demonstrated by Zhou et al. [1992].

Combining Eqs (4.147) and (4.149) provides the variation of the modal velocity as

$$v^*(\tau) = \frac{3}{1+3\gamma}(\gamma v_0 - \tau) \tag{4.151}$$

Expression (4.151) indicates that the response time in the modal solution is the same as that in the complete solution; that is $\tau_f^* = \gamma v_0 = p_f = \tau_f$ (see Eq. (4.118)). As sketched in Fig. 4.32, the variation of the modal velocity given by (4.151)

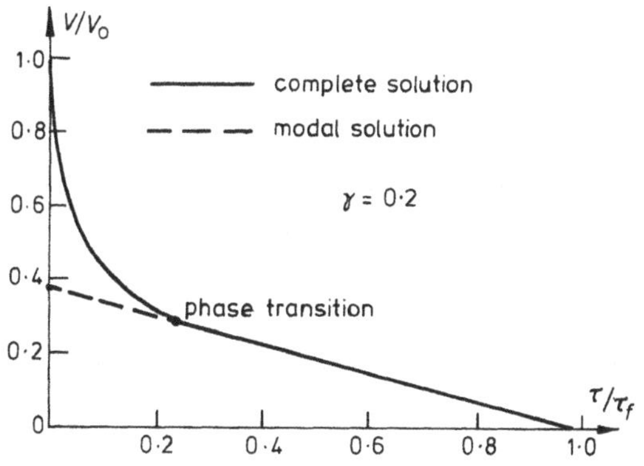

Fig. 4.32. Tip velocity variation with time for complete and modal solutions; mass ratio $\gamma = 0.2$.

exactly coincides with that given by phase II of the complete solution, although the initial kinetic energy of the modal solution is less than that in the original problem. In fact,

$$K_0 - K_0^* = \Delta_0 = \frac{1}{2}GV_0^2 - \frac{1}{2}\left(G + \frac{1}{3}\rho L\right)(V_0^*)^2 = \frac{K_0}{3\gamma + 1} \tag{4.152}$$

where K_0 and K_0^* are the initial kinetic energies for the complete solution and the modal solution, respectively. As the mass ratio γ increases, the time when the transient and modal solutions converge becomes a smaller part of the total response time.

Furthermore, the final deformed shape predicted by the modal solution can be found from Eq. (4.148) as

$$w_f^*(x) = w^*(x, \tau_f) = \frac{3}{2}\gamma v_0^2 \frac{1 - x}{1 + 3\gamma} \tag{4.153}$$

Hence, the final tip deflection predicted by the modal solution is

$$\delta_f^* = w_f^*(0) = w^*(0, \tau_f) = \frac{3}{2}\gamma v_0^2 \frac{1}{1 + 3\gamma} \tag{4.154}$$

This approximation of δ_f^* is plotted in Fig. 4.27 for comparison with the complete solution. As the mass ratio γ increases, δ_f^* approaches δ_f. A detailed comparison of results using the two methods is shown in Table 4.5; this illustrates that the modal solution is a good approximation for the final tip deflection if the mass of the colliding missile is larger than the mass of the uniform cantilever.

Table 4.5 Comparison of final tip deflections predicted by complete and modal solutions

γ	δ_f / e_0	δ_f^* / e_0	Error (%)
0.05	0.1902	0.1304	− 31.4
0.1	0.2944	0.2308	− 21.6
0.2	0.4293	0.3750	− 11.5
0.5	0.6288	0.6000	− 4.58
1	0.7628	0.7500	− 1.68
2	0.8617	0.8571	− 0.53
5	0.9384	0.9375	− 0.10
10	0.9680	0.9677	− 0.03

References

Krajcinovic, D. [1972]. On approximate solutions for rigid–plastic structures subjected to dynamic loading. *Int. J. Non-Linear Mech.* **7**, 571–575.

Lee, E.H. and Symonds, P.S. [1952]. Large deformation of beams under transverse impact. *ASME J. Appl. Mech.* **19**, 308–314.

Meirovitch, L. [1986]. *Elements of Vibration Analysis*, 2nd Edition. McGraw-Hill.

Parkes, E.W. [1955]. The permanent deformation of a cantilever struck transversely at its tip. *Proc. Roy. Soc. Lond.* **A228**, 462–476.

Youngdahl, C.K. [1970]. Correlation parameters for eliminating the effect of pulse shape on dynamic plastic deformation. *ASME J. Appl. Mech.* **39**, 744–752.

Youngdahl, C.K. [1971]. Influence of pulse shape on the final plastic deformation of a circular plate. *Int. J. Solids Struct.* **7**, 1127–1142.

Zhou, Q., Zhang, T.G. and Yu, T.X. [1992]. The correlation between the min Δ_0 technique and the conservation of the moment-of-momentum. *ASME J. Appl. Mech.* **60**, 677–679.

Zhu, G., Huang, Y.G., Yu, T.X. and Wang, R. [1986]. Estimation of the plastic structural response under impact. *Int. J. Impact Engng.* **4**, 271–282.

Chapter 5

Second-Order Effects On Dynamic Response

In the preceding chapter the dynamic response of a rigid–perfectly plastic cantilever to pulse and impact loads was studied to show how inertia influences the distribution of plastic deformation. With solely transverse loads applied to a straight beam, the rigid–perfectly plastic idealization results in a simple two degree-of-freedom displacement mechanism that clearly describes the main features of dynamic deformation by impact. However, this analysis can only be regarded as a first approximation; in some cases, strain-rate, strain hardening, shear force, rotary inertia, elasticity and geometrical change due to large deformation have significant effects. These effects modify the displacement field obtained for a rigid–perfectly plastic beam; in this respect they can be considered to be of second order. In the present chapter, dynamic analyses of major second-order effects are considered one-by-one in order to find circumstances wherein each effect is significant.

5.1 Strain-Rate Effect

5.1.1 Impulsive Load on Viscoplastic Cantilever

The rate–independent perfectly plastic behavior described in Chap. 1 and used in Chap. 4 is representative of materials such as heat-treated aluminum alloys. For other materials, however, a rate–dependent material characterization is more representative of observed or measured structural response. Most notably this is true of mild steel at moderate to large nominal strain-rates, $\mathrm{d}\varepsilon/\mathrm{d}t > 1\,\mathrm{s}^{-1}$. The effect of strain-rate on dynamic deformation of a cantilever can be incorporated via the Cowper–Symonds equation described in Sect. 1.4. The rate–dependent impact response was first analyzed by Bodner and Symonds [1960, 1962] and Ting [1964]. Their analyses are adopted below with some minor modifications.

The rate–dependent or viscoplastic analysis for flexural response of a slender member to a transverse impact is based on the following assumptions:

1. A uniform cantilever has a particle with mass G affixed at the tip.
2. Initially the cantilever is at rest while the massive particle has a transverse component of velocity V_0.

3. The rotation of every element remains small; i.e. small deflections.
4. Inertia is negligible in the plastically deforming segment adjacent to the root.
5. The quasistatic yield moment M_p is independent of the axial force.

In addition to these assumptions, the rate–dependent bending moment at a plastically deforming section is obtained from the Cowper–Symonds relation according to Eq. (1.15). Hence the rate of change of curvature $d\kappa/dt$ and bending moment M are related by

$$\frac{d\kappa}{dt} = \frac{2\dot{\varepsilon}_{or}}{h}\left(\frac{M}{M_p} - 1\right)^r \tag{5.1}$$

where

$$\dot{\varepsilon}_{or} \equiv \dot{\varepsilon}_o\left(1 + \frac{1}{2r}\right)^r \tag{5.2}$$

In order to maintain consistency with the previous rigid–plastic analysis of Sect. 4.5, the following nondimensional parameters are introduced:

$$\begin{gathered} T_o \equiv L\sqrt{\frac{\rho L}{M_p}}, \quad \tau \equiv \frac{t}{T_o}, \quad (\dot{\ }) \equiv \frac{d}{d\tau}(\ ), \quad v \equiv \frac{V}{L/T_o} \\ p \equiv \frac{P}{T_o M_p/L}, \quad q \equiv \frac{Q}{M_p/L}, \quad m \equiv \frac{M}{M_p}, \quad w \equiv \frac{W}{L} \\ k \equiv \kappa L, \quad \eta \equiv \frac{2\dot{\varepsilon}_{or}}{h}LT_o = \frac{2\dot{\varepsilon}_o L^2}{h}\left(\frac{\rho L}{M_p}\right)^{1/2}\left(1 + \frac{1}{2r}\right)^r \end{gathered} \tag{5.3}$$

Consequently, Eq. (5.1) can be recast as

$$\dot{k} \equiv \frac{dk}{d\tau} = \eta(m-1)^r \tag{5.4}$$

The inclusion of a rate–dependent yield moment into the analysis for the cantilever impact problem (depicted in Fig. 4.22) completely changes the kinematics of the system. Instead of two separate phases, the cantilever response becomes a single continuous motion. This is a consequence of the shear reaction at the root that develops instantaneously at impact due to the necessary variation of strain-rate and yield moment along the cantilever. In order to satisfy the equation of motion, yield condition and flow rule for a *rate–dependent* cantilever with an impulse load, the plastic deforming region must initially extend over the full length of the cantilever. This is in contrast to the rigid–perfectly plastic material model where plastic deformation was concentrated at a hinge. During the response of a viscoplastic structure, the plastically deforming region shrinks and an unloading region extends from the tip where the load is applied. In the unloaded region the bending moment is decreasing from a previously larger value. Assumption (4) implies that the bending moment varies linearly in the plastically deforming segment adjacent to the root.

The deformation mechanism for a viscoplastic relation (5.4) at an intermediate stage of motion is shown in Fig. 5.1a. It consists of two regions: (1) a rigid outer segment AH of length λL where the bending moment is decreasing ($m < 1$ and $\dot{k} = 0$); and (2) a plastically deforming inner segment HB of length $(1 - \lambda)L$ where the bending moment is increasing ($m > 1$ and $\dot{k} > 0$). These regions are separated by an interface H located at $\Lambda(t) = \lambda(T_o\tau)L$ where the bending moment

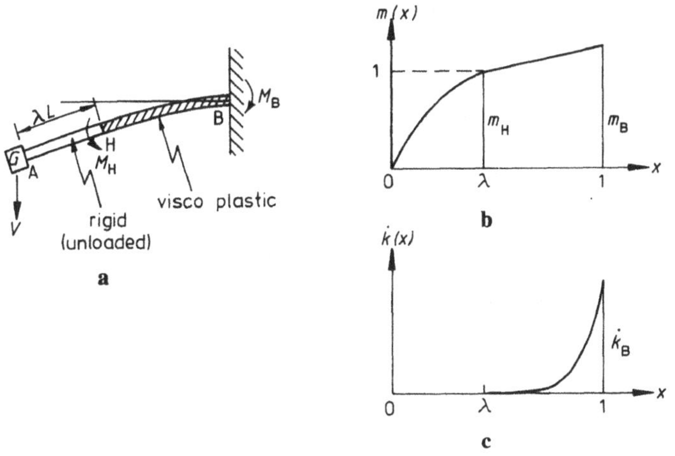

Fig. 5.1 (a) Dynamic deformation mechanism of viscoplastic cantilever; (b) distribution of bending moment; (c) distribution of curvature–rate, $r = 5$.

is equal to the plastic bending moment M_p . Hence, the bending moments at the transition section H and the root B are

$$m_H = 1, \quad m_B = 1 + q_B(1-\lambda) \tag{5.5}$$

where q_B denotes the nondimensional shear force (reaction) at the root B, and λ is an undetermined function of time.

Denoting the nondimensional curvature–rate at the root B as $\dot{k}_B = \dot{k}(1)$, Eq. (5.4) leads to

$$\dot{k}_B = \eta(m_B - 1)^r \tag{5.6}$$

Combining Eqs (5.5) and (5.6), it follows that

$$q_B = \frac{m_B - 1}{1-\lambda} = \frac{1}{1-\lambda}\left(\frac{\dot{k}_B}{\eta}\right)^{1/r} \tag{5.7}$$

Assume that inertia forces in plastically deforming segment HB are negligible so that the shear force $q_B = q_H$. This results in a linear variation of bending moment in this segment; i.e.

$$m(x) = \frac{1 - x + (x-\lambda)m_B}{1-\lambda}, \qquad \lambda \le x \le 1 \tag{5.8}$$

This bending moment distribution is shown in Fig. 5.1b. Using Eqs (5.4) and (5.6), we obtain the rate-of-change of curvature,

$$\dot{k}(x) = \dot{k}_B\left(\frac{x-\lambda}{1-\lambda}\right)^r, \qquad \lambda \le x \le 1 \tag{5.9}$$

This distribution of curvature–rate is shown in Fig. 5.1c for a typical value $r = 5$. For $r > 3$, most deformation is concentrated in a small segment near the root.

At any section the angular velocity $\dot{\theta}$ can be obtained by integrating the curvature–rate in the deforming section. The angular velocity of the rigid segment adjacent to the tip depends on the rate of curvature throughout the deforming segment.

$$\dot{\theta}_H = \dot{\theta}(\lambda) = \int_\lambda^1 \dot{k}(x)\,dx = \int_\lambda^1 \dot{k}_B\left(\frac{x-\lambda}{1-\lambda}\right)^r dx = \dot{k}_B\frac{1-\lambda}{r+1} \tag{5.10}$$

The transverse velocity in the deforming segment HB can be expressed as

$$\dot{w}(x) = \int_x^1 (\xi - x)\,d\dot{\theta} = \int_x^1 (\xi - x)\dot{k}(\xi)\,d\xi \tag{5.11}$$

$$= \frac{\dot{k}_B}{(1-\lambda)^r}\left\{\frac{(1-x)(1-\lambda)^{r+1}}{r+1} - \frac{(1-\lambda)^{r+2}-(x-\lambda)^{r+2}}{(r+1)(r+2)}\right\}, \quad \lambda \le x \le 1$$

Thus the velocity $\dot{w}_H$ at the interface is

$$\dot{w}_H = \dot{w}(\lambda) = \dot{k}_B\frac{(1-\lambda)^2}{r+2} \tag{5.12}$$

Since segment AH rotates about H as a rigid body, the velocity in segment AH is

$$\dot{w}(x) = \dot{w}_H + \dot{\theta}_H(\lambda - x) = \dot{k}_B\left[\frac{1-\lambda}{r+2}\left(1+\frac{\lambda}{r+1}\right) - \frac{1-\lambda}{r+1}x\right], \quad 0 \le x \le \lambda \tag{5.13}$$

and the velocity at the tip is

$$v = \dot{w}(0) = \dot{w}_H + \dot{\theta}_H\lambda = \dot{k}_B\frac{1-\lambda}{r+2}\left(1+\frac{\lambda}{r+1}\right) \tag{5.14}$$

With the velocity distribution defined by Eq. (5.11), the transverse momentum can be calculated as a function of the length λ by considering the transverse momentum and the moment of momentum about the root B. This results in the following two equations of motion:

$$p_o - \int_0^\tau q_B\,d\tau = \gamma v + \int_0^\lambda \dot{w}(x)\,dx \tag{5.15}$$

$$1 \cdot p_o - \int_0^\tau m_B\,d\tau = 1 \cdot \gamma v + \int_0^\lambda \dot{w}(x)(1-x)\,dx \tag{5.16}$$

where $\gamma = G/\rho L$ is the mass ratio and $p_o = \gamma v_o$ is the nondimensional transverse impulse applied at the tip at the initial instant. Notice that in Eqs (5.15) and (5.16) the momentum of segment HB is excluded since it has been assumed to be negligible. Substituting q_B from (5.7), m_B from (5.6) and $\dot{w}(x)$ from (5.13) into these equations, one obtains

$$\gamma v_o - \int_0^\tau \frac{1}{1-\lambda}\left(\frac{\dot{k}_B}{\eta}\right)^{1/r} d\tau = \dot{k}_B\Gamma_1(\lambda) \tag{5.17}$$

$$\gamma v_o - \int_0^\tau \left[1+\left(\frac{\dot{k}_B}{\eta}\right)^{1/r}\right] d\tau = \dot{k}_B\Gamma_2(\lambda) \tag{5.18}$$

with

$$\Gamma_1(\lambda) \equiv \frac{1-\lambda}{r+2}\left\{\gamma + \left[1+\frac{\gamma}{r+1}\right]\lambda - \frac{r}{2(r+1)}\lambda^2\right\}$$

$$\Gamma_2(\lambda) \equiv \frac{1-\lambda}{r+2}\left\{\gamma + \left[1+\frac{\gamma}{r+1}\right]\lambda - \frac{2r+1}{2(r+1)}\lambda^2 + \frac{2r+1}{6(r+1)}\lambda^3\right\}$$

If Eq. (5.18) is subtracted from (5.17), the resulting equation has a right hand side proportional to

$$\Gamma_1(\lambda) - \Gamma_2(\lambda) \equiv \frac{1-\lambda}{r+2}\lambda^2 \left\{\frac{1-\lambda}{2} + \frac{r+2}{6(r+1)}\lambda\right\} \tag{5.19}$$

Hence the difference between Eqs (5.17) and (5.18) is independent of mass ratio γ and always positive for $0 < \lambda < 1$.

Thus, for any specified impulse at the tip of a cantilever, the analysis has been reduced to solving for two variables $\dot{k}_B$ and λ that are both functions of time τ. This solution can be obtained by iterating Eqs (5.17) and (5.18) at each successive interval of time starting from the initial conditions at $\tau = 0$

$$\dot{k}_B(0) = \frac{(r+2)p}{\gamma} = (r+2)v_0, \qquad \lambda(0) = 0 \tag{5.20}$$

Having solved these equations for the period before the response terminates, that is for $\lambda \le 1$, the final deformed shape of the cantilever can be calculated by an integration with respect to time of the velocity given in Eqs (5.11) and (5.13).

Ting [1964] analyzed a numerical example that simulates a mild steel specimen in the impact experiments of Bodner and Symonds [1960]. The parameter values were taken as $\dot{\varepsilon}_0 = 40.4\ \mathrm{s}^{-1}$, $r = 5$, $\gamma = 1.64$ and $p_0 = 2.93$. In this example, the nondimensional response time τ_f was found to be 1.635 and the final angle at the tip 59.2° or 1.033 radians. The main results are illustrated in Figs 5.2–5.4.

In order to improve their analysis, Bodner and Symonds [1962] examined methods which bring inertia forces of the plastic segment HB into the equations of motion. It should be noted, however, that if the momentum in the plastic region is included in the analysis, the problem will inevitably require a solution for a fourth-order nonlinear parabolic equation with a floating boundary. Alternatively, a linear approximation for the velocity distribution results in additional terms, Γ_1^* on the right-hand side of Eq. (5.15) and Γ_2^* on the right-hand side of Eq. (5.16). With

$$\dot{w}(x) = \dot{w}_H \frac{1-x}{1-\lambda} = \dot{k}_B \frac{(1-\lambda)(1-x)}{r+2}$$

these terms are found to be

$$\Gamma_1^* = \int_\lambda^1 \dot{w}(x)\,\mathrm{d}x = \dot{k}_B \frac{(1-\lambda)^3}{2(r+2)}$$

$$\Gamma_2^* = \int_\lambda^1 \dot{w}(x)(1-x)\,\mathrm{d}x = \dot{k}_B \frac{(1-\lambda)^4}{3(r+2)}$$

If this modified rate–dependence is incorporated into Eqs (5.15) and (5.16), the initial values given in (5.20) should also be modified to result in

$$\dot{k}_B = \frac{3(r+2)p}{1+3\gamma} = \frac{3\gamma}{1+3\gamma}(r+2)v_0, \qquad \lambda = 0, \qquad \text{at } \tau = 0 \tag{5.21}$$

Numerical examples show, however, that this modification does not significantly alter the results. In other words, the inertia of the plastically deforming segment HB does not significantly influence the results in practice and thus it can be neglected.

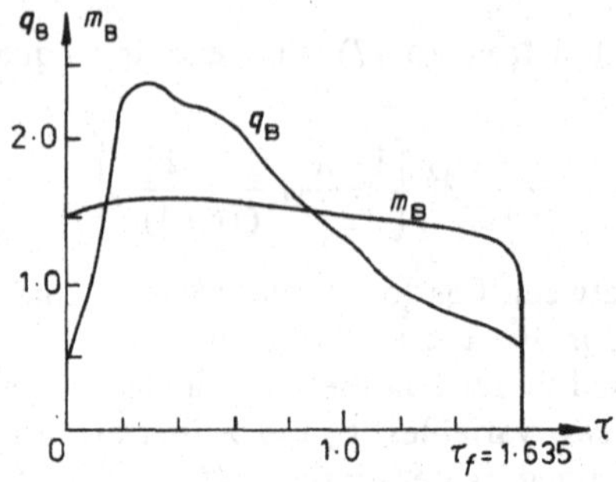

Fig. 5.2 Shear force and bending moment at root B as functions of time.

One notable feature of the viscoplastic response is that the position λ of the interface between the plastic and rigid regions varies very little during most of the response period. This is illustrated in Fig. 5.3. Consequently, most plastic deformation takes place in a portion of length $(1-\bar{\lambda}L)$ near the root, where $\bar{\lambda}$ denotes the average value of λ during the response period.

This example also shows that the bending moment at the root, m_B, varies little with time (see Fig. 5.2) and so it can be approximated by its initial value $m_B(0)$; i.e. using $\dot{k}_B$ from (5.21),

$$m_B \approx 1+\left(\frac{\dot{k}_B}{\eta}\right)^{1/r} = 1+\left[\frac{3\gamma}{1+3\gamma}\frac{(r+2)v_0}{\eta}\right]^{1/r} \tag{5.22}$$

Assuming that this bending moment is constant during the response, one finds the response time τ_f from the moment of momentum equation (5.16) with $v(\tau_f)=\dot{w}(x,\tau_f)=0$

$$\tau_f \approx \frac{p_0}{1+\left(\dfrac{\dot{k}_B}{\eta}\right)^{1/r}} = \frac{\gamma v_0}{1+\left[\dfrac{3\gamma}{1+3\gamma}\dfrac{(r+2)v_0}{\eta}\right]^{1/r}} \tag{5.23}$$

Similarly when $\dot{k}_B = 0$, from (5.15) one can obtain an estimate for the location of the boundary $\bar{\lambda}$ between deformed and undeformed regions at time τ_f

$$\bar{\lambda} \approx \frac{1}{1+\left[\dfrac{3\gamma}{1+3\gamma}\dfrac{(r+2)v_0}{\eta}\right]^{1/r}} \tag{5.24}$$

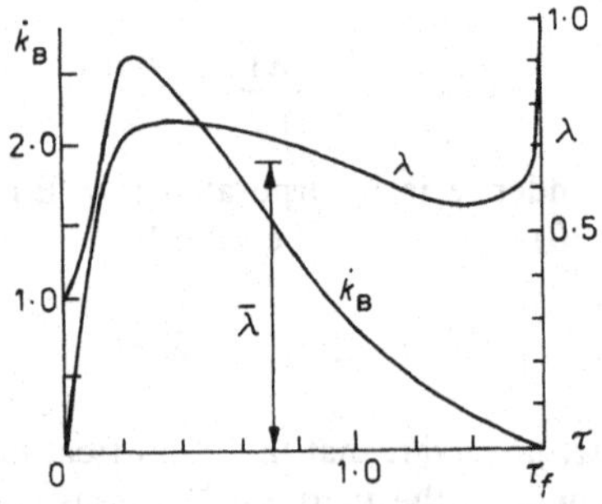

Fig. 5.3 Interface location λ between plastic and rigid regions, and curvature–rate $\dot{k}_B$ at root as functions of time.

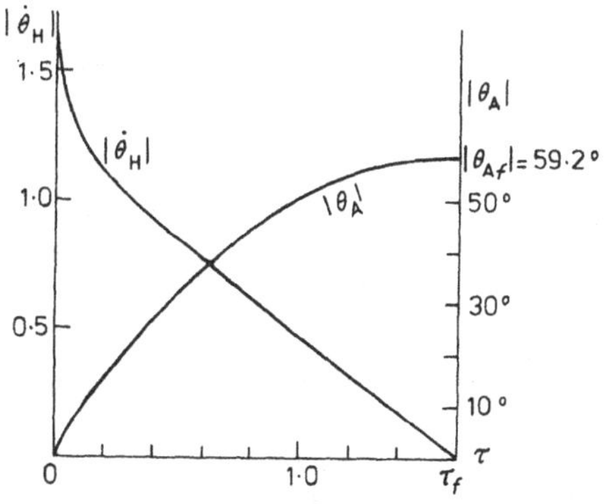

Fig. 5.4 Angular velocity $\dot{\theta}_H$ at interface, and rotation angle θ_A at tip as functions of time.

Noting that the angular acceleration $\ddot{\theta}_A$ is roughly constant, the preceding approximations provide an estimate for the final angle of rotation at the tip of the cantilever,

$$\theta_{Af} \approx -\frac{3p^2\bar{\lambda}}{2(1+3\gamma)} \tag{5.25}$$

The basic viscoplastic model at the beginning of this section neglects inertia in the plastically deforming segment adjacent to the root. The significance of this approximation has been examined by developing a modified model in which inertia of the entire beam is accounted for, but the bending moment $m(x)$ does not vary during the response; this modified model invokes a kinematic approximation that does not satisfy the yield condition at all times. Calculations by Ting [1964] have shown that the results of these two models are practically the same. For the final rotation of the tip θ_{Af}, the modified model gives rotations that are less than 10% smaller in magnitude than those of the first model; this difference is largest for a light mass at the tip. So unless the tip mass is very light, the inertia of the plastically deforming segment is negligible.

The theoretical predictions of the viscoplastic analysis above are compared with relevant experimental measurements in Chap. 7. Before making that comparison, however, it is worthwhile to point out that a prominent result from this viscoplastic analysis is that the final curvature increases from the tip to the root; consequently, the portion near the tip has only small plastic deformation. This result for a viscoplastic structure is essentially different from the results obtained from the elementary rigid–plastic theory in Sect. 4.5, where final curvature decreased with distance from the tip.

5.1.2 Elementary Estimates of the Effect of Strain-Rate on Final Deformation

In view of the laborious numerical work involved in the viscoplastic analysis above, some elementary estimates of the effect of the strain-rate on the final deformation in the cantilever impact problem could be useful both for comparisons with experiments and for preliminary calculations of dynamic response. The elementary rigid–plastic analysis in Sect. 4.5 separated the dynamic response to impact loading into two phases. Here we take the mean value of the curvature–

rate for each phase of the rigid–plastic response, and calculate the decrease in the final curvature for each phase that corresponds to the rate-effect.

For *phase I*, using Eqs (4.111), (4.117) and (4.132), the mean nondimensional curvature–rate is found to be

$$\bar{\dot{k}}_1 = \frac{|\theta_{Af}|-|\theta_{Bf}|}{\tau_1} = \frac{v_0}{2}\frac{8\gamma+3}{2\gamma+1} \tag{5.26}$$

Thus, in phase I the mean curvature–rate shows fairly weak dependence on the mass ratio γ; this curvature–rate only varies by a factor of 4/3 when γ increases from zero to infinity.

In *phase II* the cantilever rotates about a stationary hinge at the root, so the curvature–rate can be infinite for a hinge of zero length. Thus the rate-effect for the model depends on *an effective length of a plastic hinge* L_h. As discussed by Nonaka [1967], Jones [1989] and Shu [1990], this effective length is usually taken to be two to five depths of the beam, i.e.

$$L_h = (2 \sim 5)h \tag{5.27}$$

Hence, the mean curvature–rate in phase II can be calculated as

$$\bar{\dot{k}}_2 = \frac{1}{(\tau_f-\tau_1)}\frac{|\theta_{Bf}|}{L_h/L} = \frac{v_0}{2\left(1+\dfrac{1}{2\gamma}\right)}\frac{L}{L_h} \tag{5.28}$$

For a cantilever made of rate-sensitive material, the plastic bending moment increases with curvature–rate. By recalling the Cowper–Symonds relation (5.1), it is estimated that the plastic bending moment m will increase by factors

$$\mu_1 = 1+\left(\frac{\bar{\dot{k}}_1}{\eta}\right)^{1/r} = 1+\left[\frac{v_0}{2\eta}\frac{8\gamma+3}{2\gamma+1}\right] \tag{5.29}$$

$$\mu_2 = 1+\left(\frac{\bar{\dot{k}}_2}{\eta}\right)^{1/r} = 1+\left[\frac{v_0}{2\eta(1+1/2\gamma)}\frac{L}{L_h}\right]^{1/r} \tag{5.30}$$

for phase I and phase II, respectively. It follows that the deformation parameters obtained from the rigid–plastic analysis can be modified accordingly. For instance, the final rotation at the root θ_{Bf} and that at the tip θ_{Af} given by Eqs (4.117) and (4.132) can be modified as

$$\hat{\theta}_{Bf} = \frac{\theta_{Bf}}{\mu_2} = -\frac{\gamma v_0^2}{2}\frac{1+\dfrac{1}{3\gamma}}{\left(1+\dfrac{1}{2\gamma}\right)^2}\frac{1}{\mu_2} \tag{5.31}$$

$$\hat{\theta}_{Af} = -\frac{\gamma v_0^2}{2}\left\{\left[1-\frac{1+\dfrac{1}{3\gamma}}{\left(1+\dfrac{1}{2\gamma}\right)^2}\right]\frac{1}{\mu_1}+\frac{1+\dfrac{1}{3\gamma}}{\left(1+\dfrac{1}{2\gamma}\right)^2}\frac{1}{\mu_2}\right\} \tag{5.32}$$

while the following modification incorporates the effect of strain-rate on final deflection at the tip of the cantilever

$$\hat{\delta}_f = \frac{\delta_{f1}}{\mu_1} + \frac{\delta_{f2}}{\mu_2} \tag{5.33}$$

where $\delta_{f1} = \delta_f - \delta_{f2}$ with δ_f being given by Eq. (4.134) and $\delta_{f2} = \left|\theta_{Bf}\right|$.

For a heavy striker $\gamma >> 1$, the response is dominated by phase II (see Sect. 4.5), so that

$$\hat{\delta}_f \approx \left|\hat{\theta}_{Af}\right| \approx \left|\hat{\theta}_{Bf}\right| \approx \frac{\gamma v_o^2}{2}\frac{1}{\mu_2} \approx \frac{\gamma v_o^2}{2}\frac{1}{\left[1+\left(\frac{v_o}{2\eta}\frac{L}{L_h}\right)^{1/r}\right]} \tag{5.34}$$

On the other hand, for a light striker $\gamma << 1$, the response is dominated by phase I, so that

$$\left|\hat{\theta}_{Bf}\right| \approx 0,\ \left|\hat{\theta}_{Af}\right| \approx \frac{\gamma v_o^2}{2}\frac{1}{\mu_1} \approx \frac{\gamma v_o^2}{2}\frac{1}{\left[1+\left(\frac{3v_o}{2\eta}\right)^{1/r}\right]} \tag{5.35}$$

$$\hat{\delta}_f \approx \frac{2}{3}\gamma^2 v_o^2 \ln\left(\frac{1}{2\gamma}\right)\frac{1}{\mu_1} \approx \frac{2}{3}\gamma^2 v_o^2 \ln\left(\frac{1}{2\gamma}\right)\frac{1}{\left[1+\left(\frac{3v_o}{2\eta}\right)^{1/r}\right]} \tag{5.36}$$

These estimates are much simpler to calculate than the involved viscoplastic analysis presented in Sect. 5.1.1.

Indeed, Parkes [1955] proposed a correction to the plastic bending moment according to the mean strain-rate in the rigid–plastic analysis; however, he only accounted for strain-rate in phase II. Hence, his estimates are good for heavy strikers but they may fail for light strikers. The corrections for rate-effects given above certainly improve the accuracy of estimates for the final deformation.

The following examples provide a rough idea of the magnitude of the correction factors. Consider two different specimens subjected to the same impulsive load:

Size of specimens: $L = 120$ mm, $h = 3$ mm, $b = 10$ mm

Materials: mild steel and aluminum alloy, both with static yield stress Y = 300 N/mm^2 and dynamic properties as listed in Table 1.2

Colliding mass: heavy, $\gamma = G/\rho L = 10$, $V_o = 5\ \mathrm{m\,s^{-1}}$ or light, $\gamma = G/\rho L = 0.2$, $V_o = 250\ \mathrm{m\,s^{-1}}$

Length of plastic hinge at root: $L_h = 2h$

The calculated results for effective bending moments μ_1 and μ_2 are listed in Table 5.1. It is seen that (1) the strain-rate effect is larger for mild steel than for

Table 5.1 Examples of correction factors accounting for strain-rate effect

Material	η	T_o (s)	γ	v_o	μ_1	μ_2
Mild steel	40.4	0.0078	10	0.325	–	1.60
Mild steel	40.4	0.0078	0.2	16.25	1.92	–
Aluminum alloy	3780	0.0045	10	0.188	–	1.15
Aluminum alloy	3780	0.0045	0.2	9.38	1.25	–

aluminum alloy; (2) the strain-rate effect is larger for high speed impact; and (3) for both cases the strain-rate effect is not negligible.

5.2 Strain Hardening (or Strain Softening) Effect

5.2.1 Introduction

Most engineering materials display some strain hardening after initial yielding; that is, for strain in excess of the yield strain the flow stress increases with increasing strain. A few materials, such as rocks and concrete that suffer internal damage as a consequence of deformation, may display strain softening in certain ranges of strain. In Chap. 3, static analysis of flexural plastic deformation in beams showed that strain hardening diffuses the stationary plastic hinges, so the plastically deforming region is spread through a finite volume of material. Although the effect of strain hardening on static deformation of elastic–plastic beams is well established, there has been little progress in explaining the influence of strain hardening on the transient phase of dynamic structural response.

Conroy [1952] was first to consider the dynamic deformation of strain hardening beams; she commented that 'an infinite number of localized plastic regions form along the beam'. Later analyses of impulsively loaded beams by Florence and Firth [1965] and Forrestal and Sagartz [1978] obtained solutions which circumvent the more difficult transient phase of deformation; i.e. they considered strain hardening during only the modal stage. In the above investigations, an approximation for the modal phase of dynamic response was obtained by assuming that strain hardening does not influence the spatial distribution of velocity. Jones [1967] extended this kinematic approximation to include the transient stage of deformation by considering a travelling plastic hinge distributed over a predetermined short length of beam. These approximations for plastic deformation in strain hardening beams do not satisfy the yield condition at all times; nevertheless, they provide reasonable estimates for final deflections if the strain hardening modulus is small.

In this section, the effect of strain hardening on the dynamic plastic deformation of a uniform cantilever is examined. First a rough estimate for this effect is given in a manner suggested by Jones [1967]; this scheme also takes strain-rate into account. Then a complete analysis of the dynamic response of strain hardening cantilevers is presented in Sect. 5.2.3; this analysis also applies to strain softening cantilevers.

The analysis in this chapter assumes that a section is rigid if the bending moment is less than yield. If the section is plastically deforming so the rate-of-curvature $\dot{\kappa} > 0$, the deformation is represented by a linear moment–curvature relation. The moment–curvature relation for pre-yielding and post-yielding states can be expressed as

$$\left\{\begin{array}{lll} \text{pre–yielding} & M \le M_p, & \kappa = 0 \\ \text{post–yielding and loading} & M = M_p + \alpha_m \kappa, & \mathrm{d}\kappa/\mathrm{d}t > 0 \\ \text{post–yielding and unloading} & M < M_p + \alpha_m \kappa, & \mathrm{d}\kappa/\mathrm{d}t = 0 \end{array}\right\} \quad (5.37)$$

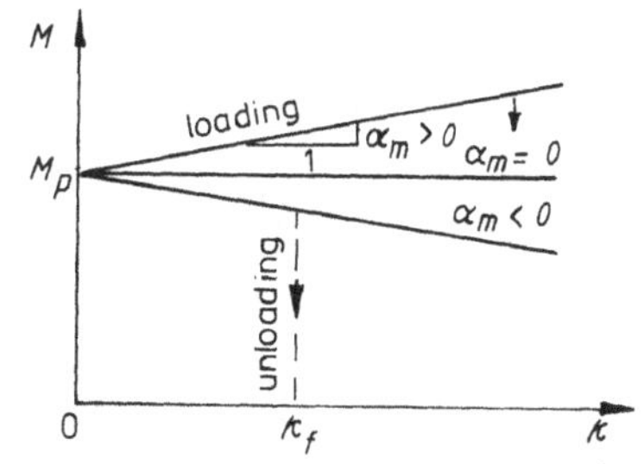

Fig. 5.5 The moment–curvature relations for rigid–linearly strain hardening and rigid–linearly strain softening materials.

These relations are illustrated in Fig. 5.5. With strain hardening, a cross-section remains rigid if the bending moment applied is less than M_p, but increases in curvature if M exceeds M_p. Here M_p is taken to be the *initial plastic bending moment* or the *initial yield moment* for the beam section. After initial yielding in a section, *loading* and *unloading* are distinguished according to the curvature-rate $d\kappa/dt$. The solid lines on Fig. 5.5 represent loading paths where $d\kappa/dt > 0$, while the dashed line is an unloading path where $d\kappa/dt = 0$. Coefficient α_m in Eq. (5.37) is either a hardening parameter if $\alpha_m > 0$, or a softening parameter if $\alpha_m < 0$.

The post-yield and loading equation from (5.37) can be rewritten in a nondimensional form as

$$m \equiv \frac{M}{M_p} = 1 + \tilde{\alpha}k \tag{5.38}$$

with $\tilde{\alpha} \equiv \alpha_m / LM_p$ and $k = \kappa L$

5.2.2 Elementary Effect of Strain Hardening on Final Deformation

Using amplification factors for stress resultants similar to those in Sect. 5.1.2, the effect of strain hardening on final deformation can be estimated by using one or more correction factors to modify the initial yield moment. This factor (or factors) can be found by substituting the mean curvature calculated without hardening into the moment–curvature relation.

For the cantilever impact problem analyzed in Sect. 4.5, the mean nondimensional curvature produced by the travelling hinge during phase I is equal to

$$\bar{k}_1 \equiv \bar{\kappa}_1 L = \left|\theta_{Af}\right| - \left|\theta_{Bf}\right| = \frac{\gamma v_0^2}{6} \frac{(8\gamma + 3)}{(2\gamma + 1)^2} \tag{5.39}$$

Introducing the effective length of a plastic hinge L_h as given by Eq. (5.27), the mean nondimensional curvature for phase II can be estimated as

$$\bar{k}_2 \equiv \bar{\kappa}_2 L = \frac{\left|\theta_{Bf}\right| L}{L_h} = \frac{\gamma v_0^2}{6} \frac{\left(1 + \dfrac{1}{3\gamma}\right)}{\left(1 + \dfrac{1}{2\gamma}\right)^2} \frac{L}{L_h} \tag{5.40}$$

If these mean curvatures are substituted into the constitutive relation (5.37), the nominal plastic bending moment during phases I and II can be expressed as

$$\overline{M}_1 = M_p + \alpha_m \overline{\kappa}_1 = M_p + \frac{\alpha_m}{L} \frac{\gamma v_0^2}{6} \frac{(8\gamma + 3)}{(2\gamma + 1)^2} \tag{5.41}$$

$$\overline{M}_2 = M_p + \alpha_m \overline{\kappa}_2 = M_p + \frac{\alpha_m}{L_h} \frac{\gamma v_0^2}{2} \frac{\left(1 + \frac{1}{3\gamma}\right)}{\left(1 + \frac{1}{2\gamma}\right)^2} \tag{5.42}$$

Alternatively, one may define *correction factors* for the nondimensional fully plastic moment during phases I and II as

$$\mu_1 \equiv \frac{\overline{M}_1}{M_p} = 1 + \frac{\alpha_m \overline{\kappa}_1}{M_p} = 1 + \tilde{\alpha} \frac{\gamma v_0^2}{6} \frac{(8\gamma + 3)}{(2\gamma + 1)^2} \tag{5.43}$$

$$\mu_2 \equiv \frac{\overline{M}_2}{M_p} = 1 + \frac{\alpha_m \overline{\kappa}_2}{M_p} = 1 + \tilde{\alpha} \frac{L}{L_h} \frac{\gamma v_0^2}{2} \frac{\left(1 + \frac{1}{3\gamma}\right)}{\left(1 + \frac{1}{2\gamma}\right)^2} \tag{5.44}$$

where $\tilde{\alpha} = \alpha_m / LM_p$. Then all the deformation quantities found in the rigid–perfectly plastic analysis (Sect. 4.5) also apply here if they are divided by these correction factors, as was done in Eqs (5.31)–(5.36). For example, using the parameters listed at the end of Sect. 5.1.2, and taking the hardening parameter $\tilde{\alpha}$ to be small ($\tilde{\alpha} = 0.01$), one finds from Eqs (5.43) and (5.44) that $\mu_2 = 1.10$ for the heavy striker ($\gamma = 10$) and $\mu_1 = 1.21$ for the light striker ($\gamma = 0.2$).

Symonds [1965] and Perrone [1966] suggested that this elementary estimate can represent the effects of both strain hardening and strain-rate. This is possible if the effects are separable,

$$\frac{\hat{Y}}{Y} = f_1\left(\frac{d\varepsilon}{dt}\right) f_2(\varepsilon) \tag{5.45}$$

In general the material strain hardening and strain-rate effects are not decoupled but Eq. (5.45) provides a rough estimate of how these effects modify the response. In (5.45), function f_1 can take the form of the Cowper–Symonds relation (see Sect. 1.3) while function f_2 can be represented by a linear function of strain.

The notion of separable functions can be extended to the moment–curvature or rate of change of curvature relation in order to represent the influences of strain hardening and strain-rate. The ratio between the modified bending moment $\hat{M}$ and the fully plastic moment M_p at any section is

$$\mu \equiv \frac{\hat{M}}{M_p} = \Phi_1\left(\frac{d\kappa}{dt}\right) \Phi_2(\kappa) \tag{5.46}$$

This ratio μ is a crude correction factor for the perfectly plastic analysis; e.g. function Φ_1 can represent expression (5.4), while Φ_2 represents the right-hand side of Eq. (5.38), leading to

$$\mu \equiv \left[1 + \left(\frac{\dot{k}}{\eta}\right)^{1/r}\right](1 + \tilde{\alpha} k) \tag{5.47}$$

where material parameters η, r and $\tilde{\alpha}$ correspond to characteristic rate of change of curvature, rate index and strain hardening modulus, respectively.

5.2.3 Dynamic Analysis of Strain Hardening and Strain Softening Cantilevers

Based on an analysis by Stronge and Yu [1989], the present section begins with equations of motion and shows how a hardening (softening) moment–curvature relation changes the final distribution of curvature in a basic manner. This is a constitutive effect that is in contrast to the modification factor μ that was developed in the preceding section for mean curvature based on the perfectly plastic material model. The following analysis examines the influence of strain hardening and strain softening on the transient stage of dynamic plastic deformation. For a transverse collision at the tip of a rigid–strain hardening (softening) cantilever, the entire beam is suddenly loaded to the initial yield moment M_p at the instant of collision. At this instant deformation begins throughout the cantilever; the deformations are accompanied by transverse accelerations that are large at sections near the tip and insignificant near the root. These accelerations soon reduce the bending moment near the tip to less than the current yield moment so a rigid (unloaded) segment quickly emerges from the tip. This previously deformed but currently rigid segment spreads from the tip until it envelops most of the cantilever. The curvature at any section develops *before* it becomes a part of the rigid segment; all sections in the rigid segment are unloading from a previously deformed state. Hence, at all sections that are not very near the root the bending moment first increases and then decreases. At each section the curvature developed during this cycle is uniquely defined for strain hardening beams. The picture is not as clear for strain softening structural elements. At a deforming section in a strain softening cantilever, the change in the bending moment with time must be negative for both loading and unloading paths, see Fig. 5.5. The loading and unloading paths are associated with increasing curvature and no further change in curvature, respectively. Hence, the sign of rate of change in bending moment cannot be used as a criterion to distinguish loading and unloading. The present theory solves this dilemma by separating the cantilever into two regions: (1) a rigid segment of increasing length that spreads from the tip, and (2) a deforming segment that shrinks towards the root. We assume that the root segment is loading and the tip segment is unloading, i.e. that all deformation develops in the root segment. Furthermore, it is assumed that inertia forces in the slowly moving root segment are negligible; this assumption is the same as that employed in Sect. 5.1.2. The present theory for plastic deformation of both strain hardening and strain softening cantilevers results in dynamics for the transient stage that converge to Parkes' rigid–perfectly plastic solution in the limit as the hardening parameter α becomes negligible.

Let us consider again the problem of transverse impact at the tip of a cantilever as shown in Fig. 4.22, but now assume that the cantilever is either strain hardening or softening with a moment–curvature relation of the form of Eq. (5.37). Impulsive loading of the massive particle at the tip at the initial instant produces a transverse motion; propagation of this disturbance away from the tip depends on inertia and the constitutive relation. At any instant of time t the cantilever is composed of two segments: a rigid segment AH near the tip that is unloading ($0 < X < \Lambda$) and

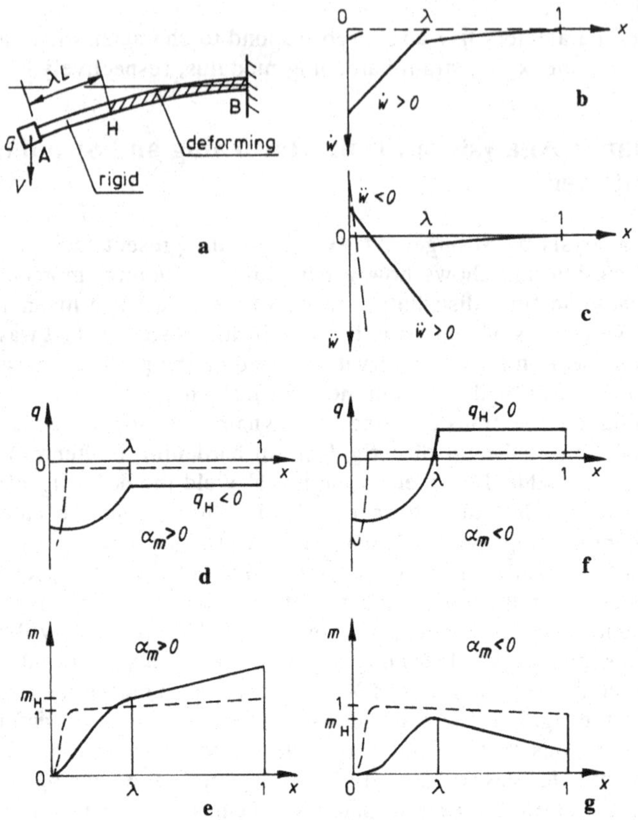

Fig. 5.6 (a) Dynamic deformation mechanism of a strain hardening (or strain softening) cantilever; (b) distribution of velocity; (c) distribution of acceleration; (d) distribution of shear force in case of strain hardening; (e) distribution of bending moment in case of strain hardening; (f) distribution of shear force in case of strain softening; (g) distribution of bending moment in case of strain softening.

the remainder of the beam HB that is plastically deforming ($\Lambda < X < L$). These two segments are shown in Fig. 5.6a. The interface $\Lambda(t)$ between the segments moves from the tip towards the root; a section located at coordinate X begins to deform on the loading path at $t = 0$ and ceases deforming when $\Lambda(t) = X$. Since segment AH is rigid while the velocity in segment HB is assumed to be negligible, the transverse velocity distribution can be expressed in nondimensional form as

$$\dot{w} = v\left(1 - \frac{x}{\lambda}\right), \qquad 0 \le x \le \lambda \tag{5.48}$$

where nondimensional quantities w, $\dot{w}$, v and λ are all as defined in Eq. (5.3). The inertia of deforming segment HB is assumed to be negligible since the velocities in this segment are small in comparison with velocities in segment AH.

Distributions of velocity, acceleration, shear force and bending moment along the cantilever at two intermediate times are sketched in Fig. 5.6. Figure 5.6d, e pertains to the case of strain hardening ($\alpha_m > 0$) whereas 5.6f, g relates to strain softening ($\alpha_m < 0$). Ahead of the interface the distribution of shear force $q(x,\tau)$ is uniform in the plastically deforming segment $\lambda < x < 1$ when the inertia is

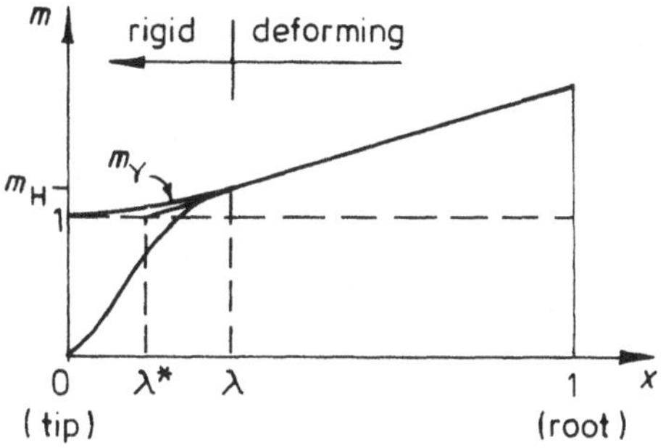

Fig. 5.7 Distribution of bending moment along cantilever and definition of λ^*.

neglected. At the interface H the shear force $q_H = q(\lambda,\tau)$ either increases or decreases monotonously with time depending on the sign of α_m. Consequently, depending on the sign of α_m, the bending moment $m(x,\tau)$ either increases or decreases linearly with x. The bending moment at the interface m_H is determined by the curvature $k_H = k(\lambda,\tau) = w''(\lambda,\tau)$ that develops before $\lambda(\tau)$ reaches any location

$$m_H = 1+\tilde{\alpha}k_H \tag{5.49}$$

where $\tilde{\alpha} = \alpha_m / LM_p$. The linear variation of the bending moment in segment HB can be expressed as

$$m(x) = 1-q_H(x-\lambda^*), \qquad \lambda \le x \le 1 \tag{5.50}$$

where $\lambda^*(\tau)$ is the intersection of $m(x)$ with m=1. Figure 5.7 shows typical distributions for bending moment m and the residual yield moment m_Y, at an instant during the transient stage.

A comparison of (5.50) with the moment–curvature relation (5.38) gives the curvature in segment HB as a function of $\lambda^*(\tau)$

$$\tilde{\alpha}w''(x) = -q_H(x-\lambda^*), \qquad \lambda \le x \le 1 \tag{5.51}$$

By integrating (5.51) with respect to x and applying the boundary condition $w'(1,\tau)=0$ at the root, the inclination in segment HB is found to be

$$w'(x,\tau) = -q_H(2\lambda^*-1-x)(1-x)/2\tilde{\alpha}, \qquad \lambda \le x \le 1 \tag{5.52}$$

Thus, the angular velocity in segment HB is

$$\dot{w}'(x,\tau) = -[\dot{q}_H(2\lambda^*-1-x)+2q_H\dot{\lambda}^*](1-x)/2\tilde{\alpha}, \qquad \lambda \le x \le 1 \tag{5.53}$$

and at the interface H,

$$\dot{w}'_H = \dot{w}'(\lambda,\tau) = -[\dot{q}_H(2\lambda^*-1-\lambda)+2q_H\dot{\lambda}^*](1-\lambda)/2\tilde{\alpha} \tag{5.54}$$

An additional relationship is required at the interface to ensure that the distribution of final curvature in the rigid segment is an analytic function. Therefore from (5.51),

$$0 = \frac{\mathrm{d}m(\lambda,\tau)}{\mathrm{d}\lambda^*} = q_H-(\lambda-\lambda^*)\frac{\mathrm{d}q_H}{\mathrm{d}\lambda^*} \tag{5.55}$$

Substituting q_H from (5.55) into Eq. (5.54), this constitutive constraint results in an expression for the angular velocity at the interface,

$$\dot{w}'_H = \dot{q}_H(1-\lambda)^2/2\tilde{\alpha} \tag{5.56}$$

It is confirmed later that at every rigid section, $0 < x < \lambda$, the moment is always less than the magnitude of m_H when the section entered the rotating segment; hence, the section is unloading. This rigid segment AH rotates with an angular velocity $d\dot{w}_H/dx$ so the transverse velocity at the tip is

$$v = -\dot{q}_H\lambda(1-\lambda)^2/2\tilde{\alpha} \tag{5.57}$$

If the small transverse velocity at the moving interface H is negligible, the velocities and accelerations in the segment AH are given by Eq. (5.48) and

$$\ddot{w}(x,\tau) = \dot{v}(1-x/\lambda)+vx\dot{\lambda}/\lambda^2, \qquad 0 \le x \le \lambda \tag{5.58}$$

respectively. These accelerations cause a shear resultant at the interface,

$$q_H = (\gamma+\lambda/2)\dot{v}+v\dot{\lambda}/2 \tag{5.59}$$

Likewise, the moment at the interface is

$$m_H = -\lambda q_H + \lambda(\dot{v}\lambda+2v\dot{\lambda})/6 \tag{5.60}$$

The continuity of the bending moment at the interface requires that Eqs (5.50) and (5.60) result in the same moment m_H at $x=\lambda$, so that

$$1+\lambda^* q_H = \lambda(\dot{v}\lambda+2v\dot{\lambda})/6 \tag{5.61}$$

If the shear resultant at the interface $q_H = 0$, these relations for the rotating rigid segment adjacent to the tip are the same as those of Parkes' rigid–perfectly plastic solution.

Equations (5.55), (5.57), (5.59) and (5.61) are rewritten to provide nondimensional equations describing the dynamics of the system

$$\dot{q}_H(\lambda-\lambda^*) = q_H\dot{\lambda}^* \tag{5.62}$$

$$\dot{q}_H\lambda(1-\lambda)^2 = -2\tilde{\alpha}v \tag{5.63}$$

$$\frac{d}{d\tau}(2\gamma v+\lambda v) = 2q_H \tag{5.64}$$

$$\frac{d}{d\tau}(\lambda^2 v) = 6(1+\lambda^* q_H) \tag{5.65}$$

These are first-order differential equations containing four unknown functions of time τ; namely λ, λ^*, q_H and v. Rearranging the equations in a form that is convenient for integration we obtain

$$\dot{\lambda}^* = -\frac{2\tilde{\alpha}v(\lambda-\lambda^*)}{q_H\lambda(1-\lambda)^2} \tag{5.66}$$

$$\dot{\lambda} = \frac{2\{3(2\gamma+\lambda)+q_H[3\lambda^*(2\gamma+\lambda)-\lambda^2]\}}{\lambda(4\gamma+\lambda)v} \tag{5.67}$$

$$\dot{q}_H = -\frac{2\tilde{\alpha}v}{\lambda(1-\lambda)^2} \tag{5.68}$$

$$\dot{v} = -\frac{2\{3+q_H(3\lambda^*-2\lambda)\}}{\lambda(4\gamma+\lambda)} \tag{5.69}$$

For compatibility with the moment distribution shown in Fig. 5.6, the initial conditions for this problem are taken as

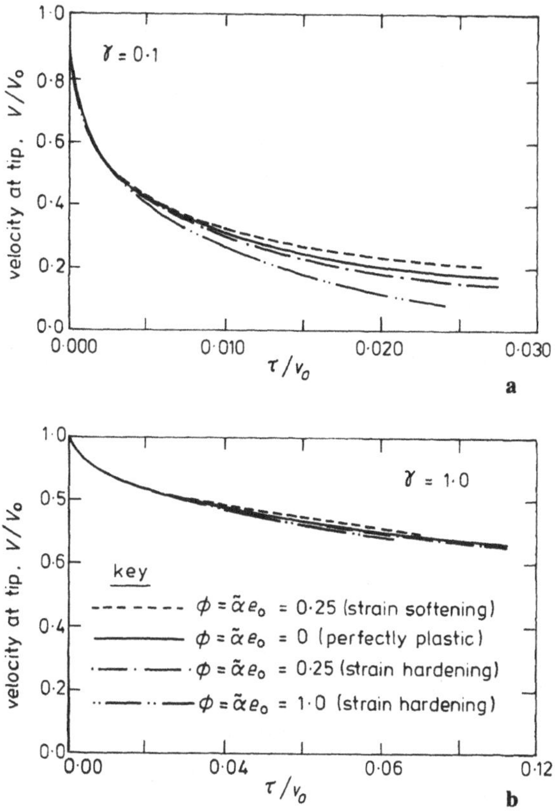

Fig. 5.8 Tip velocity as function of time during transient phase of motion: (a) mass ratio $\gamma = 0.1$; (b) mass ratio $\gamma = 1.0$.

$$\lambda = \lambda^* = q_H = 0, \quad v = v_0, \quad \text{at } \tau = 0 \tag{5.70}$$

With these initial conditions, the differential equations for the system are singular. Nevertheless, the interface speed $\dot{\lambda}$ immediately after impact can be approximated by

$$\dot{\lambda} = 3/\lambda v_0, \qquad \text{for } \tau \ll 1 \tag{5.71}$$

that is, the interface moves away from the tip with a speed that decreases with distance. It follows that

$$\lambda = \sqrt{6\tau/v_0}\,, \qquad \text{for } \tau \ll 1 \tag{5.72}$$

This early time approximation gives

$$v = v_0\left(1 - \frac{1}{2\gamma}\sqrt{\frac{6\tau}{v_0}}\right), \; \lambda^* = \frac{1}{2}\sqrt{\frac{6\tau}{v_0}}, \; q_H = -\frac{2}{3}\tilde{\alpha} v_0 \sqrt{6\tau v_0}\,, \text{ for } \tau \ll 1 \tag{5.73}$$

Numerical calculations verify that the structural deformations depend on the impact energy e_0 and the hardening parameter α; the deformation always depends on a nondimensional group $\phi \equiv \tilde{\alpha} e_0 = \tilde{\alpha}\gamma v_0^2/2$. For light and moderate mass ratios

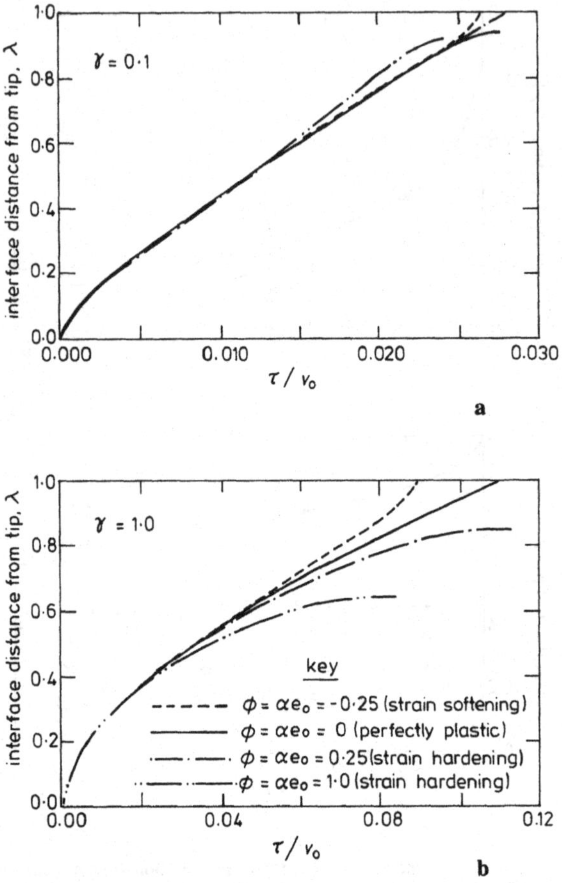

Fig. 5.9 Interface position as function of time during transient phase of motion: (a) mass ratio $\gamma = 0.1$; (b) mass ratio $\gamma = 1.0$.

γ, Fig. 5.8 shows the variation of the tip velocity with time during the transient phase of motion for impulsively loaded cantilevers; results are plotted for values of ϕ equal to $-$ 0.25, 0, 0.25 and 1.0. The rate of decrease in tip speed during this phase is primarily influenced by the mass ratio γ. If $\gamma \ll 1$, most of the initial velocity disappears and energy is dissipated during phase I; in contrast, if the mass ratio is moderate to large $\gamma \geq 1$, strain hardening mostly affects the length of the deforming segment HB for the modal phase of deformation. This is shown in Fig. 5.9 where the variation of the interface position with time is illustrated. It is also seen that hardening only slightly increases the deceleration of the tip.

For strain hardening cantilevers, the transient phase of deformation terminates when $\dot{\lambda} = 0$. The interface terminus λ_1 is some distance from the root, $\lambda_1 < 1$; this distance can be determined from Eq. (5.67),

$$\frac{\lambda_1^2}{(2\gamma + \lambda_1)} = 3\left[\lambda_1^* + \frac{1}{q_{H1}}\right] \tag{5.74}$$

with τ_1 being the time at which the transient phase terminates, $\lambda_1^* = \lambda^*(\tau_1)$, $q_{H1} = q_H(\tau_1)$. The terminal position depends on the hardening parameter $\tilde{\alpha}$ and

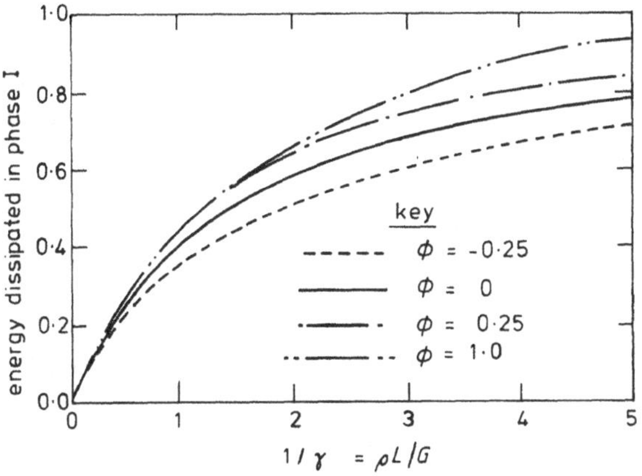

Fig. 5.10 Energy dissipated during the transient phase of motion is influenced by mass ratio and strain hardening (or softening) coefficient.

mass ratio γ. The subsequent phase of motion has deformation only in the region $\lambda_1 < x < 1$. This region is more widespread with increases in either the hardening parameter $\tilde{\alpha}$ or the initial kinetic energy e_o.

For a strain softening material, $\tilde{\alpha} < 0$, the interface always transits the entire cantilever. Hence, the transient phase terminates when the interface reaches the root, $\lambda = 1$. The subsequent modal phase of motion has all deformation concentrated in the section at the root of the cantilever.

At the end of phase I, the residual kinetic energy is $E_2 = E(t_1)$, where subscript 2 indicates the energy dissipated in phase II. The part of the initial kinetic energy $K_o = GV_o^2/2$ that is dissipated during phase II is given by

$$\frac{E_2}{K_o} = \left(1 + \frac{\lambda_1}{3\gamma}\right)\left(\frac{v_1}{v_o}\right)^2 \tag{5.75}$$

Figure 5.10 shows that the energy dissipated during phase I is remarkably insensitive to the hardening parameter unless the colliding mass G is very light. For a small mass ratio γ where most of the kinetic energy is dissipated by bending away from the root, strain hardening increases the part of the total energy dissipated during phase I. Strain softening, on the other hand, reduces the part of the initial kinetic energy that is dissipated by distributed bending during phase I.

After the transient phase (phase I), the remaining kinetic energy E_2 is dissipated in a stationary mode of deformation during the period $\tau_1 < \tau < \tau_2$ (phase II). The final response time τ_2 is defined as the time when velocity at the tip vanishes, $v(\tau_2) = 0$. During phase II deformation continues to develop in the segment $\lambda_1 < x < 1$ adjacent to the root of a strain hardening cantilever. Since in this region inertia is assumed to be negligible, the shear resultant is uniform and the bending moment increases linearly with distance from the interface terminus λ_1. At the end of phase I the moment at the terminus $m_{H1} = 1 - (\lambda_1 - \lambda_1^*)q_{H1}$ can be used to express the bending moment $m_1(x, \tau_1)$ within the root segment at the beginning of phase II, that is

$$m_1(x, \tau_1) = m_{H1} - (x - \lambda_1)q_{H1}, \qquad \lambda_1 \le x \le 1 \tag{5.76}$$

where $q_{H1} = q_H(\lambda_1, \tau_1)$. During phase II the moment at any section increases monotonously from an initial value m_1 to the final value $m_2 = m_2(x,\tau_2)$. In order to satisfy the equations of motion in the rigid segment, the moment $m(\lambda,\tau)$ and the shear resultant $q_H(\lambda_1,\tau)$ at the stationary interface must vary proportionately during phase II. Thus in the deforming segment, when motion finally ceases the bending moment is

$$m_2 = m_1 q_{H2}/q_{H1} \tag{5.77}$$

During phase II sections adjacent to the root suffer a change in curvature $k_2 - k_1 = (m_2 - m_1)/\tilde{\alpha}$ because of the increase in bending moment during this phase. The energy dissipated by this plastic deformation is equal to the kinetic energy E_2 given in (5.75). Thus, the final shear resultant at the terminus H is found to be

$$q_{H2} = \left\{ q_{H1}^2 + \tilde{\alpha} v_1^2 \frac{(3\gamma + \lambda_1)}{(1-\lambda_1)^3} \left[1 - \frac{m_{H1}}{q_{H1}(1-\lambda_1)} + \frac{m_{H1}^2}{q_{H1}^2 (1-\lambda_1)^2} \right]^{-1} \right\}^{1/2} \tag{5.78}$$

At the end of phase II, the final bending moment m_2 is obtained from this shear resultant. Hence the final curvature is completely determined by the initial conditions for phase II because during this phase the deformation is a separable function of spatial and temporal variables.

In this analysis the hypotheses that prescribe the initial and boundary conditions for phase II are consistent with the hypotheses that separate the cantilever into deforming and rigid segments. A consequence of this approximation is a discontinuity in bending moment at λ_1. This discontinuity develops during phase II with strain hardening. The possibility of a reversal in the direction of interface travel during phase II was considered by Stronge and Yu [1989]; this reversal, however, is not possible since it would result in a negative rate of energy dissipation.

A fundamental problem arises in finding the motion of a strain softening cantilever during phase II, after the deforming region shrinks to a point. The deforming region has no length so any further deformation has indefinitely large curvature and no energy dissipation. Other investigations have circumvented this problem by specifying a minimum length for the deforming region and defining a moment–curvature relation that asymptotically approaches a nonzero moment as curvature increases. Alternatively, a different constitutive relation can be defined for the localized region, i.e. a moment–rotation rather than a moment–curvature relation, see Martin [1989]. The latter technique is commonly employed at stationary hinges in rigid–perfectly plastic members.

The final curvature distributions in strain hardening cantilevers are shown in Fig. 5.11 for mass ratios $\gamma = 0.1$ and 1.0. With a heavy mass, most of the initial kinetic energy is dissipated during phase II while the small part dissipated in the transient phase is almost independent of the hardening coefficient $\tilde{\alpha}$. During phase II the length of the deforming segment at the root is influenced by the hardening coefficient; the length of this segment increases with $\tilde{\alpha}$ and this diminishes the increase in curvature during this phase. Consequently, strain hardening has a larger effect on final tip deflection if the mass ratio is large as shown in Fig. 5.12. If the tip mass is light, $\gamma < 0.5$, most of the impact energy is dissipated in distributed curvature that develops during phase I. Figure 5.10 indicates that strain hardening accentuates this effect — the fraction of impact energy dissipated in phase II decreases as hardening increases. Strain softening is exceptional only during the final part of the transient phase when the interface is

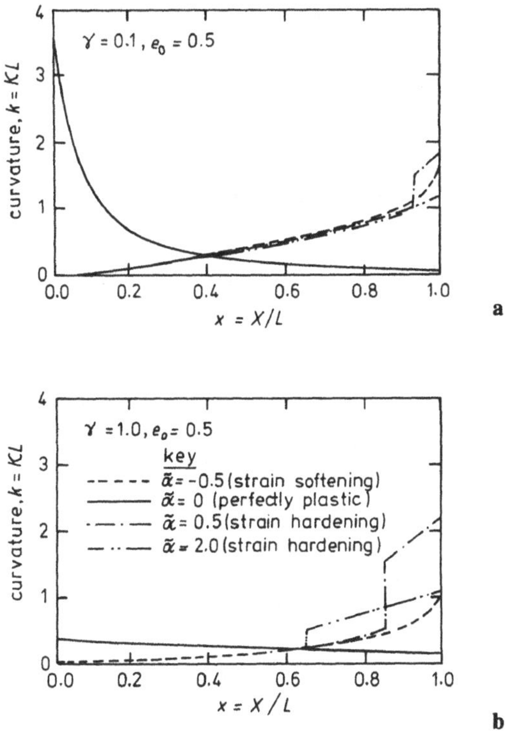

Fig. 5.11 Final curvature distributions along cantilevers: (a) $\gamma = 0.1,\ e_0 = 0.5$; (b) $\gamma = 1.0,\ e_0 = 0.5$

almost at the root. Then the shear resultant q_{H1} increases very rapidly and there is a final large increase in curvature near the root.

The discontinuities in curvature that arise at λ_1 during phase II are not apparent in the final deformed configurations, see Fig. 5.12. These deformed profiles show that strain hardening decreases the deformation; the only apparent effect on the distribution of curvature is the extent of the phase II deforming region if the mass ratio is large, $\gamma > 1$. Changes in the strain hardening coefficient $\tilde{\alpha}$ and the nondimensional initial kinetic energy $e_0 = \gamma v_0^2/2$ have exactly the same effect on the distribution of deformation for all stages of the dynamic response. Hence, the final deformation is directly proportional to e_0 and the effect of increasing e_0 on the distribution of curvature is equivalent to a proportional increase in $\tilde{\alpha}$.

As indicated in Fig. 5.11, the present analysis yields a distribution of curvature which increases with distance from the tip whereas the rigid–perfectly plastic theory (Sect. 4.5) indicates that curvature increases with distance from the root. These distinctly different behaviors result from similar hypotheses; the difference is a consequence of the assumed moment–curvature relation. With the rigid–plastic idealization, all deformation occurs at a hinge which travels away from the tip. The segment ahead of the hinge does not deform because it does not carry any shear force. On the other hand for an *elastic-hardening-softening M-κ relationship*, a pulse-loaded cantilever has finite length regions of hardening and softening that initiate and develop from an interior point some distance away from the tip, Reid, Yu and Yang [1995].

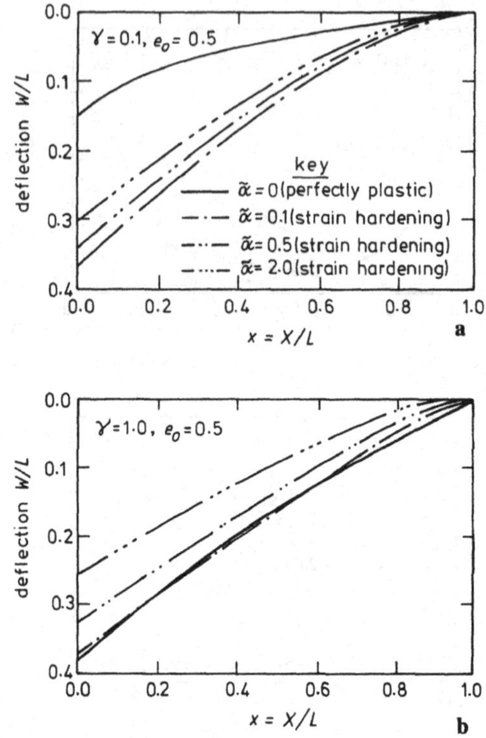

Fig. 5.12 Final deflections of cantilevers: (a) $\gamma = 0.1$, $e_0 = 0.5$; (b) $\gamma = 1.0$, $e_0 = 0.5$.

The final distribution of curvature obtained from the present analysis is distinctly different from that for a perfectly plastic material obtained in Chap. 4. The analysis of dynamic deformation in a strain hardening cantilever neglects inertia in the deforming segment whereas in the perfectly plastic analysis, this same segment of the member is at yield but not deforming. Experiments discussed in Chap. 7 show that in practice the final deformation of strain hardening structures is somewhere between the predictions of these two models; hence in comparison with the perfectly plastic analysis, the curvature near the tip is reduced by strain hardening while in a region near the root, the curvature is increased during the final or modal phase of deformation.

5.3 Effects of Transverse Shear and Rotary Inertia

5.3.1 Interface Conditions for Concentrated Mass

In Chap. 4, the dynamic response of a rigid–plastic cantilever to a transverse force $F(t)$ or impact is analyzed by assuming that shear strain and the effect of shear force on yield are negligible. These assumptions are valid for statically loaded beams, frames and thin plates since the shear stresses are usually much smaller than the normal stress. For instance, when a cantilever is loaded by a static force F at the tip, the largest shear force $Q_B = F$ is at the root; the largest bending moment $M_B = FL$ is at the same cross-section. Hence, when the force reaches the static collapse force $F_c = M_p/L$, the largest shear force is $Q_{max} = M_p/L$. For a rigid–perfectly plastic material the fully plastic shear force Q_p for a cross-section can be estimated as the product of the shear yield stress τ_Y and the cross-sectional area A, i.e.

$$Q_p = \tau_Y A \approx YA/2 \tag{5.79}$$

where Y is the uniaxial yield stress in tension. Thus

$$\frac{Q_{max}}{Q_p} = \frac{2M_p}{YAL} \tag{5.80}$$

For a rectangular cross-section, $M_p = Ybh^2/4$ and $A = bh$, so

$$\frac{Q_{max}}{Q_p} = \frac{h}{2L} << 1 \tag{5.81}$$

Thus in a slender member the resultant shear force is much smaller than the fully plastic shear force unless the applied force is very large, $F >> F_c$.

For rigid–perfectly plastic structures, however, theoretical and experimental studies (e.g. Symonds [1968], Nonaka [1977], Jones and Song [1986] and Liu and Jones [1987]) have shown that shear is more important for an impact or blast load than it is for a static load. This point can be verified by inspecting the dynamic solutions given in Chapter 4. For example, Sect. 4.1 explains the dynamic deformation of a cantilever subjected to a suddenly applied steady force F at its tip. The solution shows that the largest shear force is immediately adjacent to the loading point where shear force is exactly equal to the applied force, F. The applied force can be arbitrarily large and consequently, the ratio Q_{max}/Q_p is no longer necessarily small. The shear stress in a small region near the tip could be as large as the yield stress. This situation is significant for the cases of impact or impulsive loading where the force is indefinitely large. As shown in Sect. 4.5.3, if a transverse impulse is applied to a particle the shear force at a section of the beam adjacent to the particle is indefinitely large. Hence, in a small region near the particle at the tip of the cantilever the deformation must be predominantly shear rather than flexure. The effect of shear on yield during dynamic deformation is often significant; consequently, shear deformation and the possibility of shear failure have to be considered in addition to flexural deformation.

The flexural analysis presented in Chap. 4 also neglects the effect of rotary inertia. Jones and de Oliveira [1979], de Oliveira and Jones [1979] and de Oliveira [1982] examined a series of problems to study the effect of rotary inertia on the dynamic response of uniform beams. Their results indicate that rotary inertia of the beam is not of practical significance unless the beam is extremely short. In the following, therefore, rotary inertia is not included in the equations of motion. However, when the mass attached at the tip of the cantilever is a block rather than a particle, the effect of rotary inertia of the block can be of great importance. If an impulse is applied to a block attached to the tip and the moment of inertia for the block is taken into account, then there can be both a transverse velocity discontinuity and an angular velocity discontinuity at the junction between the block and the tip of the cantilever. These two discontinuities cause simultaneous shear and flexural deformation at the junction. In the following section, shear forces at the interface where a heavy particle is attached to the tip of a rigid–perfectly plastic cantilever are taken into account. Then in Sect. 5.3.3 the effects of both shear and rotary inertia of a finite size block at the tip are brought into the analysis. Thus, the validity of the 'heavy particle' idealization is examined. Finally, shear failure or separation is discussed in Sect. 5.3.4 based on the preceding models of rigid–plastic response. All of these modeling considerations are set in the framework of a cantilever impact problem (Fig. 4.22 in Sect. 4.5).

5.3.2 Shear Deformation Adjacent to Colliding Particle

From the rigid–perfectly plastic analysis for impact of a particle at the tip of a cantilever (Sect. 4.5), the shear force q_A at the tip is given by

$$q_A(\tau) \equiv \frac{Q_A L}{M_p} = -f = -\frac{6\gamma}{\lambda(\lambda+4\gamma)} \tag{5.82}$$

while the bending moment $m_A(\tau)=0$. Immediately after impact, Q_A exceeds the fully plastic shear stress Q_p since $\lambda \to 0$ when $\tau \to 0$. Let the nondimensional fully plastic shear force q_p be defined as

$$q_p \equiv \frac{Q_p L}{M_p} \tag{5.83}$$

This limiting shear force depends on the shape of the cross-section and the size of the beam. For instance, for a beam of rectangular cross-section with depth h and length L,

$$q_p = \frac{2L}{h} \tag{5.84}$$

The limiting shear force represents a ratio of shear to bending strength.

Deformation Mechanism A deformation mechanism with boundary condition (5.82) and limiting force (5.84) is shown in Fig. 5.13a. At the tip of the cantilever a limiting shear stress resultant Q_p opposes the initial relative motion of the heavy particle at the tip of the cantilever. Thus, a downward shear force of magnitude Q_p is exerted on the tip of the cantilever; this force is steady during an initial phase of motion. The steady force Q_p accelerates a segment of length $\lambda_0 L$ that is rotating about a stationary hinge; this case is examined in Sect. 4.1. Taking the nondimensional force $f = q_p$ for the present case, and requiring that $q_p = 2L/h > 3$, one obtains from (4.29) that the distance from the tip to the stationary hinge is

$$\lambda_0 = \frac{3}{q_p} = \frac{3h}{2L} \tag{5.85}$$

The magnitude of the shear force decreases linearly from $|Q| = Q_p$ at the tip to $Q=0$ at $\Lambda_0 = \lambda_0 L = 3h/2$ (Fig. 5.13b). Hence, expression (5.85) implies that shear deformation only takes place in a localized region near the tip, with the length of the sheared region being of the order of the depth. It should be noted, however, that this length depends on the shape of the cross-section. For wide-flanged I–beams the parameter q_p may be quite small even for relatively long span-to-depth ratios (see Jones and de Oliveira [1979]).

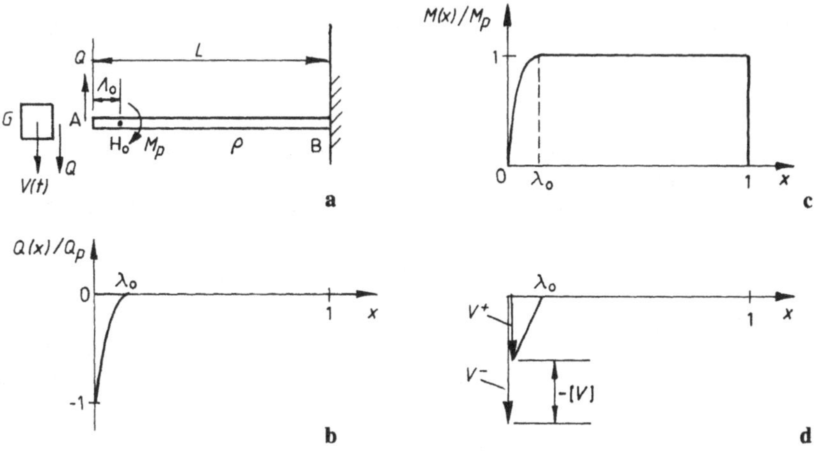

Fig. 5.13 (a) Dynamic deformation mechanism in which shear yielding occurs at a section of a cantilever adjacent to an impulsively loaded particle; (b) distribution of shear force; (c) distribution of bending moment; (d) distribution of transverse velocity.

Yield Criterion Hodge [1959] obtained an *interactive yield criterion* that combines shear force and bending moment at rectangular cross-sections of slender beams. This can be approximated by Eq. (1.37),

$$\left|\frac{Q}{Q_p}\right|^2 + \left|\frac{M}{M_p}\right|^2 = 1 \tag{5.86}$$

For a suddenly applied force at the tip of a cantilever, the following expressions for shear force and bending moment in segment AH_0 are given in Chap. 4.

$$\left|\frac{Q}{Q_p}\right| = \left|\frac{q}{q_p}\right| = \left(1-\frac{x}{\lambda_0}\right)^2, \ \left|\frac{M}{M_p}\right| = |m| = 1-\left(1-\frac{x}{\lambda_0}\right)^3, \quad 0 \le x \le \lambda_0 \tag{5.87}$$

The shear force and bending moment distributions are shown in Fig. 5.13b and c, respectively. These results are based on an assumed pattern of deformation. While this pattern is suggested by the distribution of stress resultants, for a complete solution we must ensure that the stress resultants finally satisfy the yield condition throughout the structure. Substitution of (5.87) into (5.86) indicates that the interactive yield condition (5.86) is not violated in segment $0 < x < \lambda_0$, and the fully plastic yield condition is satisfied at only two locations: (1) the particle junction at the tip where $Q_A = -Q_p$, $M_A = 0$ and (2) the interior plastic hinge H_0 where $x = \lambda_0$, $Q_H = 0$, $M_H = M_p$. Although an interaction between shear force and bending moment is considered, all cross-sections in segment AH_0 remain rigid and shear and flexural deformations occur only at these distinct locations as long as the tip force $F = Q_p$ is maintained constant. Hence, the solution is complete since at every cross-section in the cantilever the proposed deformation field gives stress resultants that satisfy the yield condition.

Equations of Motion When the transverse shear force at a cross-section equals the fully plastic shear force, the flow rule for plastically deforming materials requires that the associated deformation at this section is pure shear. This corresponds to a discontinuity in transverse velocity at this section,

$$[V] \equiv \lim_{\varepsilon \to 0}\{V(0+\varepsilon) - V(0-\varepsilon)\} \equiv V^+ - V^- \tag{5.88}$$

where for any dependent variable, [ ] denotes the change in value at a discontinuous section. This velocity discontinuity is depicted in Fig. 5.13d. Associated with this discontinuity, the energy dissipation due to shear D_q has a rate of change

$$\frac{\mathrm{d}}{\mathrm{d}t}D_q = Q_p\left|[V]\right| \tag{5.89}$$

In order to calculate the jump in vertical velocity at the junction between the particle and the tip of the cantilever, the acceleration at the tip of the uniform cantilever subjected to a suddenly applied force Q_p is calculated using Eq. (4.30),

$$\dot{v}^+ = \frac{2}{3}q_p^2 = \text{const} \tag{5.90}$$

Here $v \equiv VT_o / L$ is the nondimensional vertical velocity, $\tau \equiv t/T_o$ and $(\dot{\ }) \equiv \mathrm{d}(\ )/\mathrm{d}\tau$. Consequently, the transverse velocity at the tip of the uniform cantilever is

$$v^+ = \frac{2}{3}q_p^2\tau \tag{5.91}$$

On the other hand, the equation of motion for the particle attached at the tip gives

$$G\frac{\mathrm{d}V^-}{\mathrm{d}t} = -Q_p \tag{5.92}$$

or in nondimensional form,

$$\gamma\dot{v}^- = -q_p \tag{5.93}$$

With the initial condition $v = v_o$ at $\tau = 0$, this gives the velocity of the particle

$$v^- = v_o - \frac{q_p}{\gamma}\tau \tag{5.94}$$

At the junction there is a discontinuity in transverse velocity that can be evaluated from Eqs (5.91) and (5.94)

$$[v] \equiv v^+ - v^- = -\left\{v_o - q_p\left(\frac{1}{\gamma} + \frac{2}{3}q_p\right)\tau\right\} \tag{5.95}$$

This velocity jump decreases linearly with time τ; the shearing terminates at an instant $\tau = \tau_o$ determined by

$$\tau_o = \frac{v_o}{q_p\left(\dfrac{1}{\gamma} + \dfrac{2}{3}q_p\right)} \tag{5.96}$$

At the interface where the heavy particle is connected to the uniform cantilever, a displacement discontinuity develops with time as a consequence of shear; this occurs if the material is rigid–perfectly plastic. The relative displacement at the interface can be found by integrating the velocity jump with respect to time; i.e.

$$[w] \equiv w^+ - w^- = \int_0^\tau [v]\,\mathrm{d}\tau = -\left\{v_o\tau - \frac{q_p}{2}\left(\frac{1}{\gamma} + \frac{2}{3}q_p\right)\tau^2\right\} \tag{5.97}$$

so that the discontinuity at $\tau = \tau_o$ is finally

$$\delta_q \equiv [w]_{\tau=\tau_0} \equiv -\frac{\gamma v_0^2}{2} \frac{3}{q_p(3+2\gamma q_p)} \tag{5.98}$$

Shear Rupture Adjacent to Particle Shear rupture can occur if shear deformation at the interface between the heavy particle and the uniform cantilever exceeds a critical value. An estimate for conditions that cause rupture can be based on the magnitude of the displacement discontinuity at this interface. This kind of *shear failure* will be discussed in Sect. 5.3.5. Here we simply assume that shear rupture occurs at the particle junction if relative displacement becomes as large as the depth of the cantilever, i.e. $|\delta_q| = h/L$.

Flexural Deformation after Shear Terminates If $|\delta_q| < h/L$ for $\tau \le \tau_0$, then at $\tau = \tau_0$ the plastic shear deformation terminates while the velocity of the particle and the tip are both equal to

$$v(\tau_0) = \frac{2\gamma q_p}{3+2\gamma q_p} v_0 \tag{5.99}$$

Thereafter shear yielding and relative sliding at the junction disappear, and the shear force at the tip, $q(0)$, simply depends on acceleration of the particle,

$$|q(0)| = |\gamma \dot{v}| < q_p \tag{5.100}$$

The equations of motion for $\tau > \tau_0$ are the same as those in Sect. 4.5, that is Eqs (4.103) and (4.104). Integrating these equations and using (5.85) and (5.99) as initial conditions, one obtains the same solutions given in (4.105)–(4.118). This means that after the initial *shear phase* there are transient and modal phases of response that are represented completely by the rigid–plastic flexural theory in Sect. 4.5. The additional effect of shear on structural response is illustrated in Fig. 5.14 for the case of $\gamma = 0.2$ and $L/h = 6$ The flexural theory (Sect. 4.5) is shown as dashed curves while the solid lines pertain to the present analysis which takes shear yielding into account. The initial shear phase has no effect on the subsequent deformation that takes place after τ_0.

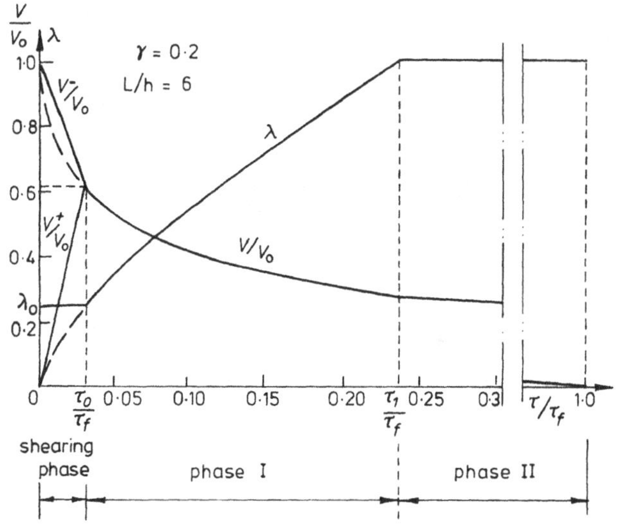

Fig. 5.14 Tip velocity v and position of the plastic hinge λ as functions of time for $\gamma = 0.2$ and $L/h = 6$.

For solutions with and without shear deformation at the junction, the velocity curves are slightly different only in the short period $0 \le \tau \le \tau_0$. Since the area underneath the curve of $v(\tau)$ in Fig. 5.14 represents the final deflection at the tip, this figure indicates that shear yielding has very little effect on the final tip deflection. For example, if $\gamma = 0.2$ and $L/h = 6$, the final tip deflection is changed by 1.2% by shear deformation, while if $\gamma = 0.2$ and $L/h = 30$, this change is less than 0.02%. Obviously, the effect on final deflection is very small and practically negligible. This effect may, of course, be more important for other beam support conditions.

Distribution of Energy Dissipation During the period of increasing deflection, plastic deformation dissipates all of the kinetic energy initially imparted to the system. Now let us examine the effect of shear deformation on the distribution of the energy dissipation. If shear rupture does not occur, i.e. $|\delta_q| < h/L$, then during the initial *shear phase* $0 \le \tau \le \tau_0$, the energy dissipation d_0 consists of two parts: dissipation due to shearing, d_q, and dissipation due to bending at H_0, d_{om}.

$$d_q = |\delta_q| q_p = \frac{\gamma v_0^2}{2} \frac{1}{1 + \frac{2}{3}\gamma q_p} \tag{5.101}$$

$$d_{om} = |\theta(\lambda_0)| = \frac{\delta_0^+}{\lambda_0} = \frac{\gamma^2 v_0^2 q_p}{(3 + 2\gamma q_p)^2} \tag{5.102}$$

Hence, total dissipation d_0 during this first phase of dynamic response to impact is the sum of these separate parts,

$$d_0 = d_q + d_{om} = \frac{\gamma v_0^2}{2} \frac{9 + 8\gamma q_p}{(3 + 2\gamma q_p)^2} \tag{5.103}$$

Since the response during $\tau > \tau_0$ is exactly the same as that in the rigid–plastic bending-only theory, the energy dissipation in the sequential phases, I and

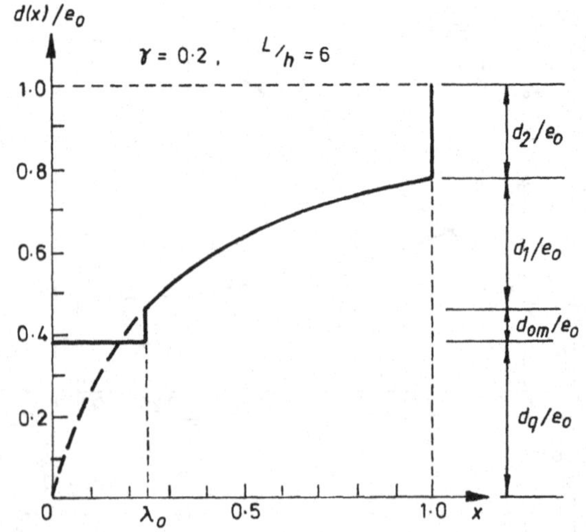

Fig. 5.15 The part of the impact energy dissipated before hinge transits any section x, for $\gamma = 0.2$ and $L/h = 6$.

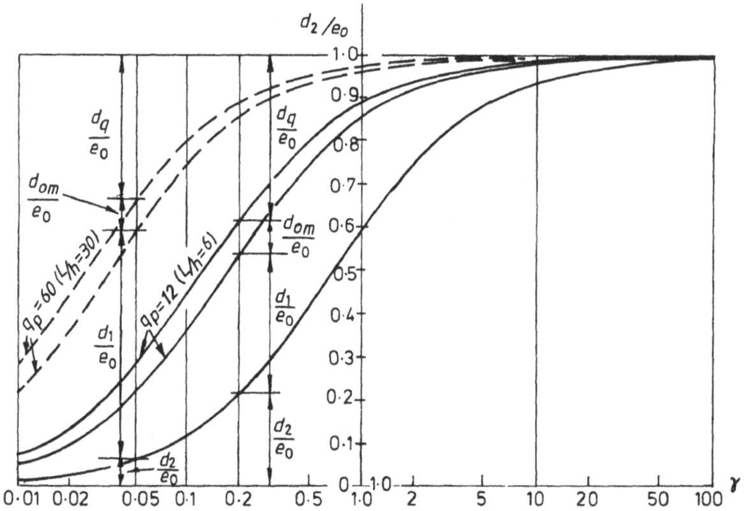

Fig. 5.16 The part of the impact energy dissipated in various phases of response depends on mass ratio $\gamma = G/\rho L$.

II, is given by

$$d_1 = e_o - d_o - d_2 \tag{5.104}$$

$$d_2 = \left|\theta_{Bf}\right| = e_o \frac{4\gamma(1+3\gamma)}{3(1+2\gamma)^2} \tag{5.105}$$

As an example, for a stubby beam $L/h = 6$ and a relatively light particle $\gamma = 0.2$, Fig. 5.15 shows the part of the impact energy dissipated before the travelling hinge transits any position x. This function has a discontinuity at the tip due to shear deformation. Again, shear deformation at the junction makes little or no difference to the final deformation throughout most of the cantilever.

To provide a broader picture of the influence of shear deformation upon the distribution of energy dissipation, Fig. 5.16 demonstrates the proportions of energies dissipated in various response phases as functions of mass ratio $\gamma = G/\rho L$. It is clear that the shear deformation is most significant for a light colliding mass and a stubby cantilever (i.e. small slenderness ratio L/h). These conditions are just where assumptions of this model are most at odds with the practical situation. For any value of q_p the distribution of energy can be obtained by simply shifting the original curve in Fig. 5.16 (for which shear yielding is not considered) to the left along the γ-axis a distance of $q_p/3$.

5.3.3 Shear and Rotary Inertia of Finite Size Colliding Missile

Most previous studies of structural impact have neglected the effect of size or rotary inertia of a colliding body that embeds in the beam — the analysis is notably simplified by assuming that the mass of this body is concentrated in a particle. Shu et al. [1992] took size and rotary inertia of a block at the tip into account when analyzing oblique impact on a rigid–plastic cantilever. Here a more complete analysis is given; this considers both transverse shear force at the junction

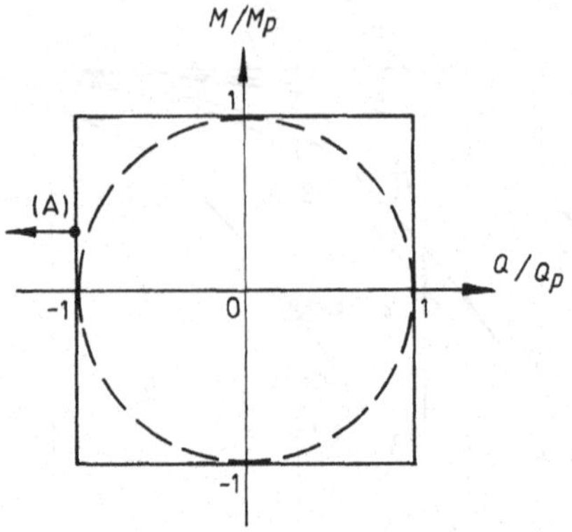

Fig. 5.17 Interactive and independent yield conditions in the shear force–bending moment plane: – – – interactive yield condition (5.86); ——— independent yield condition (5.106); (A) refers to Eq. (5.125).

where a block is attached to the tip of a cantilever and also the rotary inertia of the block.

In this example we employ a *separated* or *independent yield condition* rather than the interactive yield condition (5.86) used previously. With this independent yield condition a complete solution is obtained that always satisfies yield at every cross-section. As shown in Fig. 5.17, this yield condition is represented by a square in the shear force–bending moment plane:

$$\begin{cases} \left|\dfrac{Q}{Q_p}\right| = 1, & \left|\dfrac{M}{M_p}\right| \leq 1 \\[2ex] \left|\dfrac{M}{M_p}\right| = 1, & \left|\dfrac{Q}{Q_p}\right| \leq 1 \end{cases} \tag{5.106}$$

Although shear force and bending moment are independent of each other in the yield condition, the coupling between them still enters the formulation through the equations of motion.

When rotary inertia of the block at the tip is taken into account, the bending moment at the junction between the block and the tip of the uniform beam is no longer equal to zero. Changes in the rate of rotation at the tip require a bending moment at the junction; this moment depends on the angular acceleration and the polar moment of inertia for the colliding mass at the tip of the cantilever. However, the magnitude of the bending moment acting at the junction depends on the size and rotary inertia of the block. If both shear force and bending moment at the junction are significant, the consequent deformation associated with these generalized forces is simultaneous shear and flexure of the plastic hinge at the junction.

Equations of Motion Consider the deformation mechanism shown in Fig. 5.18, where both shear and flexural deformation are assumed to be present simultaneously at the junction between the block and the tip of the cantilever. For an independent yield condition, the transverse shear force and the bending moment at that cross-section are equal to Q_p and M_p, respectively. In the cantilever, the suddenly applied steady load of Q_p and M_p at the junction results in a rotating segment of

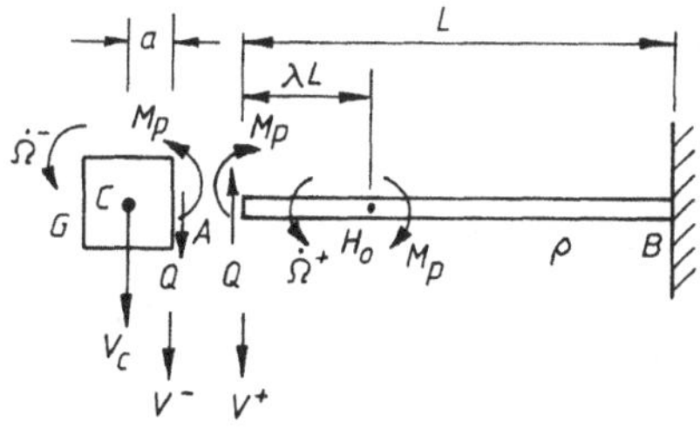

Fig. 5.18 Dynamic deformation mechanism with both shear and flexural yielding at a section adjacent to impulsively loaded block.

constant length $\Lambda = \lambda L$ adjacent to the tip. During an initial period wherein the shear force at the tip $Q(0) = Q_p$, this segment rotates about a second stationary plastic hinge H_0 that is between the ends. By using previously defined non-dimensional parameters, the equations of motion for segment AH_0 are found to be

$$\frac{1}{2}\lambda\dot{v}^+ = q_p \tag{5.107}$$

$$\frac{1}{3}\lambda^2\dot{v}^+ = q_p\lambda - 2 \tag{5.108}$$

where $\dot{v}^+$ denotes the transverse acceleration at the tip of the cantilever while the shear resultant $q_p = Q_p L / M_p$ as used in Sect. 5.3.2. These equations give the segment length λ and tip velocity $\dot{v}^+$,

$$\lambda = \frac{6}{q_p} = \text{const} \tag{5.109}$$

$$\dot{v}^+ = \frac{1}{3}q_p^2 \tag{5.110}$$

The segment of length λ adjacent to the junction at the tip has an angular acceleration

$$\dot{\omega}^+ = \frac{\dot{v}^+}{\lambda} = \frac{1}{18}q_p^3 \tag{5.111}$$

The associated dimensional angular velocity of this segment is $\Omega^+ \equiv \omega^+ / T_0$ with a positive sense indicated in Fig. 5.18.

With the independent yield condition the equations of motion for the block at the tip are separated from those for the cantilever. The transverse and rotational equations for this block with a distance a between the center-of-mass and the junction are

$$-\gamma\dot{v}_c = q_p \tag{5.112}$$

$$\gamma\left(\frac{r_g}{L}\right)^2\dot{\omega}^- = 1 + q_p\frac{a}{L} \tag{5.113}$$

where $\dot{v}_c$ is the transverse acceleration of the center-of-mass of the rigid block, $\dot{\omega}^-$ is the angular acceleration of the block with a positive sense indicated in Fig. 5.18 and r_g is the radius of gyration of the rigid block. From Eq. (5.113), the angular acceleration of the rigid block can be determined,

$$\dot{\omega}^- = \frac{1 + q_p\frac{a}{L}}{\gamma(r_g / L)^2} \tag{5.114}$$

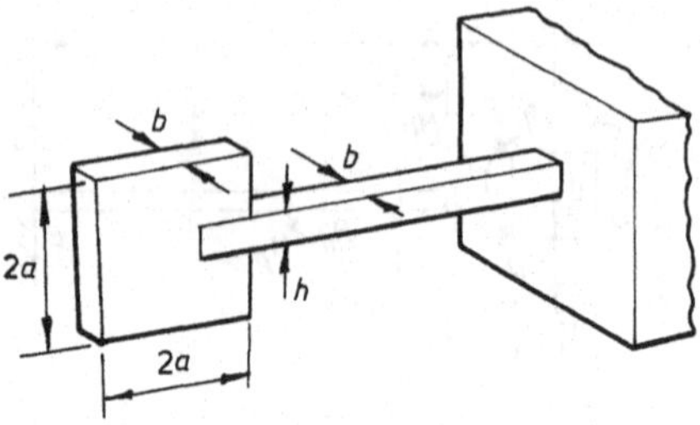

Fig. 5.19 Square block attached at tip of cantilever; the cantilever cross-section is rectangular while the width and mass density of block are identical to those of cantilever.

Consequently, the transverse acceleration of the block at the junction is

$$\dot{v}^- = \dot{v}_c - \dot{\omega}^-\left(\frac{a}{L}\right) = -\frac{q_p}{\gamma} - \frac{\frac{a}{L}\left(1+q_p\frac{a}{L}\right)}{\gamma(r_g/L)^2} \tag{5.115}$$

Note that at the instant of impact both ω^+ and ω^- are zero, and the angular accelerations $\dot{\omega}^+$ and $\dot{\omega}^-$ are both constant. While $M_A = M_p$ there is a plastic hinge at junction A and the flow rule requires relative rotation across this hinge. Since hinges at A and H_0 must have opposite signs, flexural yield at the junction requires $\dot{\omega}^+ > \dot{\omega}^-$ or

$$\frac{\gamma}{18}\left(\frac{r_g}{L}\right)^2 q_p^3 > 1+q_p\frac{a}{L} \tag{5.116}$$

Impulse at Center-of-Mass of Block at Tip of Cantilever The following example is developed specifically for a cantilever with a rectangular cross-section and a square block at the tip. The block has side length $2a$, while the depth and material density for cantilever and block are identical, as shown in Fig. 5.19. For this system, it follows that

$$\gamma\left(\frac{r_g}{L}\right)^2 = \frac{G}{\rho L}\left(\frac{r_g}{L}\right)^2 = \frac{4\rho_v ba^2}{\rho_v bhL^3}\frac{2a^2}{3} = \frac{8}{3}\frac{a^4}{hL^3} \tag{5.117}$$

where ρ_v is the material density per volume. Substituting (5.117) and the shear force $q_p = 2L/h$ from (5.84) into the flexural yield condition (5.116) results in

$$\frac{2}{27}\left(\frac{2a}{h}\right)^4 - \frac{2a}{h} - 1 > 0 \tag{5.118}$$

Thus, the deformation mechanism with a combined plastic hinge and shear deformation at the junction (see Fig. 5.18) requires that

$$\frac{2a}{h} > 2.65 \tag{5.119}$$

The transverse velocities on either side of the junction can be obtained by integrating the accelerations given in Eqs (5.110) and (5.115):

$$v^+ = \frac{1}{3}q_p^2\tau \tag{5.120}$$

$$v^- = v_0 - \left\{ \frac{q_p}{\gamma} + \frac{\frac{a}{L}\left(1 + q_p \frac{a}{L}\right)}{\gamma\left(\frac{r_g}{L}\right)^2} \right\} \tau \qquad (5.121)$$

Shear Rupture at Junction with Block Transverse relative displacement at the junction terminates at the instant τ_0 given by $v^+(\tau_0) = v^-(\tau_0)$; therefore,

$$\tau_0 = v_0 \left\{ \frac{q_p}{\gamma} + \frac{\frac{a}{L}\left(1 + q_p \frac{a}{L}\right)}{\gamma\left(\frac{r_g}{L}\right)^2} + \frac{1}{3} q_p^2 \right\}^{-1} \qquad (5.122)$$

In particular, for the square block of side length $2a$ mentioned above,

$$\tau_0 = v_0 \left(\frac{a}{L}\right)^2 \left\{ \frac{5}{4} + \frac{3}{4}\frac{h}{2a} + \frac{1}{3}\left(\frac{2a}{h}\right)^2 \right\}^{-1} \qquad (5.123)$$

During the *shear phase* $0 \le \tau \le \tau_0$, the total sliding distance between two sides of the interface is

$$|\delta_q| = \frac{1}{2} v_0 \tau_0 = \frac{v_0^2}{2}\left(\frac{a}{L}\right)^2 \left\{ \frac{5}{4} + \frac{3}{4}\frac{h}{2a} + \frac{1}{3}\left(\frac{2a}{h}\right)^2 \right\}^{-1} \qquad (5.124)$$

Shear rupture will occur at a critical relative displacement; a detailed discussion of this point is presented in Sect. 5.3.5. For now suppose that shear rupture occurs when $|\delta_q| = h/L$. If the block does not shear off during the shear phase $0 \le \tau \le \tau_0$, the shear force vanishes at time τ_0. Thereafter solely flexural deformation of the cantilever continues. After the initial period $0 \le \tau \le \tau_0$ the magnitude of the shear force at the junction decreases $|Q_A| < Q_p$, while the remaining momentum makes the plastic hinge travel from H_0 towards the root. The analysis for the period when $\tau > \tau_0$ is omitted here since it is similar to that given by Shu et al. [1992] and the previous flexural theory in Sect. 4.5.

Junction with Shear Deformation but no Flexure If the polar moment-of-inertia of the block at the tip is small there can be shear without flexure. A small polar moment-of-inertia does not satisfy conditions (5.116) and (5.119). In this case a different model of junction deformation is required instead of the mechanism shown in Fig. 5.18. At the junction the bending moment M_A is smaller than the yield moment M_p, so there is shear deformation but no flexure; i.e. in the shear force–bending moment plane of a separated or independent yield condition the resultant 'force' is on the side $Q_A = -Q_p$ and $M_A < M_p$, which is marked by (A) in Fig. 5.17. The 'forces' at the junction that correspond to this state of deformation must satisfy

$$q_A \equiv \frac{Q_A}{Q_p} = -1, \qquad |m_A| \equiv \left|\frac{M_A}{M_p}\right| < 1 \qquad (5.125)$$

Since Q_A is constant during an initial phase of motion, all accelerations and hence the moment M_A are also constant. On either side of the junction the angular accelerations are found to be

$$\dot{\omega}^+ = \frac{2q_p^3}{9(1+m_A)^2}, \qquad \dot{\omega}^- = \frac{m_A + aq_p/L}{\gamma(r_g/L)^2} \tag{5.126}$$

where r_g is the radius of gyration for the center-of-mass of the block. In this case there is no flexural plastic hinge at the junction; therefore, there is no relative angular acceleration across the junction; i.e.

$$\dot{\omega}^+ = \dot{\omega}^- \tag{5.127}$$

and consequently,

$$(1+m_A)^2\left(m_A + q_p\frac{a}{L}\right) = \frac{2}{9}q_p^3\gamma\left(\frac{r_g}{L}\right)^2 \tag{5.128}$$

If the block is square with side length $2a$ and it has the same width and mass density as the cantilever, then condition (5.128) boils down to a cubic equation for m_A if the beam cross-section is rectangular.

$$(1+m_A)^2\left(m_A + \frac{2a}{h}\right) = \frac{8}{27}\left(\frac{2a}{h}\right)^4, \qquad 1 \le \frac{2a}{h} < 2.65 \tag{5.129}$$

At the junction which is shearing, the bending moment m_A is a function of $(2a/h)$. This expression is based upon $q_p = 2L/h$ or $Q_p = Ybh/2$, which is representative if $2a > h$.

If the side length of the block is less than the depth of the cantilever, $2a < h$, then the cross-sectional area at the junction is reduced to $2ab$ instead of bh, so it is reasonable to take $Q_p = Yab$ and correspondingly $q_p = Q_pL/M_p = 4aL/h^2$. This leads to a modified equation for determining m_A as a function of $(2a/h)$,

$$(1+m_A)^2\left\{m_A + \left(\frac{2a}{h}\right)^2\right\} = \frac{8}{27}\left(\frac{2a}{h}\right)^7, \qquad \frac{2a}{h} < 1 \tag{5.130}$$

By combining the solutions of Eqs (5.129) and (5.130), we obtain the variation of bending moment m_A for the entire range of relative size for the block, $(2a/h)$. This is shown in Fig. 5.20, where the solid line is the solution of Eq. (5.129) and the dashed line is the solution of (5.130).

For a very small block, Fig. 5.20 shows that fully plastic shearing results in

$$|m_A| < 0.45, \qquad \text{if } \frac{a}{h} < 1 \tag{5.131}$$

When the side length of the colliding block is less than twice the depth of the cantilever, the moment applied at the junction is relatively small so the mass can be regarded as a particle. Consequently, the boundary condition at the junction is $M_A = M(0) \approx 0$.

5.3.4 Shear Rupture due to Impact

As mentioned in Sect. 5.3.2, if the total sliding distance during the *shear phase* reaches or exceeds the depth of the beam, then *shear failure* or *rupture* occurs at the junction where shear deformation develops. Initially we assume rupture occurs if

$$|\delta_q| \ge h/L \tag{5.132}$$

If rotary inertia of the colliding mass is negligible, at the junction the deformation is only shear (see Sect. 5.3.2). Assuming that the shear force at this section remains constant while the relative displacement increases, an estimate can be

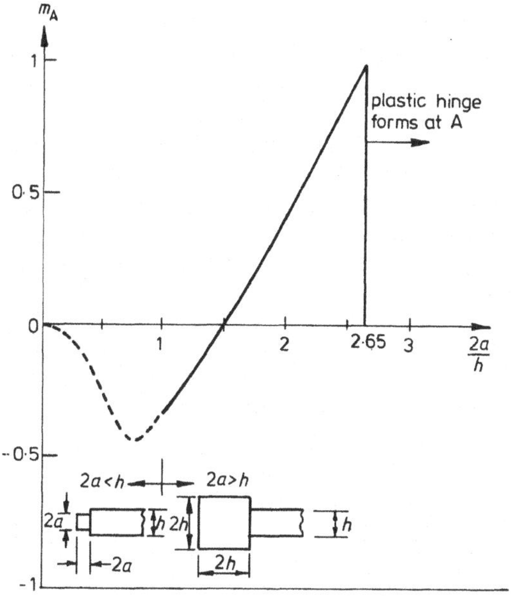

Fig. 5.20 The bending moment m_A at the junction (section A) varies with the size of the square block.

obtained for the impact energy that results in rupture. By substituting Eqs (5.84) and (5.98) into the rupture criterion (5.132) we obtain the energy criterion for a rectangular cross-section,

$$\frac{1}{2}\gamma v_o^2 \geq 2 + \frac{4}{3}\gamma q_p \tag{5.133}$$

or

$$e_o \equiv \frac{K_o}{M_p} = \frac{GV_o^2/2}{M_p} \geq 2 + \frac{8}{3}\frac{G}{\rho L}\frac{L}{h} \tag{5.134}$$

where $K_o = GV_o^2/2$ is the initial kinetic energy or impact energy of the colliding mass. The rupture and nonrupture regions illustrated in Fig. 5.21 are separated by a straight line, Eq. (5.134). Rupture depends on the ratio of the initial kinetic

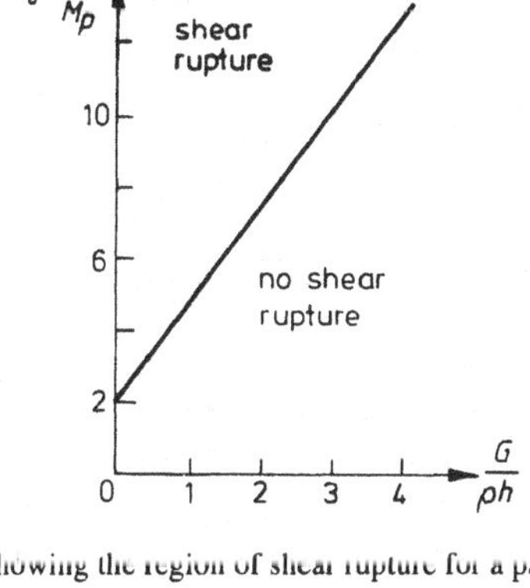

Fig. 5.21 Map showing the region of shear rupture for a particle at the tip.

energy to the fully plastic bending moment of the cantilever, in comparison with the product of mass and slenderness ratios $G/\rho h$. Note that the rupture criterion (5.134) and the map showing conditions which favor shear rupture (Fig. 5.21) both vary with the shape of the cross-section.

To compare the rupture criterion for pure shear with that due to combined shear and flexure, consider the special case adopted previously. That is, assume a square block is attached to the tip; the block has side length $2a$ and the same mass density and width as the cantilever. By using the results for shear displacement obtained in Sects 5.3.2 and 5.3.3, the rupture criterion based on these two models can be cast in a unified form

$$e_0 \equiv \frac{K_0}{M_p} \geq e_r \tag{5.135}$$

where the rupture energy e_r is a function of $(2a/h)$ as given below.

1. *Pure shear rupture* (Fig. 5.13a, Sect. 5.3.2):

$$e_r = 2 + \frac{8}{3}\left(\frac{2a}{h}\right)^2, \qquad \text{if } \frac{2a}{h} \geq 1 \tag{5.136a}$$

$$e_r = \frac{2a}{h}\left\{2 + \frac{8}{3}\left(\frac{2a}{h}\right)^3\right\}, \qquad \text{if } \frac{2a}{h} < 1 \tag{5.136b}$$

2. *Shear–flexure rupture* (Fig. 5.18, Sect. 5.3.3):

$$e_r = 5 + 3\left(\frac{h}{2a}\right) + \frac{4}{3}\left(\frac{2a}{h}\right)^2, \qquad \text{if } \frac{2a}{h} \geq 2.65 \tag{5.137a}$$

$$e_r = 5 + 3m_A\left(\frac{h}{2a}\right) + \frac{8}{3(1+m_A)}\left(\frac{2a}{h}\right)^2, \qquad \text{if } \frac{2a}{h} < 2.65 \tag{5.137b}$$

where m_A is determined by Eq. (5.129) if $1 \leq 2a/h < 2.65$, or by Eq. (5.130) if $0 \leq 2a/h < 1$.

These criteria can be summarized as a map, Fig. 5.22, where the broken line describes rupture due to pure shear deformation at the junction while the solid line describes rupture due to shear–flexure deformation. For each model, the curve divides the space of $(2a/h, e_0)$ into a region of shear rupture and a region where there is no rupture at the tip. If the block is large, more kinetic energy is required to cause shear rupture. From (5.135)–(5.137) we note that the shear rupture criteria are independent of length of the cantilever; that is, shear rupture is a local phenomenon that depends on impact energy and the relative size of the block.

5.3.5 Measurements of Energy for Shear Rupture

The preceding analysis considered a discontinuity in transverse displacement that developed due to shear between an impulsively loaded block and the uniform cantilever; the discontinuity at the junction increases with time while shear force $Q(0)$ at this section remains equal to the fully plastic shear force $Q_p = Ybh/2$ Relative motion at the junction stops when the shear force becomes less than that

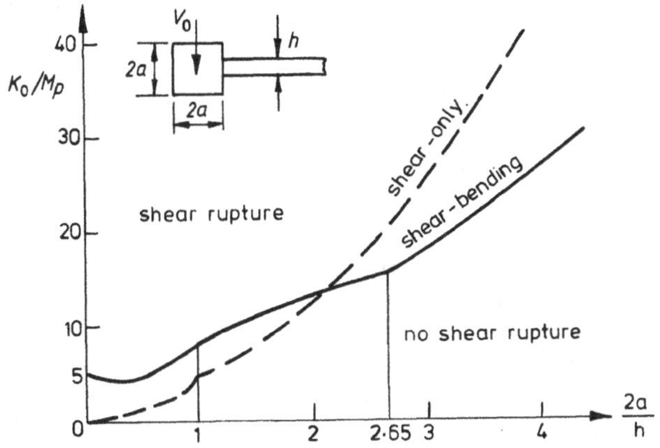

Fig. 5.22 Map showing the region of shear rupture for simple shear and shear–bending theories if the cantilever has a square block at the tip.

required for plastic deformation, $Q(0) < Q_p$. This model is used to calculate conditions that result in rupture where the impulsively loaded block is sheared from the cantilever. The rupture criterion is that the relative displacement Δ_q equals the depth h; i.e. $|\Delta_q| \equiv |\delta_q| L = h$. This criterion provides an upper bound on the impact energy that results in shear failure; the shear force–relative displacement relation for this model is shown as the solid line in Fig. 5.23.

For ductile materials with small strain hardening, a better model for rupture energy is one that considers that fully plastic shear force decreases with increasing relative displacement between the beam and the transversely moving block at the tip. If relative displacement at rupture is equal to the depth h, then a typical value for the shear resultant at the junction is

$$Q_p = \frac{1}{2} Ybh \left(1 - \frac{|\Delta_q|}{h} \right) \tag{5.138}$$

This is shown as the broken line in Fig. 5.23.

Both models assume that shearing consists of a transverse component of relative displacement at one section. In practice shear strain is distributed over a short

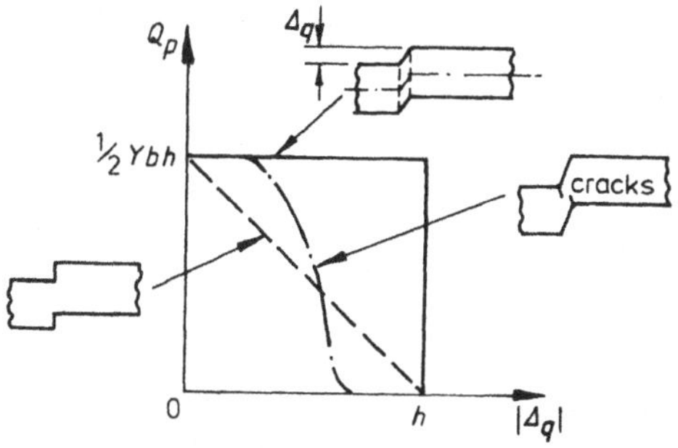

Fig. 5.23 Variation of the plastic shear force at junction during shearing.

Table 5.2 Material Properties and Measured Rupture Energy D_A for Impact Cropping Experiments (Jouri and Jones [1988], Atkins [1990])

Material	Y (MNm^{-2})	σ_u (MNm^{-2})	n	ε_{fr}	R/τ_Y (m)	$2D_A/Ybh^2$
Al. alloy	115	320	0.223	0.49	5.9×10^{-3}	1.54
Mild steel	245	485	0.224	0.90	3.0×10^{-3}	1.04

length and shear rupture is the culmination of a fracture process that initiates at a critical strain. Jouri and Jones [1988] performed impact cropping experiments on sharply notched aluminum alloy and mild steel specimens of various depths. For notched specimens the highly deformed shear zone width was 10–15% of the plate depth h while narrower zones were associated with thicker plates (5 mm $< h$ $<$ 40 mm). The properties of these materials are listed in Table 5.2; the relevant properties are yield stress Y, ultimate stress σ_u, work-hardening index n and true tensile strain at fracture ε_{fr}. The ratio of fracture toughness R to yield stress τ_Y for pure shear was obtained by Atkins [1990]. In Table 5.2, D_A denotes the energy dissipated at the junction A, which includes energy dissipation due to both shear deformation and the initiation and propagation of fracture.

For notched specimens with depth h in the range 5mm $< h <$ 40mm, Jouri and Jones found the following surface displacement at rupture δ_f :

$$\begin{aligned} \frac{\delta_f}{h} &= 0.16 + \frac{1.19}{h\,(\text{mm})} \qquad \text{for Al. alloy} \\ \frac{\delta_f}{h} &= 0.27 + \frac{0.94}{h\,(\text{mm})} \qquad \text{for mild steel} \end{aligned} \tag{5.139}$$

The impact energy D_A that produced rupture in these tests went into local shear deformation in a diffuse shear band and fracture energy. For 5 mm thick specimens the fracture energy was roughly twice the energy dissipated by plastic deformation in the shear band, but this ratio depends on the extent of shear band diffusion; i.e. on beam depth, impact speed and displacement constraints that focus shear band development (see Johnson and Travis [1968]). Because of diffuse plastic deformation the impact energy for rupture will usually be larger than that measured with a notched specimen; nevertheless, this energy is not very different from the work D_A done by the fully plastic shear force during rupture; i.e. $D_A \approx Ybh^2/2$.

In these tests the diffuse shear band was well developed and had propagated through most of the depth h before fracture initiated at the periphery of the contact region. The fracture at the junction with the colliding block began on the impact surface where the junction is both sheared and stretched due to flexure of the beam. After initiation the fracture propagated through the depth while penetration continued; the propagation was quite rapid in aluminum but more prolonged in mild steel. Accordingly, the variation of shear force Q_p during the fracture process can be sketched as the chain line in Fig. 5.23.

Despite the crude model of shear rupture, the energy requirement $2D_A/Ybh^2$ shown in Table 5.2 indicates that this approximation for rupture energy is reasonable for these tests. Of course if the beam were not notched, the relative extent of the deforming zone would increase with depth and the nonfracture part of the dissipated energy at rupture would also somewhat increase.

5.4 Effect of Large Deflection

5.4.1 General Considerations

The preceding analysis has been based on the assumption that the deflections of the structure remain infinitesimally small. This means that the equations of motion are written once for the initial configuration and that the geometric changes are small enough that the configuration does not change noticeably. For deflections that become large, however, the equations of motion must reflect the current configuration at each instant of motion. When rotations of some sections become large, it is no longer sufficient merely to consider the transverse momentum distribution — the axial momentum also comes into play. It should be recognized, however, that the rigid–perfectly plastic theory is applicable only if the curvatures are much larger than the elastic curvature. In order to apply the rigid–perfectly plastic constitutive model to practical engineering problems, it is important to take the effect of large deflections into account. Some experiments on dynamically loaded cantilevers are discussed in Chap. 7. In these experiments the final tip deflection is of the order of the length of the specimen, while the final rotation of the tip is as large as 60° or more. Notable discrepancies between the experimental measurements of final tip deflection and rigid–plastic predictions have been reported for these specimens that have undergone large deflections. The neglect of the geometrical changes owing to large deformations is partly responsible for these discrepancies together with the effects of strain-rate and strain hardening.

In Sect. 5.6 it will be seen that the rigid–perfectly plastic idealization is valid only if the ratio R of input energy to the maximum elastic deformation energy which can be stored in the structure is much larger than unity. For an impulsively loaded cantilever this requires

$$R = \frac{1}{2}GV_0^2\frac{2EI}{M_p^2 L} >> 1 \tag{5.140}$$

For cantilevers of rectangular cross-section, the moment-of-inertia $I = bh^3/12$ and the fully plastic moment $M_p = Ybh^2/4$. Here, the total elastic capacity is represented by the bending moment M_p. For both small and large deflection analysis the magnitude of the final rotation of the tip of the cantilever $|\theta_{Af}|$ equals the ratio e_0 between the impact energy and the fully plastic bending moment, i.e.

$$|\theta_{Af}| = e_0 = \frac{GV_0^2}{2M_p} \tag{5.141}$$

Consequently, the condition for the rigid–plastic idealization (5.140) requires

$$R = |\theta_{Af}|\left(\frac{2Eh}{3YL}\right) = |\theta_{Af}|\frac{4}{3k_Y} >> 1 \tag{5.142}$$

where the nondimensional curvature at yield $k_Y \equiv \kappa_Y L$ is given by

$$k_Y = \frac{M_Y L}{EI} = \frac{2M_p L}{3EI} = \frac{2YL}{Eh} \tag{5.143}$$

Typically for a ductile metal cantilever, $E/Y = 500$, so that (5.142) results in

$$|\theta_{Af}| = 0.003(L/h)R$$

where $R >> 1$. For a fairly modest slenderness ratio $h/L = 1/30$ and an impact energy ratio $R = 5$, this results in a tip rotation $|\theta_{Af}| = 0.45$ rad $\approx 26^{\circ}$, which is a large deflection. This rough estimate indicates that for a slender cantilever, validity of the rigid–perfectly plastic idealization implies that the final deflections are large.

5.4.2 Large Deflection of Impulsively Loaded Cantilever

The large deflection analysis in this section follows the method presented by Ting [1965], but for consistency, the nondimensionalization adopted is that used in the remainder of the book. In this analysis, all the assumptions in Sect. 4.5 are applicable other than assumption (5) on small deflection; in particular, the axial elongation and axial force components are neglected.

Consider a uniform rigid–perfectly plastic cantilever with a particle of mass G attached at the tip; the particle is instantaneously given velocity V_0 at time $t = 0$ (see Fig. 4.22). Thereafter a typical state of the cantilever at any time t is as shown in Fig. 5.24a; all deformation is localized in a discrete plastic hinge at a distance $\Lambda = \lambda L$ from the tip. Figure 5.24b shows the nondimensional arc length $s = S/L$ measured from the tip A towards the root B, and coordinates ξ, η that have an origin at hinge H. Since axial elongation is neglected, the hinge H is located at $s = \lambda$, and the root B occurs at $s = 1$. As in Sect. 4.5, the following nondimensional quantities can be defined

$$T_0 \equiv L\sqrt{\frac{\rho L}{M_p}}, \quad \tau \equiv \frac{t}{T_0}, \quad v_0 \equiv \frac{V_0}{L/T_0}, \quad \gamma \equiv \frac{G}{\rho L}$$

The position of the travelling hinge, λ, is a function of time τ. If $\lambda(\tau)$ is a monotonously increasing function then λ and τ are uniquely related; in this case the spatial variable λ can be taken as the independant variable rather than τ. Therefore, when the hinge is at λ the inclination angle $\theta(s,\lambda)$ at s is (see Fig. 5.24b)

$$\theta(s,\lambda) = -\int_s^{\lambda} k(z)\,\mathrm{d}z \tag{5.144}$$

where $k(s) = \kappa L$ is the nondimensional curvature of the cantilever at s. Within the framework of rigid–perfectly plastic theory, at any instant the deformation in the cantilever takes place only at a plastic hinge. Therefore, at a given section s, $k(s) = 0$ for $s > \lambda$.

Consider a point in segment AH between the tip and the hinge ($0 \le s \le \lambda$). Referring to the Cartesian coordinates ξ, η shown in Fig. 5.24b, one finds

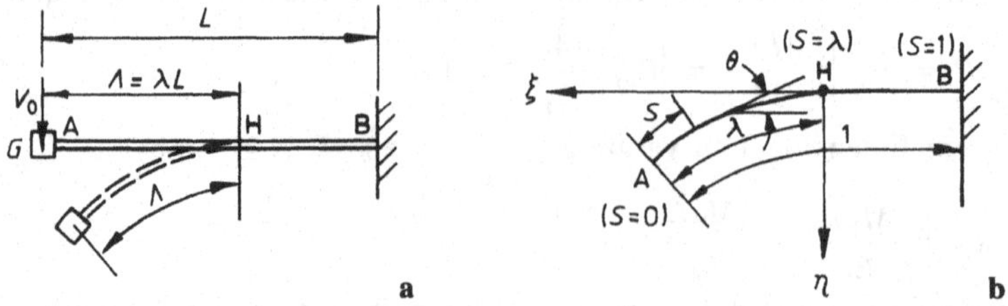

Fig. 5.24 Large deformation mechanism; (a) hinge location; (b) coordinate $\xi H \eta$.

$$\xi(s,\lambda) = \int_s^{\lambda} \cos(-\theta(z,\lambda))\,dz \tag{5.145}$$

$$\eta(s,\lambda) = \int_s^{\lambda} \sin(-\theta(z,\lambda))\,dz$$

Then, the first moment of mass for the deformed segment AH about the ξ- and η-axes, and the second moment of mass about the plastic hinge H are respectively,

$$I_{\xi}(\lambda) \equiv \int_0^{\lambda} \xi(s,\lambda)\,ds + \gamma\xi(0,\lambda) \tag{5.146a}$$

$$I_{\eta}(\lambda) \equiv \int_0^{\lambda} \eta(s,\lambda)\,ds + \gamma\eta(0,\lambda) \tag{5.146b}$$

$$I_{H}(\lambda) \equiv \int_0^{\lambda} \{\xi^2(s,\lambda) + \eta^2(s,\lambda)\}\,ds \ + \ \gamma\{\xi^2(0,\lambda) + \eta^2(0,\lambda)\} \tag{5.146c}$$

where the mass ratio $\gamma = G/\rho L$.

Using Eqs (5.146), the nondimensional equations of conservation of transverse momentum and moment of momentum about the original position of the tip for segment AH are,

$$-I_{\xi}(\lambda)\dot{\theta}_{A}(\lambda) = \gamma v_0 \tag{5.147}$$

$$-I_{H}(\lambda)\dot{\theta}_{A}(\lambda) = \tau - \gamma v_0 \lambda \tag{5.148}$$

The angular velocity $\dot{\theta}_A(\lambda)$ of the deformed segment about the plastic hinge is derived from Eq. (5.144)

$$\dot{\theta}_{A}(\lambda) \equiv \frac{d\theta(0,\lambda)}{d\tau} = \frac{d\theta(x,\lambda)}{d\tau} = -k(\lambda)\dot{\lambda}, \quad 0 \le x \le \lambda \tag{5.149}$$

In addition, the current kinetic energy plus the energy that has been dissipated at the hinge equal the impact energy; that is,

$$\frac{1}{2}\dot{\theta}_{A}^{2}(\lambda) I_{H}(\lambda) + \int_0^{\lambda} k(z)\,dz = \frac{\gamma v_0^2}{2} \tag{5.150}$$

where $\gamma v_0^2/2 = e_0 = K_0/M_p$ is the nondimensional initial or impact energy. To eliminate one variable, $\dot{\theta}_A$, the equation of translational motion (5.147) can be substituted into the expression for conservation of energy (5.150),

$$\frac{\gamma v_0^2}{2}\frac{I_{H}(\lambda)}{I_{\xi}^{2}(\lambda)} + \int_0^{\lambda} k(z)\,dz = \frac{\gamma v_0^2}{2} \tag{5.151}$$

Direct differentiation of first and second moments of mass about H, Eqs (5.146), with respect to λ gives

$$\begin{aligned} dI_{\xi}(\lambda)/d\lambda &= \lambda + \gamma - k(\lambda) I_{\eta}(\lambda) \\ dI_{\eta}(\lambda)/d\lambda &= k(\lambda) I_{\xi}(\lambda) \\ dI_{H}(\lambda)/d\lambda &= 2 I_{\xi}(\lambda) \end{aligned} \tag{5.152}$$

Using Eq. (5.152), the curvature $k(\lambda)$ at the plastic hinge can be obtained by differentiating Eq. (5.151) with respect to λ; i.e.

$$k(\lambda) = \gamma^2 v_0^2 \frac{(\lambda+\gamma) I_{H}(\lambda) - I_{\xi}^{2}(\lambda)}{I_{\xi}^{3}(\lambda) + \gamma^2 v_0^2 I_{H}(\lambda) I_{\eta}(\lambda)} \tag{5.153}$$

for $0 \le \lambda < 1$.

Similar to the small deflection rigid–plastic analysis (Sect. 4.5), the transient or travelling hinge phase is called *phase I*. During this phase, the relation between time and the hinge position can be obtained by eliminating the angular velocity from Eqs (5.147) and (5.148):

$$\tau = \gamma v_0 \left(\lambda - \frac{I_H(\lambda)}{I_\xi(\lambda)} \right) \tag{5.154}$$

When $\lambda = 1$, the travelling plastic hinge reaches the root of the cantilever, and phase I terminates. According to Eq. (5.154) this happens at an instant

$$\tau_1 = \gamma v_0 \left(1 - \frac{I_H(1)}{I_\xi(1)} \right) \tag{5.155}$$

After this instant, the cantilever rotates about a stationary hinge at the root until the remained kinetic energy is completely dissipated by the fully plastic moment that does work in resisting this rotation. This stationary hinge or modal phase is called *phase II*.

The transition from phase I to phase II occurs when $\lambda = 1$; at this instant the energy balance gives

$$\frac{\gamma^2 v_0^2}{2} \frac{I_H(1)}{I_\xi^2(1)} + \int_0^1 k(z) \mathrm{d}z = \frac{\gamma v_0^2}{2} \tag{5.156}$$

The remaining kinetic energy at time τ_1 is given by the first term on the left-hand side. This remaining kinetic energy is dissipated by deformation at the root during phase II. Hence, when motion ceases, there is a final rotation angle at the root,

$$\theta_{Bf} = -\frac{\gamma^2 v_0^2}{2} \frac{I_H(1)}{I_\xi^2(1)} \tag{5.157}$$

According to (5.147), phase II begins with the entire cantilever rotating at an angular velocity

$$\dot{\theta}_A(1) = -\frac{\gamma v_0}{I_\xi(1)} \tag{5.158}$$

Hence, the duration of phase II is found to be

$$\tau_2 - \tau_1 = \frac{2|\theta_{Bf}|}{\gamma v_0 / I_\xi(1)} = \gamma v_0 \frac{I_H(1)}{I_\xi(1)} \tag{5.159}$$

The total response time τ_f is obtained as the sum of Eqs (5.155) and (5.159)

$$\tau_f = \tau_2 = \gamma v_0 = p_f \tag{5.160}$$

where $p_f = \gamma v_0$ is the impulse applied at the initial instant. Note that the response time in this large deflection analysis is exactly the same as that obtained in the small deflection analysis, Eq. (4.118).

At time τ_f, all motion terminates and the cantilever reaches its final deformed state. A further study of the energy equation (5.151) confirms that the final inclination angle at s can be expressed as

$$\theta_f(s) = -\frac{\gamma^2 v_0^2}{2} \frac{I_H(s)}{I_\xi^2(s)} \tag{5.161}$$

Thus, the final inclination angle at any point s is equal to the nondimensional kinetic energy in the cantilever when the plastic hinge reaches the section s. In particular, *the final angle at the tip is equal to the nondimensional input energy*; i.e.

$$-\theta_{Af} = e_0 \equiv \frac{K_0}{M_p} \tag{5.162}$$

which is exactly the same relation as that obtained in the small deflection analysis. Finally, note that integration of Eq. (5.161) along the arc length s provides the final deformed shape of the cantilever for the case of large deflection.

At this point the formulation of the large deflection analysis of the cantilever subjected to impact load is completed. The solution can be obtained by a numerical scheme as discussed in detail by Ting [1965]. In brief, Eqs (5.144)–(5.146) and (5.153) provide a scheme of successive approximation for the calculation of the curvature $k(\lambda)$. Simultaneously, the values of θ, I_ξ, I_η and I_H are all found for a sequence of values for λ and τ. This process is continued until the hinge reaches the root $\lambda = 1$. Thereafter energy, inclination angle and final deflection can all be determined from the expressions above.

For example, final deformed shapes of cantilevers calculated by this numerical scheme are shown in Fig. 5.25. The solid lines are the results based on the large deflection analysis, while the broken lines are obtained with the small deflection analysis. It is seen that the difference between them increases as the nondimensional input energy increases.

5.4.3 Methods of Approximating Large Deflection Effects

It is worthwhile noticing that in the small deflection analysis each section is assumed to move only vertically, so that the deformed cantilevers (the broken lines in Fig. 5.25) appear to be longer than their original length. This causes significant discrepancy in the final shapes. On the other hand, this fact implies that the small deflection analysis in Chap. 4 can be improved by taking Eq.

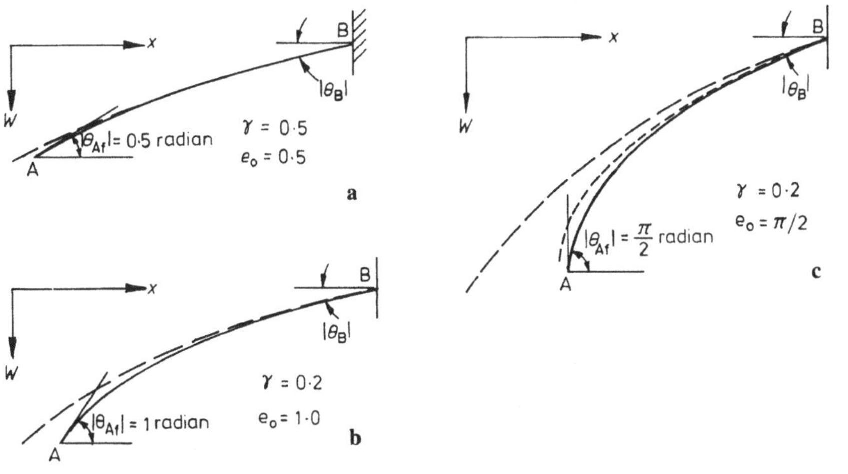

Fig. 5.25 Final deformed shapes of impulsively loaded cantilevers. ——— large deflection analysis; – – – small deflection analysis. (a) $\gamma = 0.5$, $e_0 = 0.5$; (b) $\gamma = 0.2$, $e_0 = 1.0$; (c) $\gamma = 0.2$, $e_0 = \pi/2$.

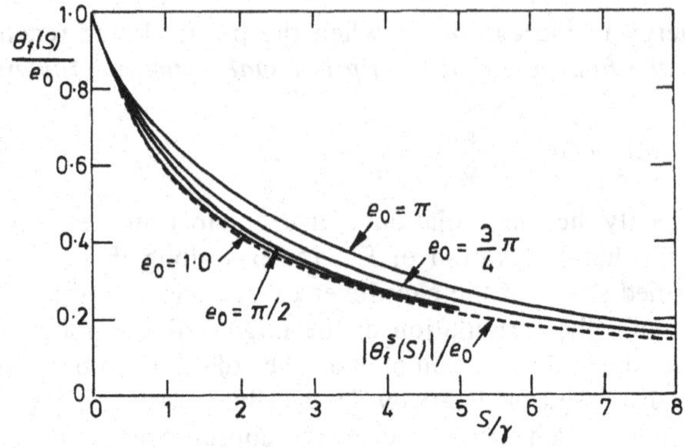

Fig. 5.26 Distribution of final inclination θ_f along cantilever. Solid lines result from large deflection analysis, with various specified values of e_0; θ_f^s results from an approximation, refer to (5.163).

(4.131) as the distribution of the final inclination angle rather than the function $w_f'(x)$ that was obtained in producing the final deformed shape of the cantilever. Taking arc length s instead of x in the small deflection expression (4.131) results in

$$\theta_f^s(s) = -\frac{v_0^2}{6}\frac{3\gamma + s}{\left(1+\dfrac{s}{2\gamma}\right)^2} \tag{5.163}$$

where superscript s designates the small deflection analysis. A numerical example shown in Fig. 5.25c for the case of $\gamma = 0.2$ and $e_0 = \pi/2$, results in the dashed line that is close to the deformed shape obtained by the large deflection analysis (the solid line).

Another way of comparing the improved small deflection solution (5.163) to the exact large deflection solution is shown in Figs 5.26 and 5.27. For small e_0,

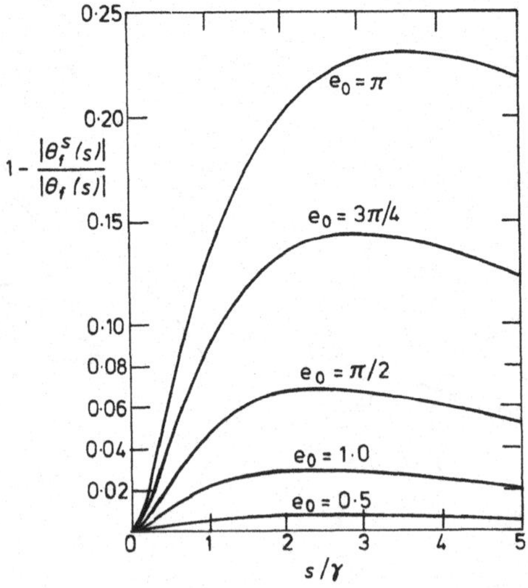

Fig. 5.27 Error of approximate θ_f^s given by (5.163) compared with large deflection result θ_f.

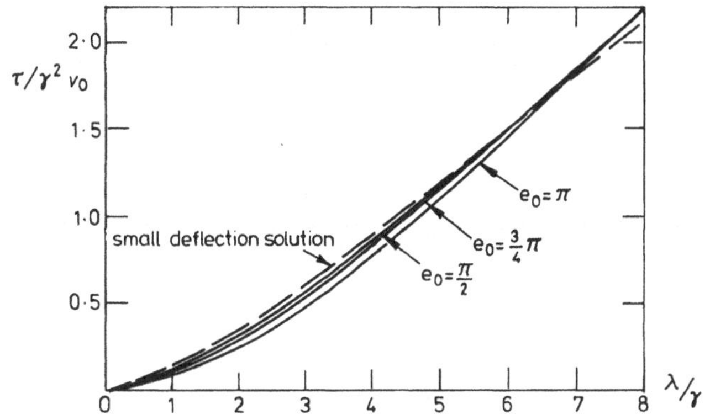

Fig. 5.28 Variation of hinge position with time. ——— large deflection solution; - - - small deflection solution.

$\theta_f^s(s)$ is a good approximation of $\theta_f(s)$. The approximation is also better for any e_0 if (s/γ) is either small or sufficiently large. From Fig. 5.27 it is seen that (1) for the case of large mass ratio $\gamma >> 1$ where $(s/\gamma) << 1$, the improved small deflection inclination $\theta_f^s(s)$ is an excellent approximation of $\theta_f(s)$; but (2) for the case of small mass ratio, $\theta_f^s(s)$ is most inaccurate in the segment $2\gamma < s < 4\gamma$.

Figure 5.28 shows the variation of the hinge position with time. Again the broken line is based on the improved small deflection solution (5.163). For a small impact energy ratio e_0, the large deflection analysis approaches the small deflection analysis in Sect. 4.5. If an initially straight member has only small deflections, then

$$\cos\theta \approx 1, \quad \sin\theta \approx \theta, \quad s \approx x \tag{5.164}$$

Hence, the coordinates of a point in segment AH become

$$\xi \approx \lambda - x, \quad \eta \approx 1 - x/\lambda \tag{5.165}$$

and the inertia parameters defined by Eqs (5.146) are recast as

$$I_\xi(\lambda) \approx \frac{1}{2}\lambda^2 + \gamma\lambda, \; I_\eta(\lambda) \approx \frac{1}{2}\lambda + \gamma, \; I_{\mathrm{H}}(\lambda) \approx \frac{1}{3}\lambda^3 + \gamma\lambda^2 \tag{5.166}$$

Employing Eqs (5.166), the equations of conservation of transverse momentum (5.147) and moment-of-momentum about the tip (5.148) lead to the same results as those obtained with the assumption of small deflections (4.105) and (4.106). The equation for curvature (5.153) asymptotically approaches (4.130) which resulted from the small deflection analysis. If the deflection is small the final deformed shape of the cantilever is given by (5.163). To sum up, the large deflection analysis presented in this section asymptotically approaches the small deflection analysis in Sect. 4.5, provided the deflections remain small.

5.4.4 Effect of Centripetal Acceleration on Bending Moment Distribution

The large deflection analysis of Sect. 5.4.2 determines the dynamic response of the cantilever under impact loading, and calculates the variation of the velocity field with time. The original analysis by Ting [1965], however, did not calculate the distribution of the bending moment in segment AH and so it did not determine whether this solution satisfies the yield condition in the entire segment AH at all times.

In the large deflection analysis there is another term — centripetal acceleration — that must be added to the acceleration field as a result of the rotation of segment AH about the plastic hinge H. The magnitude of the centripetal acceleration is determined by

$$\ddot{w}_r = \dot{\theta}_A^2 \sqrt{\xi^2 + \eta^2} \tag{5.167}$$

where ξ and η are the coordinates of a section in segment AH. This is a radial component of acceleration so it does not affect the bending moment at the plastic hinge H. Figure 5.29a illustrates the magnitude of the radial acceleration $\ddot{w}_r$ along the arc length s, while the direction of this term of acceleration is shown in Fig. 5.29b. Without further calculation, these figures show that the inertia related to the centripetal acceleration of segment AH (including the tip mass) will *diminish* the bending moment along segment AH (i.e. in $0 < s < \lambda$), but it does not change the bending moment at the tip ($M = 0$) or plastic hinge H where $M = M_p$. Thus, the distribution of the bending moment in the cantilever can be sketched as shown in Fig. 5.29c, where the broken line pertains to the small deflection analysis and the solid line pertains to the large deflection result. It is clear from this figure that the appearance of centripetal acceleration during large dynamic deflections of the cantilever does not cause a violation of yield condition $|M| \le M_p$ in any section of the cantilever.

A similar problem is related to impact and bending of a rigid–plastic fan blade; the blade may be regarded as a cantilever that is subjected to continuous acceleration in the radial direction due to the fan rotation. The blade may also be subjected to transverse impact at the tip. This problem was analyzed by Stronge and Shioya [1984] and Shioya and Stronge [1984], who took both transverse and centrifugal forces into account in determining the moment distribution and the position of the travelling hinge.

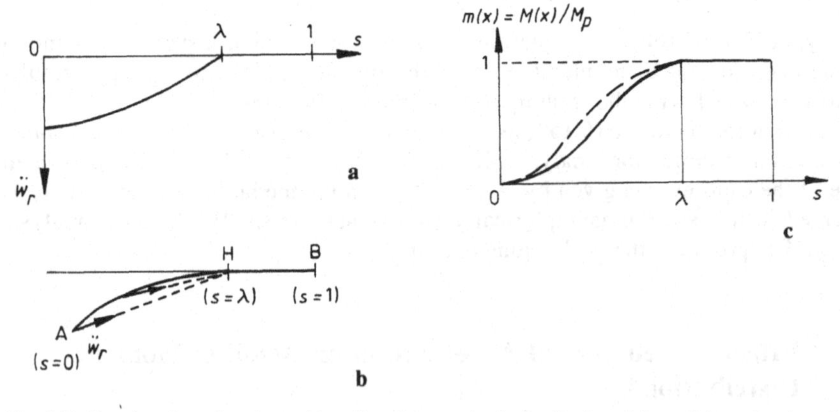

Fig. 5.29 Centripetal acceleration and its effect: (a) magnitude distribution of the radial acceleration $\ddot{w}_r$; (b) direction of $\ddot{w}_r$; (c) distribution of bending moment. – – – small deflection analysis and ——— large deflection analysis.

The conclusion that centrifugal force does not alter the basic pattern of dynamic deformation of a rigid–plastic beam is valid only for a yield condition that is independent of the axial force. In general the inertia related to centripetal acceleration of segment AH (including the tip mass) produces an axial force in segment AH; this force increases from the tip to the hinge section H. Therefore, the argument that centripetal accelerations during large dynamic deflections of a cantilever are unlikely to cause a violation of yield condition is correct only if the effect of axial force on yielding is negligible; i.e. if the axial force is small in comparison with the fully plastic yield force for the section at hinge H.

The cantilever analyzed here has large flexural deformation but it has been assumed that the deformation is inextensional. In some other structural members such as axially constrained beams, circular plates and shells the axial or membrane forces unavoidably induce stretching as the deflection becomes large. In these cases, the effects of the axial or membrane forces on the yield condition and the equations of motion must be considered. This coupling of forces by the yield condition has been discussed by many authors, e.g. Symonds and Mentel [1958], Jones [1971] and Yu and Stronge [1990].

5.5 Effect of Elastic Deformation

5.5.1 General Considerations

Most ductile metals exhibit elastic–plastic behavior rather than the rigid–plastic idealization that has been considered thus far. Although the largest elastic strains are limited (typically $\varepsilon_Y \approx 0.002$ for structural metals) the *elastic* deformation is *always* a precursor to subsequent plastic deformation. If a structure is subjected to intense dynamic loading, the first phase of response is always elastic. Initially stresses increase as deformation spreads away from the loaded region and soon some sections have stresses as large as the yield stress; plastic deformation then begins at these sections. The regions with large stresses move during dynamic deformation — sometimes these regions disappear and reappear. After plastic deformation has dissipated most of the input energy in an elastoplastic structure, the structure is left in a final state of elastic vibration. Ultimately this is also damped out and the structure comes to rest in a state of permanent deformation with an associated field of residual stress. The rigid–plastic idealization considered earlier in this book glossed over these details. It is possible that there is a synergistic effect that was overlooked.

Due to the complex intermingling of elastic and plastic behavior in the structural response, analytical methods have not made any headway in dealing with effects of inertia combined with distributed plastic deformation. Thus we turn to numerical schemes such as finite element and finite difference to gain some insight into the effect of elasticity on subsequent plastic deformations. The aim is to identify conditions where the rigid–plastic model accurately represents the response of elastoplastic structures.

The rigid–plastic idealization for structural deformation in response to impulsive loading is based on the assumption that at every section the plastic deformation is much larger than the elastic deformation. If the total input energy imparted to a structure by dynamic loading is much larger than the maximum elastic strain energy that the structure is able to store, then most input energy is dissipated by plastic deformation of the structure. The maximum elastic energy capacity $U_e^{\max}$ can be compared with the total input energy imparted by dynamic loading E_{in}; thus the specific input energy is defined as an *input energy ratio, R,* where

$$R \equiv \frac{E_{in}}{U_e^{\max}} \tag{5.168}$$

If $R >> 1$ the rigid–plastic structural idealization usually provides a good approximation for deformation of a structure subjected to a brief but intense loading pulse. By brief we mean short in comparison with the rigid–plastic structural response period. Quantitative analysis of the accuracy of the rigid–perfectly plastic approximation will be discussed in Sect. 5.6.

In the present section the problem of an elastic–plastic cantilever struck at the tip by a particle with mass G is examined to assess how elastic effects modify dynamic response to impulsive loading. The reasons for choosing this example are:

1. the solution can be compared with that of a rigid–perfectly plastic cantilever (see Sect. 4.5 or Parkes [1955]).
2. a few experimental results are available for comparison (see Chap. 7).
3. no axial force is induced by large deflections.

For a cantilever of length L, the input energy ratio R becomes

$$R = \frac{K_0}{U_e^{\max}} = \frac{GV_0^2 EI}{M_p^2 L} \tag{5.169}$$

where V_0 is the initial velocity at the tip, EI is the elastic stiffness and M_p is the fully plastic moment.

5.5.2 Mass–Spring Finite Difference Structural Model (MS–FD)

For elastoplastic structures, equations of motion (and of static equilibrium) are nonlinear because of both the material properties and the effects of large deflection. Methods of calculating dynamic response of nonlinear continua (e.g. beams, plates, shells) can be based on either a finite difference or a finite element discretization of the structure. Here a mass–spring finite difference model (MS–FD) is developed for transient response of a cantilever to impact; this is used to investigate effects of elastic deformations on the subsequent plastic deformations. The results are compared with an alternative finite element model of the same problem. This comparison elucidates effects due to short wave length flexural elastic waves that are neglected by the finite difference model formulation.

MS–FD Model and Its Formulation At an arbitrary section of a beam, the effect of elastic deformation preceding any plastic deformation can be considered by discretizing the beam into a number of rigid links connected by elastic–plastic hinges. To calculate elastic–plastic deformation of impulsively loaded cantilevers, Hou et al. [1995] adopted a finite difference discretization based on equations of motion for *small deflection*. The constitutive equation for this model is an elastic strain hardening relation between bending moment and relative rotation angle.

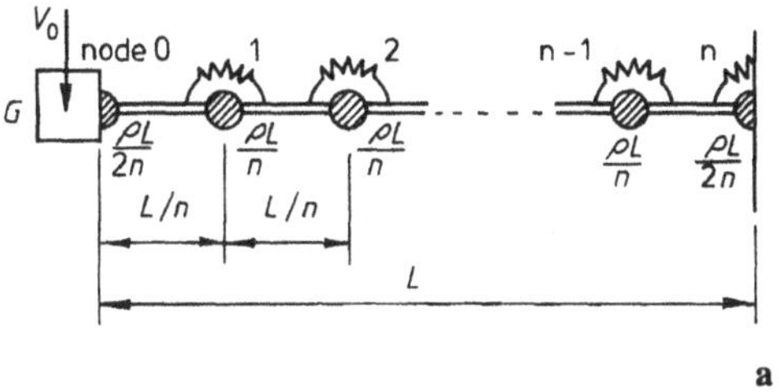

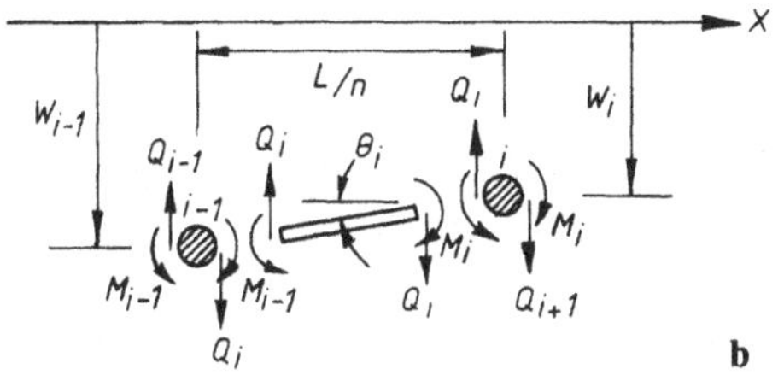

Fig. 5.30 Mass-spring finite difference (MS–FD) structural model: (a) discretization of a cantilever; (b) positive sense of deflection, shear force and bending moment.

Strictly speaking, this idealization is correct for an *ideal sandwich Euler–Bernoulli beam*, but generally it is a useful approximation for compact cross-sections. This phenomenological modeling is convenient for a semi-analytical approach because it naturally brings forth nondimensional variables that represent the system. The deformation approximations, however, limit the present formulation to small deflections in response to transverse loads. A more elaborate multilayered beam discretization by Hashmi et al. [1972] is more versatile but less easy to interpret for comparisons with the analytical results in the preceding chapters.

The MS–FD model is sketched in Fig. 5.30a. The present discretization separates the cantilever into n rigid elements of equal length connected by flexible joints or node points. The mass of the cantilever ρL is concentrated at the nodes; each node has mass $\rho L/n$. The nodes are numbered consecutively, beginning from the tip. The colliding mass G is attached to the zeroth node at the tip. Hence, the model consists of $(n+1)$ lumped masses connected by n massless rigid links of length L/n. Since both shear and axial deformations are neglected, only flexure is admissible; this is represented by relative rotation between adjacent rigid links. At each node the transverse nodal displacement $W_i(t)$ $(i=0,1,\cdots,n)$ is a function of time; the tip has deflection $W_0(t)$. If the links are also labelled so the ith link joins nodes $i-1$ and i, then under the assumption of small deflection, rotation θ_i of each link is given by

$$\theta_i = n(W_{i-1} - W_i)/L, \qquad i=1,2,\cdots,n \tag{5.170}$$

Thus at the ith joint the *relative* rotation ψ_i between the adjacent links i and $(i+1)$ is

$$\psi_i \equiv \theta_i - \theta_{i+1} = n(W_{i+1} - 2W_i + W_{i-1})/L, \qquad i=1,2,\cdots,n-1 \tag{5.171a}$$

$$\psi_n = \theta_n \tag{5.171b}$$

This relative rotation between each pair of links is resisted by elastic–plastic stiffness at each joint, which reflects the flexural rigidity of the cantilever. This

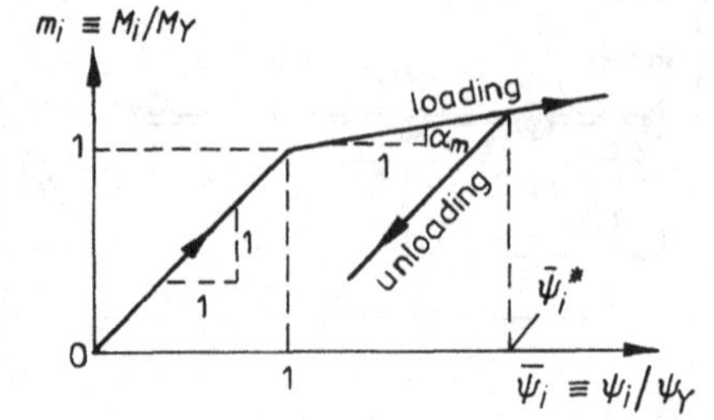

Fig. 5.31 Elastic–linear hardening relation between moment and relative rotation angle.

stiffness is portrayed in Fig. 5.30 by an elastic–plastic rotational spring at each joint. The joint is elastic for small rotations but the elastic range is limited by a 'yield' rotation angle ψ_Y. For $\psi_i > \psi_Y$ a joint exhibits linear 'strain' hardening. Thus, as shown in Fig. 5.31, for increasing relative rotation $\dot{\psi}_i > 0$, the moment M_i at a node is given by

$$M_i(\psi_i) = \begin{cases} C\psi_i, & \psi_i \le \psi_Y \\ M_Y + \alpha C(\psi_i - \psi_Y), & \psi_i > \psi_Y,\ \dot{\psi}_i > 0 \end{cases} \tag{5.172a}$$

If the largest rotation at a joint ψ_i^* exceeds the yield rotation, $\psi_i^* > \psi_Y$, and subsequently the bending moment is decreased, the joint unloads elastically. The moment–rotation relation for unloading is

$$M_i(\psi_i) = M_i(\psi_i^*) - C(\psi_i^* - \psi_i), \qquad \psi_i^* > \psi_i > \psi_i^* - 2\psi_Y \tag{5.172b}$$

where the range of applicability is limited to $\psi_i > \psi_i^* - 2\psi_Y$ by the Bauschinger effect for kinematic strain hardening.

This discretization represents the inertia by a particle with mass $\rho L/n$ at each node as shown in Fig. 5.30b; this specifically neglects the effect of rotary inertia and thereby decouples displacement variables in the equations of motion. The nodes are connected by rigid links. Equations of motion for this model give nodal shear force Q_i and bending moment M_i

$$Q_1 = \left(G + \frac{\rho L}{2n}\right)\frac{d^2 W_0}{dt^2} \tag{5.173a}$$

$$Q_i - Q_{i-1} = \frac{\rho L}{n}\frac{d^2 W_{i-1}}{dt^2}, \qquad i = 2,3,\cdots,n \tag{5.173b}$$

$$M_i - M_{i-1} = -\frac{L}{n}Q_i, \qquad i = 1,2,\cdots,n \tag{5.174}$$

For a cantilever the boundary conditions are

$$Q_0 = M_0 = 0, \qquad W_n = 0 \tag{5.175}$$

Impact by a colliding body that strikes the tip transversely at a velocity V_o results in initial conditions

$$W_0 = 0, \qquad \frac{dW_0}{dt} = \frac{V_o}{\sqrt{1 + (\rho L/2nG)}}$$
$$W_i = \frac{dW_i}{dt} = 0, \qquad i = 1,2,\cdots,n \tag{5.176}$$

where at impact the speed of the particle at the tip and the speed of the colliding mass G instantaneously change to a common speed $V_o/\sqrt{1+(\rho L/2nG)}$ obtained from conservation of transverse momentum.

Using matrix notation, the above equations may be recast in condensed form as

$$\{\theta\} = \frac{n}{L}\mathbf{A}^{\mathrm{T}}\{W\} \tag{5.170'}$$

$$\{\psi\} = \mathbf{A}^{\mathrm{T}}\{\theta\} \tag{5.171'}$$

$$\mathbf{A}\{Q\} = \frac{\rho L}{n}\mathbf{B}\frac{\mathrm{d}^2}{\mathrm{d}t^2}\{W\} \tag{5.173'}$$

$$\mathbf{A}\{M\} = -\frac{L}{n}\{Q\} \tag{5.174'}$$

where

$$\{Q\} \equiv (Q_1, Q_2, \cdots, Q_n)^{\mathrm{T}}, \qquad \{M\} \equiv (M_1, M_2, \cdots, M_n)^{\mathrm{T}}$$
$$\{W\} \equiv (W_0, W_1, \cdots, W_{n-1})^{\mathrm{T}} \tag{5.177a}$$
$$\{\theta\} \equiv (\theta_1, \theta_2, \cdots, \theta_n)^{\mathrm{T}}, \qquad \{\psi\} \equiv (\psi_1, \psi_2, \cdots, \psi_n)^{\mathrm{T}}$$

$$\mathbf{A} \equiv \begin{bmatrix} 1 & 0 & & & \\ -1 & 1 & & & \\ & & \ddots & & \\ & & & 1 & 0 \\ & & & -1 & 1 \end{bmatrix}, \quad \mathbf{B} \equiv \begin{bmatrix} n\gamma+\frac{1}{2} & 0 & & & \\ 0 & 1 & & & \\ & & \ddots & & \\ & & & 1 & 0 \\ & & & 0 & 1 \end{bmatrix} \tag{5.177b}$$

Energy Ratios and Nondimensional Formulation To nondimensionalize the formulation, define

$$T_Y \equiv L\sqrt{\frac{\rho L}{M_Y}}, \qquad \tau \equiv \frac{t}{T_Y}, \qquad (\dot{\ }) \equiv \frac{\mathrm{d}}{\mathrm{d}\tau}(\ )$$
$$\{w\} \equiv \frac{1}{L}\{W\}, \qquad \{q\} \equiv \frac{L}{M_Y}\{Q\}, \qquad \{m\} \equiv \frac{1}{M_Y}\{M\}$$
$$\psi_Y \equiv \frac{M_Y L}{nEI}, \qquad \bar{\psi} \equiv \frac{\psi}{\psi_Y}, \qquad \bar{\theta} \equiv \frac{\theta}{\psi_Y} \tag{5.178}$$
$$\gamma \equiv \frac{G}{\rho L}, \qquad e_o \equiv \frac{K_o}{M_p} = \frac{GV_o^2}{2M_p}, \qquad R \equiv \frac{K_o}{U_e^{\max}}$$

where R denotes the ratio of the initial kinetic energy to the maximum elastic deformation energy that can be stored in the cantilever, while e_o is the ratio of initial kinetic energy to fully plastic bending moment M_p at any section. These ratios relate input energy to either the global elastic strain energy capacity $U_e^{\max}$ or a local elastic capacity of a section $M_Y = M_p/\phi_m$. The latter energy ratio e_o is useful in making comparisons with rigid–perfectly plastic solutions. The cross-section shape factor for pure bending $\phi_m = M_p/M_Y$ and yield curvature κ_Y can be used to relate these energy ratios if strain hardening is negligible and the cross-section is uniform, $e_o/R = \phi_m L\kappa_Y/2 = \phi_m L\varepsilon_Y/h$. In particular, when an ideal sandwich cross-section is considered, $\phi_m = 1$ and $M_Y = M_p$.

By using these nondimensional variables, Eqs (5.170'), (5.171'), (5.173') and (5.174') are recast in nondimensional form:

$$\{\bar{\theta}\} = \frac{n}{\psi_Y}\mathbf{A}^{\mathrm{T}}\{w\} \tag{5.170"}$$

$$\{\bar{\psi}\} = \mathbf{A}^{\mathrm{T}}\{\bar{\theta}\} \tag{5.171"}$$

$$\mathbf{A}\{q\} = \frac{1}{n}\mathbf{B}\{\ddot{w}\} \tag{5.173"}$$

$$\mathbf{A}\{m\} = -\frac{1}{n}\{q\} \tag{5.174"}$$

Combining these equations gives

$$\{\ddot{\bar{\psi}}\} = -\frac{n^3}{\psi_Y}\mathbf{F}\{m\} \tag{5.179}$$

where

$$\mathbf{F} \equiv (\mathbf{A}^{\mathrm{T}})^2\mathbf{B}^{-1}(\mathbf{A})^2 = \begin{bmatrix} (n\gamma+\frac{1}{2})^{-1}+5 & -4 & 1 & & & & \\ -4 & 6 & -4 & & & & \\ 1 & -4 & 6 & & & & \\ & & & \ddots & & & \\ & & & & 6 & -4 & 1 \\ & & & & -4 & 5 & -2 \\ & & & & 1 & -2 & 1 \end{bmatrix} \tag{5.180}$$

The constitutive relation between moment and relative rotation angle, (5.172) can also be recast in nondimensional form (see Fig. 5.31)

$$m_i = \begin{cases} \bar{\psi}_i, & \bar{\psi}_i \le 1 \\ 1+\alpha(\bar{\psi}_i - 1) & \bar{\psi}_i > 1,\ \dot{\bar{\psi}}_i \ge 0 \end{cases} \tag{5.181a}$$

$$m_i = \bar{\psi}_i - (1-\alpha)(\bar{\psi}_i^* - 1), \qquad (\bar{\psi}_i^* - 2) \le \bar{\psi}_i \le \bar{\psi}_i^* \quad \dot{\bar{\psi}}_i < 0 \tag{5.181b}$$

with $i = 1, 2, \cdots, n$.

The present model consists of n nonlinear springs at the nodes, so when all springs have rotated through the yield angle the model has a maximum elastic deformation energy capacity

$$U_e^{\max} = \frac{n}{2}M_Y\psi_Y \tag{5.182}$$

The energy capacity $U_e^{\max}$ given by (5.182) can be equated to the maximum elastic deformation energy which can be stored in an elastic–plastic cantilever, $U_e^{\max} = M_p^2L/2EI$. Thus the elastic limit ψ_Y is related to compliance of a uniform cantilever by

$$\psi_Y = \frac{\phi_m M_p L}{nEI} \tag{5.183}$$

where ϕ_m is the cross-section shape factor for bending. Accordingly, the elastic spring constant $C = M_Y/\psi_Y = nEI/L\phi_m^2$.

The relationship between the energy ratios R and e_0 obtained from (5.182) is

$$\psi_Y = \frac{2U_e^{\max}}{nM_Y} = \frac{2\phi_m e_0}{nR} \tag{5.184}$$

so Eq. (5.179) can be rewritten as

$$\{\ddot{\overline{\psi}}\} = -\frac{n^4 R}{2\phi_m e_o}\mathbf{F}\{m\} \tag{5.185}$$

Equations (5.185) and (5.181) are a set of $2n$ equations for $2n$ unknown functions $\overline{\psi}_i$ and m_i $(i = 1,2,\cdots,n)$. Since a linear strain hardening relationship between moment and relative rotation angle is employed, if the hardening coefficient α tends to zero this structural representation approaches that of an elastic–perfectly plastic ideal sandwich beam. Therefore, the shape factor is taken to be $\phi_m = 1$ in all calculations.

Solution of Initial–Value Problem The nondimensional form of the initial conditions (5.176) at $\tau = 0$ can be written as

$$w_0 = 0, \qquad \dot{w}_0 = \frac{v_o}{\sqrt{1+\dfrac{1}{2n\gamma}}} = \sqrt{\frac{2e_o}{\gamma + \dfrac{1}{2n}}} \tag{5.176'}$$

$$w_i = \dot{w}_i = 0, \qquad i = 1,2,\cdots,n$$

Hence the second order differential equation (5.185) has the following initial conditions:

$$\overline{\psi}_1 = 0 \qquad \dot{\overline{\psi}}_1 = n^2 R\left[e_o(2\gamma + n^{-1})\right]^{-1/2}$$
$$\overline{\psi}_i = 0, \qquad \dot{\overline{\psi}}_i = 0, \qquad i = 2,3,\cdots,n \tag{5.186}$$

An examination of Eqs (5.181), (5.185) and the initial conditions (5.186) indicates that there are *four nondimensional parameters* γ, e_o, R and α, together with the number of links, n. These *five parameters* specify the problem. With initial conditions, a solution can be obtained by numerical integration of the initial-value problem using a Runge–Kutta procedure. Calculations confirm that when n is sufficiently large, say $n \geq 20$, then the qualitative features of the solution are insensitive to n.

5.5.3 Timoshenko Beam Finite Element Model (TB–FE)

Model and Parameters In recent years some large general-purpose finite element numerical codes have been available for calculating the dynamic deformation of elastic–plastic structures in response to arbitrary loading. Both Symonds and Fleming [1984] and Reid and Gui [1987] have reported finite element analyses of impulsively loaded elastic–plastic cantilevers. These analyses employed a finite-element code ABAQUS and assumed elastic–perfectly plastic material behavior. The principal approximations of these analyses (designated as the TB–FE model) are summarized in Table 5.3. This table compares the TB–FE and MS–FD models. The major differences between these models are:

1. The TB–FE model employs an *elastic–perfectly plastic* constitutive relation between stress σ and strain ε; while the MS–FD model employs an *elastic–linear hardening* constitutive relation between moment M and relative rotation angle ψ.

Table 5.3 Comparison of analytical finite element and finite difference models for transverse impact on cantilever

	Rigid–plastic theory E–B beam Sect. 4.5	TB–FE model Timoshenko beam Sect. 5.5.3	MS–FD model E–B beam Sect. 5.5.2
Elasticity (Sects 5.5, 5.6)	0	×	×
Strain-rate effect (Sect. 5.1)	0	0[a]	0
Strain hardening (Sect. 5.2)	0	0	×
Shear, elastic	0	×	0
Shear, plastic (Sect. 5.3)	0	0	0
Rotary inertia of beam	0	×	0
Rotary inertia of tip mass (Sect. 5.3)	0	0	0
Large geometry changes (Sect. 5.4)	0	×	0
Impulsive loading	×	×	×

Note: ×, present; 0, neglected.
[a] Except Example 3 in Reid and Gui [1987].

2. The TB–FE model includes the effects of elastic shear deformation and rotary inertia of the section; i.e. it represents a *Timoshenko beam of rectangular cross-section*. The MS–FD model, on the other hand, represents an *ideal–sandwich Euler–Bernoulli beam*.

3. The TB–FE model includes the effects of *large changes in geometry*, while the MS–FD formulation is limited to small deflections.

5.5.4 Dynamic Deformation of Elastic–Plastic Cantilever from Impact

Choosing Numerical Examples In order to explore the elastic–plastic dynamic behavior of cantilevers after impact, we look first at two typical examples; one with a moderately large mass ratio and the other with a small mass ratio. For these examples, numerical calculations of the dynamic response obtained from the TB–FE and MS–FD models will be presented and compared. Then more systematic results obtained with the MS–FD model will be discussed.

In order to compare with experimental data for a steel cantilever presented by Bodner and Symonds [1962], Symonds and Fleming [1984] calculated the deformations with the TB–FE model using the parameter values of Table 5.4. This impact test had a moderately large mass ratio for the colliding mass $\gamma = 1.64$. Symonds and Fleming calculated deformations resulting from values of R ranging from 2 to 14.8; for this cantilever, these energy ratios correspond to values of V_0 ranging from 4.7 to 12.9 ms^{-1}.

A similar TB–FE model was used by Reid and Gui [1987] for finite element calculations aimed at explaining how the presence of elastic deformation changes the response history; in particular the development of plastic deformation along the cantilever after it is hit at the tip. They calculated the response for three cases of impact: (1) conditions identical to those considered by Symonds and Fleming but with more detail on the distribution of bending moment and plastic work; (2) a high velocity bullet impact test with a small mass ratio γ that was performed by Parkes [1955]; (3) the same high-velocity impact problem but including strain-rate sensitivity in the plastic constitutive law (Cowper–Symonds). The latter analysis

Table 5.4 Parameters used to calculate elastic–perfectly plastic response to impact (Symonds and Fleming [1984], Reid and Gui [1987])

	Symonds and Fleming	Reid and Gui Example 1	Reid and Gui Example 2
$\gamma = G/\rho L = G/\rho_v bhL$	1.64	1.64	0.0228
$e_0 = K_0/M_p = GV_0^2/2M_p$	0.228–1.69	1.69	2.93
$R = 2K_0EI/M_p^2L$	2.0–14.8	14.8	51.7
R^{-1}	0.5–0.0676	0.0676	0.0193
Number of elements	28	28	24
M_p (N m)	16.5	16.5	24.8
G (kg)	0.336	0.336	0.0023
V_0 (ms^{-1})	4.7–12.9	12.9	251.5

showed that with a small mass ratio the response was not qualitatively altered by strain-rate effects.

Elastic–Plastic Behavior for Moderately Large Mass Ratio ($\gamma = 1.64$) For a small global energy ratio $R = 2$ the elastic–plastic response at early times shows little resemblance to the travelling hinge phase of response given by the rigid–perfectly plastic analysis in Sect. 4.5. Most plastic deformation develops near the root after a period wherein elastic flexural waves transmit energy away from the impact point.

For a moderately large energy ratio $R = 14.8$ and mass ratio $\gamma = 1.64$ (i.e. Example 1 in Table 5.4), Reid and Gui [1987] calculated the bending moment distribution at various times. They described the elastic–plastic response of the cantilever as the following sequence of events:

Phase 1: *elastic–plastic bending wave*. The moment distribution in this phase shows many of the characteristics of wave propagation in Timoshenko beams. The moment distribution is oscillatory in nature and the dispersion of the bending wave is evident. At the tail of the disturbed region there exists a small region wherein the bending moment is the fully plastic moment M_p; this may be regarded as a plastic hinge. As time progresses, the position of this short plastically deforming segment moves along the cantilever in a manner reminiscent of the rigid–plastic solution. In this example, the head of the bending wave reaches the root of the cantilever after 0.28 ms.

Phase 2: *reversed hinge at root — arrest of travelling hinge*. After 0.28 ms the elastic bending wave is reflected at the root; the interaction between the primary bending wave and the reflected wave arrests the progress of the travelling hinge. Thereafter the hinge tends to move back towards the tip. Meanwhile at the root the bending moment oscillates and increases in magnitude until, at approximately 1.43 ms, a hinge with reverse rotation is formed at the root.

Phase 3: *initiation of positive root rotation*. While a certain amount of plastic work is consumed in reversed bending at the root, the position of the deforming segment slowly moves back towards the tip and the moment at the root is reduced in magnitude. A positive root hinge forms after 4.9 ms. This continually changing location of plastic deformation is illustrated in Fig. 5.32.

Phase 4: *root rotation and elastic vibration*. The plastic deforming region rapidly shrinks to a hinge at the root; after approximately 62.0 ms only elastic vibrations remain in the cantilever.

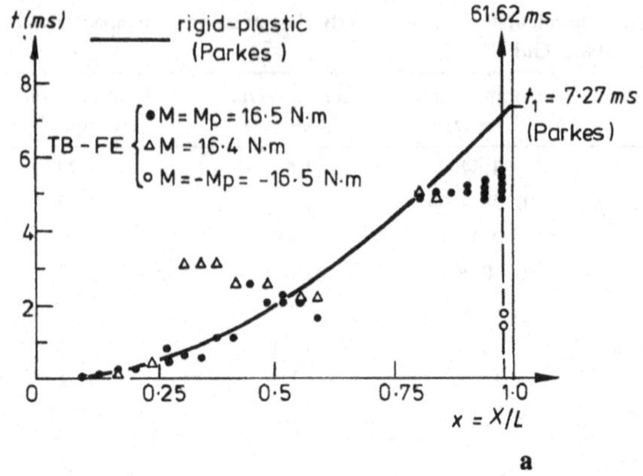

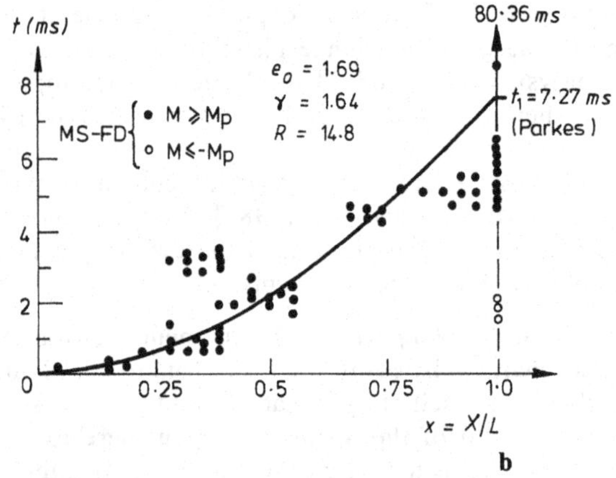

Fig. 5.32 Evolution of plastic deformation region along cantilever during dynamic response when $\gamma = 1.64$, $e_0 = 1.69$ and $R = 14.8$; (a) obtained from the TB–FE model; (b) obtained from the MS–FD model.

This example can also be calculated by means of the finite difference model (MS–FD). For comparison with the solutions obtained by the TB–FE model of a perfectly plastic cantilever, a small value of the strain hardening coefficient is taken; e.g. $\alpha = 0.001$. Apart from this, all other parameters (i.e. e_0, γ and R) were taken from Example 1 of Table 5.4.

Figure 5.32 shows the time when plastic deformation is present at locations along the cantilever. Here Fig. 5.32a was obtained from the TB–FE model, while Fig. 5.32b was calculated with the MS–FD model. The two models show almost identical behavior. Initially a focused region of plastic deformation travels away from the tip. This seems similar to a travelling plastic hinge except that this region disappears midway along the length when the bending moment at the root reverses. Thereafter plastic deformation occurs mostly near the root in a mode-like mechanism. Unlike the rigid–plastic moment distribution (Fig. 4.23c) which

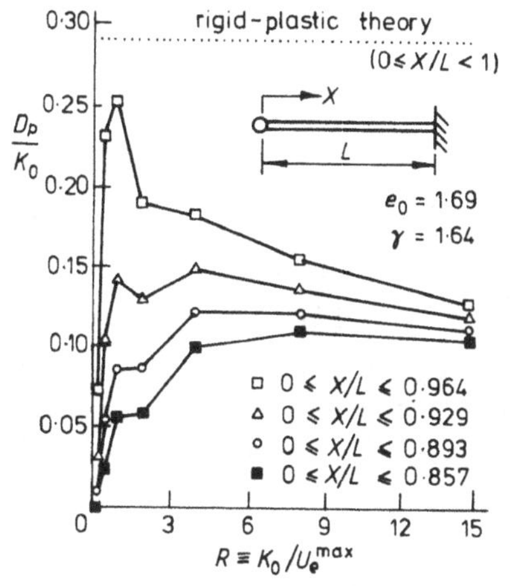

Fig. 5.33 Energy dissipation in a segment adjacent to the root for $\gamma = 1.64$, $e_0 = 1.69$; ——— results obtained from the TB–FE model; – – – energy dissipation at root according to the rigid–plastic theory.

has a uniform plastic moment M_p throughout the region between the travelling hinge and the root, here at any instant the bending moment is close to the fully plastic moment for only short segments of the cantilever.

The development of reverse bending at the root begins as a consequence of an elastic flexural wave that has travelled from the impact point; when this arrives at the root it is reflected back towards the tip. The reflected elastic wave and the focused region of plastic deformation meet near mid-length of the cantilever and this disrupts the travel of the 'plastic hinge'. The subsequent transient development of plastic deformation depends on the mass ratio γ and energy ratio R. In any case with this moderately large mass ratio $\gamma = 1.64$, the early transient stage of deformation has substantially less plastic deformation (permanent curvature) than that given by the rigid–plastic theory. Figure 5.33 shows that the total energy dissipated by curvature away from the root is substantially less than that predicted by the rigid–perfectly plastic theory.

For the present example, calculations of the changing distribution of curvature within the cantilever are shown in Figs. 5.34a, b, for the TB–FE and MS–FD models, respectively. The elastic–plastic calculations predict that almost 87% of the available kinetic energy is absorbed in the element at the root compared with a prediction of only 72% from the rigid–plastic idealization. The peak values of plastic work are found near the middle of the cantilever as a consequence of the oscillation in the position of the plastic hinge in phases 2 and 3. Nevertheless, the vast majority of the energy is dissipated in root rotation during a modal phase of deformation that dominates the behavior for moderate or large γ.

Elastic–Plastic Behavior for Small Mass Ratio (γ = 0.0228) Example 2 in Table 5.4 pertains to bullet impact on a cantilever so it has a very small mass ratio $\gamma = 0.0228$ and a high impact speed. Calculations of the response by the TB–FE model, show roughly the same phases as in Example 1. The location, duration and extent of plastic deformation for this example are plotted as a function of time in Fig. 5.35a. Again the rigid–plastic analysis shows good agreement

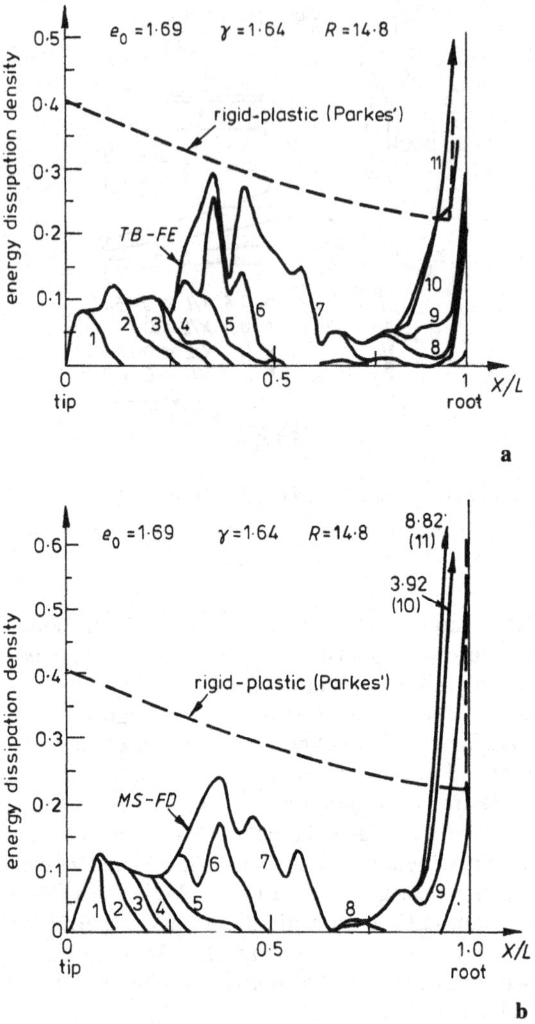

Fig. 5.34 Change of curvature distribution along cantilever during dynamic response when $\gamma = 1.64$, $e_0 = 1.69$ and $R = 14.8$; (a) obtained from the TB–FE model; (b) obtained from the MS–FD model.

with the elastic–plastic solution during the period before the travelling hinge halts midway along the beam. In the elastic–plastic solution the root rotation initiates slightly later than the 3.1ms obtained with the rigid–plastic idealization.

For a light tip mass $\gamma = 0.0228$, the majority of the energy (91.75%) is absorbed between the ends of the cantilever. As shown in Fig. 5.35b, the distribution of plastic work approaches more closely that predicted by the rigid–plastic analysis (given by the dashed line in the figure); however, a major discrepancy exists around the central part of the cantilever where a high curvature region (or 'kink' in the final deformed shape) appears. This is a result of the arrest of the travelling hinge by reflected flexural waves.

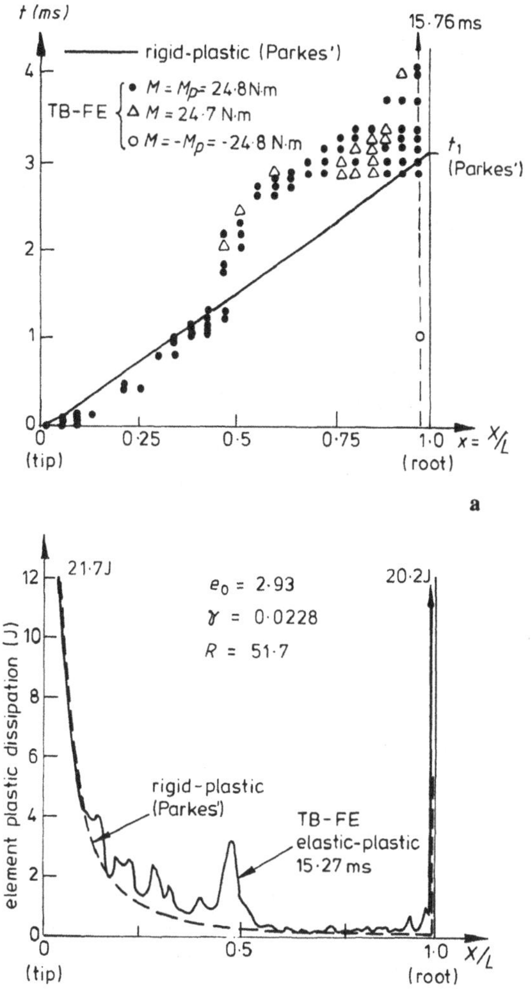

Fig. 5.35 (a) Evolution of plastic deformation region along cantilever obtained from the TB–FE model for $\gamma = 0.0228$, $e_0 = 2.93$ and $R = 51.7$; (b) distribution of plastic dissipation.

Influence of Mass Ratio γ and Energy Ratio R in Strain Hardening Cantilevers
The MS–FD model can be used to examine the influence of nondimensional parameters such as mass ratio γ and energy ratio R, on the dynamic behavior of impulsively loaded beams. Some results obtained with this model are given here.

The distributions of bending moment along a strain hardening cantilever at various instants are given in Figs 5.36 and 5.37 for mass ratios $\gamma = 5$ and 1, respectively. The case considered has a fairly small energy ratio $R = 5$; other parameters are taken to be $e_0 = 1$, $\alpha = 0.1$ and $n = 20$. Fig. 5.36 shows that for a large mass ratio ($\gamma = 5$) the bending moment at the root can reach a value well beyond the initial plastic bending moment M_p; this implies a mode-like response since the root dissipates most of the input energy. For a smaller mass ratio $\gamma = 1$,

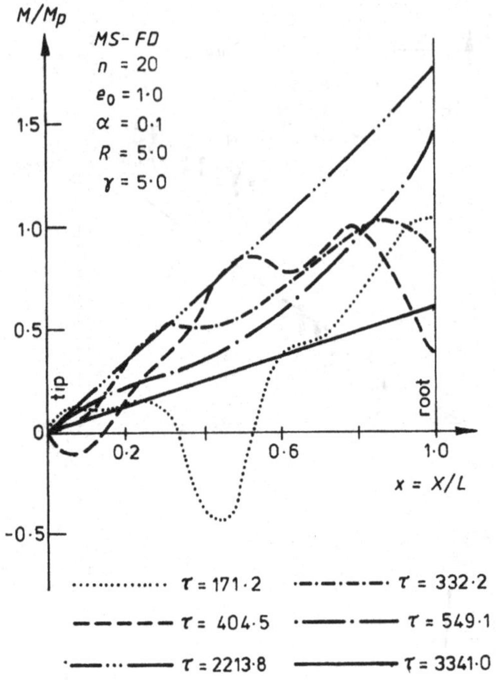

Fig. 5.36 Distribution of bending moment along strain-hardening cantilever, obtained from the MS–FD model; $\gamma=5$, $e_0=1$, $R=5$ and $\alpha=0.1$.

Fig. 5.37 shows that the largest bending moment in the transient phase is only M_p; this may be regarded as a travelling plastic hinge that propagates from the tip towards the middle part of the cantilever. This transient deformation field becomes unclear after it begins to interact with reflected elastic waves returning from the fixed end.

For a heavy colliding mass ($\gamma=5$) and a very light colliding mass ($\gamma=0.025$), Figs 5.38a, b shows the change in position of plastic regions during the response. It is seen from Fig. 5.38a that when $\gamma=5$ and $R=5$, the plastic region first appears near the middle of the cantilever for a very short time; then it reappears at a region close to the root at a time later than that predicted by the rigid–plastic solution as noted in Fig. 5.36. This behavior is more like a modal solution and differs significantly from the rigid–plastic solution. When $\gamma=5$ but $R=15$, the plastic region also first appears at the middle part of the cantilever but there are indications that the plastic region continues to move (backwards then forwards) along the cantilever until it finally reaches the root at the time predicted by the rigid–plastic solution. It is noted that both reverse yielding at the root and arrest of the travelling hinge occur only in the case of $R=15$; for smaller R, (e.g. $R=5$) the root section directly enters modal-type yielding without prior reverse yielding.

As shown in Fig. 5.38b, when the mass ratio γ is very small (e.g. $\gamma=0.025$), the travelling hinge only reaches about 0.5–0.6 of the cantilever length before it disappears. This implies that the response mode is quite similar to Pattern IIb in

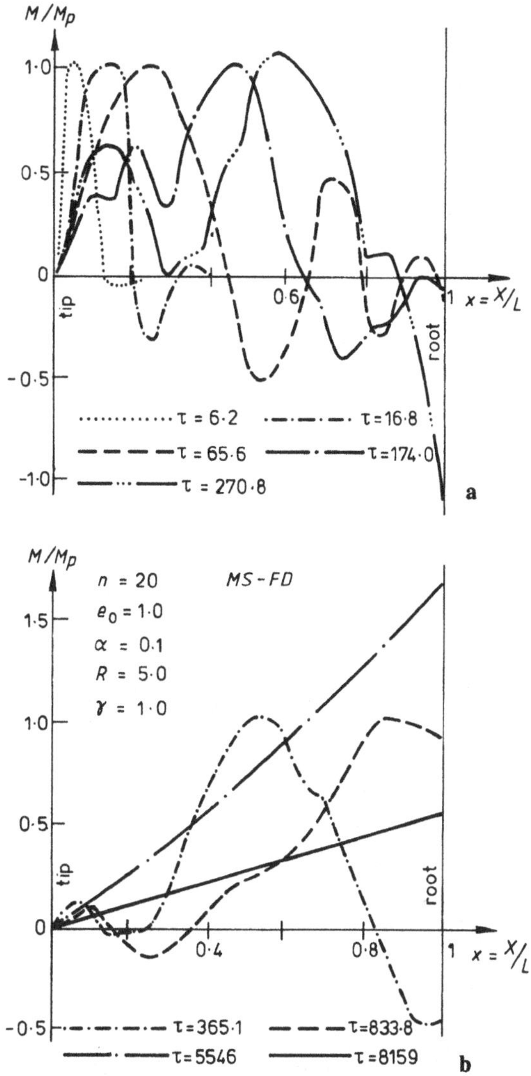

Fig. 5.37 Distribution of bending moment along strain hardening cantilever, obtained from the MS–FD model; $\gamma=1$, $e_0=1$, $R=5$ and $\alpha=0.1$

Sect. 5.5.5; that is, only a part of the cantilever undergoes plastic deformation when the colliding mass is very light or the cantilever itself is very long (i.e. those cases with a very small mass ratio γ). Also, if $\gamma \ll 1$, most of input energy must be dissipated in curvature at interior sections and only a small fraction of energy is dissipated by reverse flexure at the root.

Figure 5.39a, b demonstrates the influence of both mass ratio γ and energy ratio R, on the distribution of plastic dissipation d_p in a strain hardening cantilever. This dissipation is proportional to the final curvature of each section of the cantilever. It is seen that more energy is dissipated in the outer half of the cantilever

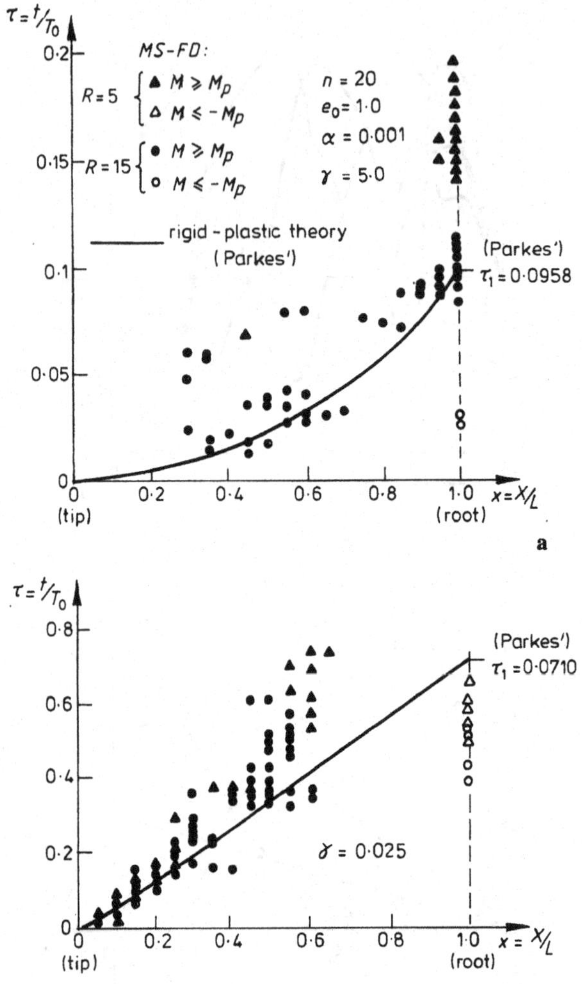

Fig. 5.38 Evolution of plastic deformation region along the cantilever during dynamic response, obtained from the MS–FD model, $e_o = 1$, $\alpha = 0.001$, $R = 5$ and 15; (a) $\gamma = 5$; (b) $\gamma = 0.025$.

(near the tip) when the mass ratio γ is smaller or elastic energy ratio R is larger.

The final deflected shape of the cantilever is insensitive to minor changes in the distribution of energy dissipation, because the deflection is calculated by a double integral of the curvature. The calculated results confirm that the final deflected shape obtained by the present elastic–plastic model almost coincides with that predicted by the rigid–plastic solution, provided mass ratio γ is not very small, see Fig. 5.40a. However, when mass ratio γ is very small, e.g. $\gamma = 0.025$, a notable difference is seen in the middle part of the cantilever as shown in Fig. 5.40b. Nevertheless the tip deflections predicted by the elastic–plastic model and the rigid–plastic solution are almost the same. In Fig. 5.40b

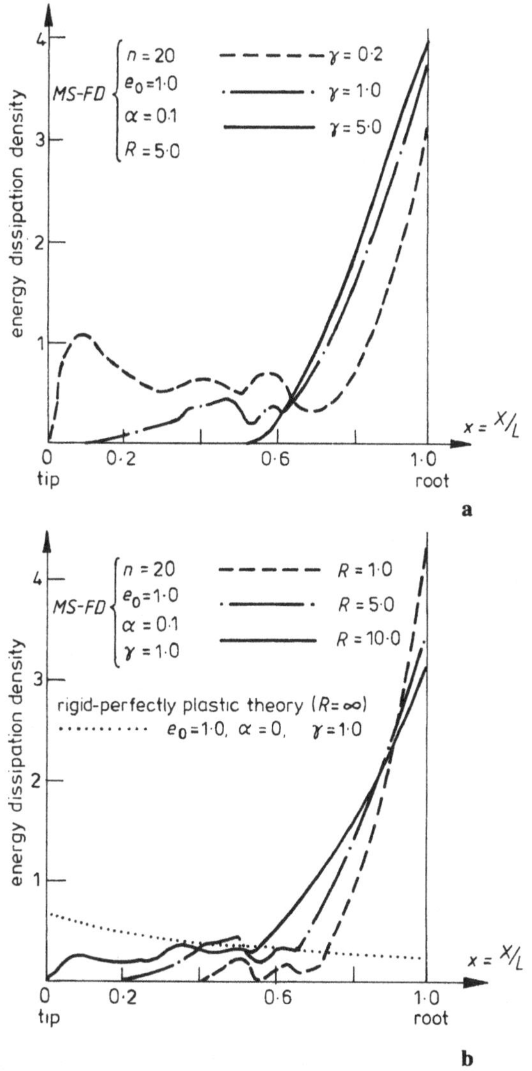

Fig. 5.39 Distribution of energy dissipation along cantilever; (a) influence of mass ratio γ ; (b) influence of energy ratio R.

there is also a clear indication of reverse rotation at the root as a result of reverse yielding during an early phase of the response.

5.5.5 Effect of Elastic Deformation at Root of Cantilever

Model with Rotational Spring at Root The preceding sections have shown that elastic deformations near the root influence the development of plastic deformation throughout the remainder of the structure if the energy ratio R is not too large; this influence is due to elastic compliance that provides some springback. It seems that elastic deformation near the root is the source of both the early termination

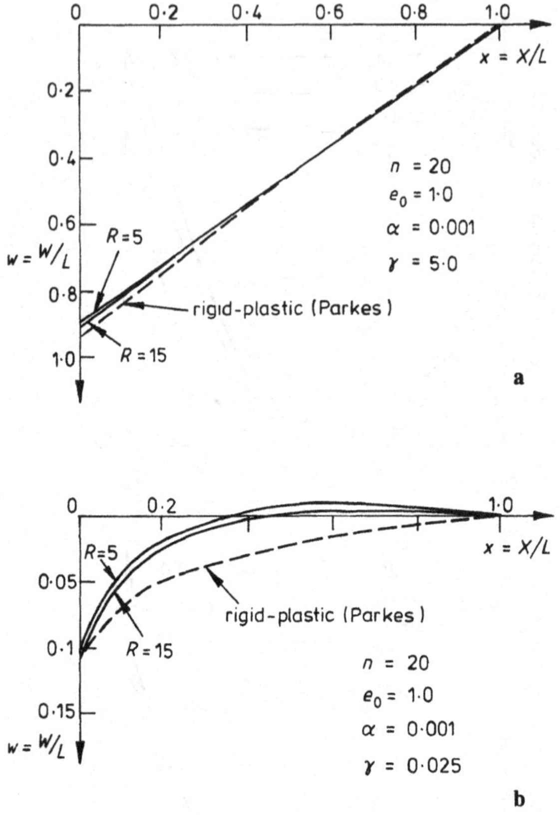

Fig. 5.40 Final deflected shape of cantilever, $e_0 = 1$, $\alpha = 0.001$, $R = 5$ and 15; (a) $\gamma = 5$; (b) $\gamma = 0.025$.

of the travelling hinge phase for cantilevers with high mass ratio and elastic springback from the final modal phase of deformation. The single degree-of-freedom mass–spring system proposed and studied by Symonds et al. (see Sect. 5.6) refers solely to the modal motion of a structure. For a dynamically loaded cantilever, this single degree-of-freedom system is in fact equivalent to setting an elastic–plastic rotational spring at the root while the cantilever itself is regarded as a rigid body. Combining this elastic–plastic rotational spring at the root and the rigid–perfectly plastic beam of most previous analyses, Wang and Yu [1991] proposed a simplified structural model which has some capacity for storing elastic energy in the modal deformation configuration although the majority of the beam is rigid–plastic. This model is certainly not as good as the previous numerical models in representing details of elastic–plastic deformation, but it is much simpler in analysis and reflects some basic features of the dynamic response of an elastic–plastic structure.

The simplified structural model proposed by Wang and Yu [1991] is shown in Fig. 5.41a; this consists of a *rigid*–perfectly plastic cantilever and an *elastic*–perfectly plastic rotational spring at the root. The material is assumed to be rate–independent, the influence of shear on yielding is neglected, the deflections are assumed to be small, and the colliding body is treated as a particle with mass G.

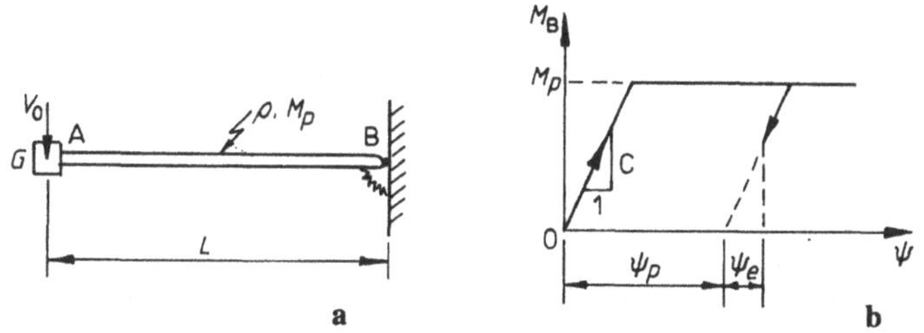

Fig. 5.41 (a) Simplified structural model with an elastic–perfectly plastic rotational spring at the root. (b) Relation between moment and rotation angle at the root.

Under these assumptions, the dynamics of this structural model are almost the same as those of the cantilever analyzed by Parkes [1955], except the bending moment at the root is related to the root rotation by a moment–rotation angle relationship for an elastic–perfectly plastic spring, see Fig. 5.41b. The spring has an elastic constant k, a root rotation angle composed of an elastic part ψ_e and a plastic part ψ_p. During loading the moment at the root M_B is elastic for rotations $\psi \le M_p/C$ and fully plastic M_p for $\psi \ge M_p/C$. For any cross-section of the cantilever the fully plastic bending moment is M_p.

Based on the idea that all of the elastic strain energy of the structure is represented by the elastic capacity of the spring, the spring constant C can be easily determined. For a cantilever of length L and flexural rigidity EI, the maximum elastic deformation energy that can be stored is $U_e^{\max} = LM_p^2/2EI$, while the maximum elastic energy capacity of the root spring is $U_e^{\max} = M_p^2/2C$. Equating the two expressions for $U_e^{\max}$ gives $C = EI/L$. In this model the energy ratio R represents the part of the input energy that can be stored elastically by the spring.

Formulation For the model described above, impact at the tip causes plastic bending of the cantilever as well as elastic–plastic deformation of the rotational spring at the root. The former implies a travelling hinge, as shown in the rigid–plastic analysis and Fig. 5.42. At any time t this plastic hinge is at an interior section H, located a distance $\Lambda(t)$ from the tip. The undeformed segment ahead of the hinge, HB, has rigid-body rotation with angular velocity $\mathrm{d}\psi/\mathrm{d}t$ about the root B, where ψ is the angle between HB and the X-axis. In segment HB, the velocity of a point X is

$$\frac{\mathrm{d}W}{\mathrm{d}t} = \frac{\mathrm{d}\psi}{\mathrm{d}t}(L - X), \qquad \Lambda \le X \le L \tag{5.187a}$$

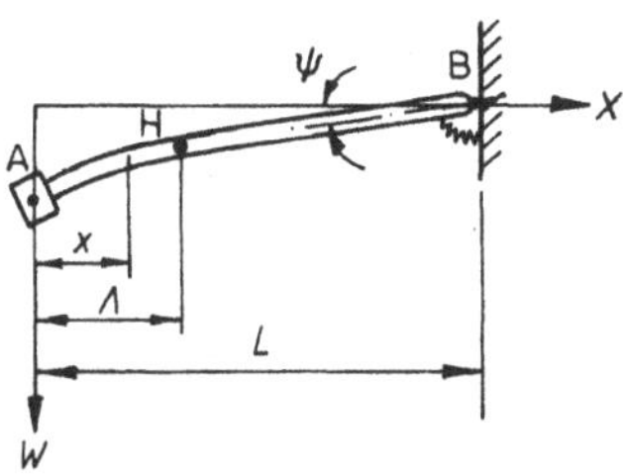

Fig. 5.42 Deformation mechanism with a travelling hinge at H and rotation at root B.

If the heavy particle at the tip has velocity V relative to a reference frame fixed on the rotating segment HB, then segment AH has a rigid-body rotation with relative angular velocity V/Λ. From the sum of rotational velocities about the travelling hinge H and the root B, point X in segment AH has transverse velocity

$$\frac{dW}{dt} = \frac{d\psi}{dt}(L-X)+V(1-X/\Lambda), \qquad 0 \le X \le \Lambda \tag{5.187b}$$

The transverse acceleration throughout the entire length of the cantilever is obtained by differentiating the velocity,

$$\frac{d^2W}{dt^2} = \begin{cases} \dfrac{d^2\psi}{dt^2}(L-X), & \Lambda \le X \le L \\ \dfrac{d^2\psi}{dt^2}(L-X)+\dfrac{dV}{dt}(1-X/\Lambda)+\dfrac{d\Lambda}{dt}VX/\Lambda^2, & 0 \le X \le \Lambda \end{cases} \tag{5.188}$$

At hinge H the bending moment has a local maximum, so shear force $Q_H = 0$. By considering segments AH and HB respectively, the equations of motion are

$$M_B - M_p + \int_\Lambda^L \rho \frac{d^2\psi}{dt^2} X^2 \, dX = 0 \tag{5.189}$$

$$G\frac{d^2W(0)}{dt^2} + \int_0^\Lambda \rho \frac{d^2W}{dt^2} \, dX = 0 \tag{5.190}$$

$$-M_p + \int_0^\Lambda \rho \frac{d^2W}{dt^2} X \, dX = 0 \tag{5.191}$$

By substituting (5.188) into (5.189)–(5.191) and integrating, one obtains three differential equations for $\psi(t)$, $V(t)$ and $\Lambda(t)$. To nondimensionalize these equations, define

$$x \equiv \frac{X}{L}, \quad w \equiv \frac{W}{L}, \quad \lambda \equiv \frac{\Lambda}{L}, \quad m_B \equiv \frac{M_B}{M_p}, \quad \gamma \equiv \frac{G}{\rho L}$$
$$T_o \equiv L\sqrt{\frac{\rho L}{M_p}}, \quad v \equiv \frac{V}{L/T_o}, \quad \tau \equiv \frac{t}{T_o}, \quad (\dot{\ }) \equiv \frac{d}{d\tau}(\ ) \tag{5.192}$$

Thus, Eqs (5.189)–(5.191) can be recast as

$$m_B - 1 + \frac{1}{3}(1-\lambda)^3\ddot{\psi} = 0 \tag{5.193}$$

$$\left(\gamma + \frac{1}{2}\lambda\right)\dot{v} + \left[\gamma + \lambda\left(1 - \frac{\lambda}{2}\right)\right]\ddot{\psi} + \frac{1}{2}v\dot{\lambda} = 0 \tag{5.194}$$

$$\frac{1}{6}\lambda^2\dot{v} + \left(\frac{1}{2} - \frac{\lambda}{3}\right)\lambda^2\ddot{\psi} + \frac{1}{3}v\lambda\dot{\lambda} - 1 = 0 \tag{5.195}$$

where

$$m_B = \begin{cases} C\psi_e/M_p, & \text{elastic or unloading} \\ \pm 1, & \text{otherwise} \end{cases} \tag{5.196}$$

The three unknown functions $\psi(\tau)$, $v(\tau)$, $\lambda(\tau)$ can be solved from Eqs (5.193)–(5.195) with initial conditions

$$\psi(0) = 0, \quad \dot{\psi}(0) = 0, \quad v(0) = v_o, \quad \lambda(0) = 0 \quad \text{at } \tau = 0 \tag{5.197}$$

In the formulation (5.193)–(5.197), some quantities such as $\dot{v}$ and $\dot{\lambda}$ are singular at $\tau = 0$; however this can be removed by starting the integration at a very small initial time τ_0 $(0 < \tau_0 << 1)$ with modified initial conditions. Three nondimensional parameters govern the solution; i.e.

$$\gamma \equiv \frac{G}{\rho L}, \quad e_0 \equiv \frac{K_0}{M_p} = \frac{2GV_0^2}{Ybh^2}, \quad R \equiv \frac{K_0}{U_e^{\max}} = \frac{GV_0^2 C}{M_p^2} \tag{5.198}$$

The first two, γ and e_0 are the only significant parameters for the rigid–perfectly plastic analysis in Sect. 4.5; while the stored energy ratio R^{-1} is the part of the input energy that can be stored in elastic deformation throughout the structure.

Patterns for Process of Dynamic Deformation If these three nondimensional parameters are specified, a solution of the present problem can be obtained by step-by-step numerical integration using a Runge–Kutta procedure. The results indicate that the introduction of elastic deformation at the root notably changes the dynamic behavior from that of the rigid–perfectly plastic idealization. The latter always has a travelling hinge that transits from the impact point to the support at the root of a cantilever; this transient phase is followed by a modal phase of motion. In contrast, the present model exhibits two distinctly different patterns of response depending on the stored energy ratio R^{-1}.

Pattern I. A travelling plastic hinge moves from the tip towards the root, but before it reaches the root, the rotation at the root exceeds the elastic range $\psi > M_p/C$. In this case the travelling hinge and the hinge at the root occur simultaneously in the cantilever. With this pattern, the travelling hinge always finally coalesces with hinge at the root of the cantilever; thereafter, deformation enters a modal phase where the remaining kinetic energy is partly dissipated and partly transformed into elastic vibration about the root.

Pattern II. A travelling plastic hinge moves from the tip towards the root, but at a certain interior section the hinge motion halts and the hinge disappears; hinge travel ceases while the root rotation is still elastic. After the disappearance of the travelling hinge the root *may* or *may not* enter into a plastic state depending on the remaining kinetic energy. Thus Pattern II can be further separated into *pattern* IIb and *pattern* IIa, depending on the largest root rotation.

Numerical examples have established that the stored energy ratio R^{-1} has the largest effect on the pattern of response. A small value of R^{-1} results in pattern I response which is similar to the rigid–perfectly plastic solution. When the stored energy ratio R^{-1} is very small, the root enters the plastic state soon after the hinge begins to move from the tip. In this case the shear force at the hinge vanishes so there is no angular acceleration of the segment between the hinge and the root; this is just Parkes' solution. On the other hand, a large value of R^{-1} results in Pattern II. Hence if the energy stored in elastic deformation at the root is relatively large, only a part of the cantilever undergoes plastic deformation.

The variation of the hinge location with time is plotted in Fig. 5.43. It shows that the value of stored energy ratio R^{-1} has only a minor effect on the speed of the hinge before it reaches $\lambda \equiv \Lambda/L \approx 0.4$; thus the elastic–plastic response of the present model is not very different from the rigid–perfectly plastic (Parkes') solution during an initial period. According to the rigid–perfectly plastic solution, the travelling hinge passes through the entire cantilever. The present results indicate, however, that if the input energy is relatively small (compared with the elastic

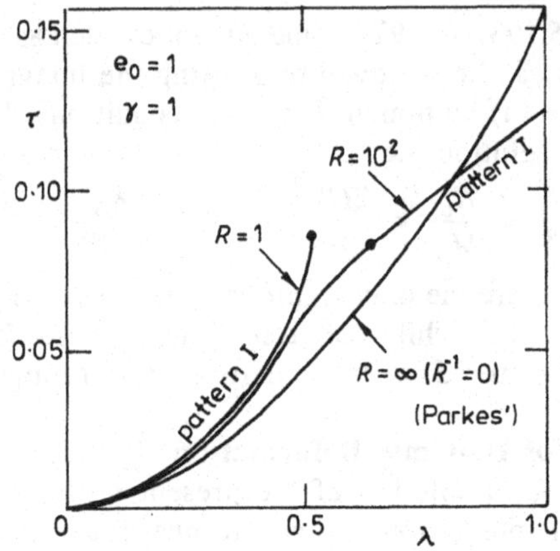

Fig. 5.43 Variation of hinge location with time for different values of R; $\gamma = 1$, $e_0 = 1$.

capacity of the cantilever), the travelling hinge passes through only a part of the cantilever before it disappears (pattern II). Thereafter, the entire cantilever simply rotates about the root. If the remaining kinetic energy is small at the beginning of the modal phase, the root deforms in the elastic range (pattern IIa); if this energy is large, the root deforms plastically (pattern IIb). Figure 5.43 also shows that for either pattern I or pattern II, the hinge slows down near the middle of the cantilever. This is similar to the results obtained by finite–element approaches; it results in a high-curvature region (or a 'kink') around the middle part of the cantilever in its final configuration.

For $e_0 = 1$, Fig. 5.44 shows a map, which illustrates the regions of occurrence for these patterns in the R^{-1}–γ plane. If the stored energy ratio $R^{-1} > 1$, the elastic capacity is more than the input energy so the response is always pattern IIa. If $R^{-1} < 1$ but the mass ratio $\gamma >> 1$, the response is always pattern IIb. In this case, most of the input energy is dissipated plastically at the root in a modal

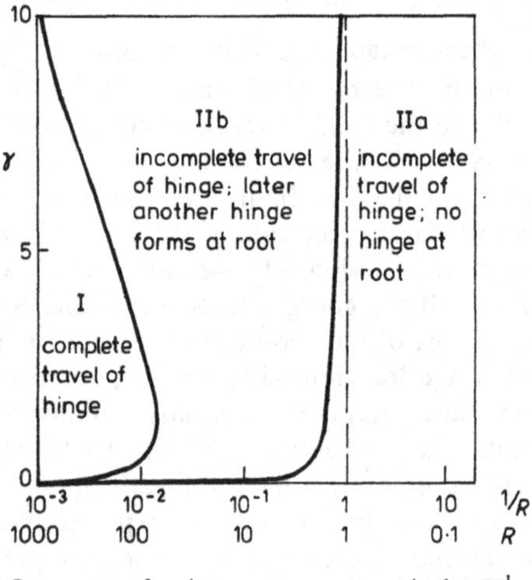

Fig. 5.44 Occurrence of various response patterns in the $R^{-1} - \gamma$ plane.

phase of deformation. Finally, if the stored energy ratio R^{-1} is very small while the mass ratio γ is moderate, the response is Pattern I. This is similar to the response of a rigid–perfectly plastic beam. For a small local energy ratio $e_o = 1$, the global energy ratio R can be large only if the beam is stubby; i.e. L/h is small.

It is interesting to compare this conclusion with the previous results of finite element analyses. For instance, the parameters adopted for Example 1 in Table 5.4 seem to fall in the region of modal response, i.e. pattern IIb.

Energy Dissipation At any instant during dynamic deformation, part of the input energy has already been dissipated, part remains as kinetic energy and part is stored as elastic strain energy. The kinetic energy of the system at any instant is determined by

$$K = \frac{1}{2}GV^2 + \frac{1}{2}\int_0^L \rho\left(\frac{\mathrm{d}W}{\mathrm{d}t}\right)^2 \mathrm{d}X \tag{5.199}$$

Substituting the velocity field (5.187) into (5.199) gives

$$\frac{K}{M_p} = \frac{1}{2}\gamma v^2 + \frac{1}{6}[\dot{\phi}^2 + \dot{\phi} v\lambda(3-\lambda) + v^2\lambda] \tag{5.200}$$

The elastic deformation energy at the root, U_e, is

$$U_e = \frac{1}{2}C\psi_e^2 \le U_e^{\max} = \frac{M_p^2}{2C} \tag{5.201}$$

where $U_e^{\max}$ denotes the maximum elastic energy that can be stored in the system. Consequently, the ratio of elastic strain energy to the local bending stiffness is

$$\frac{U_e}{M_p} = \frac{R\psi_e^2}{4e_o} = \frac{R\psi_e^2}{2\gamma v_o^2} \tag{5.202}$$

Let D_p denote the total plastic dissipation, while D_p^R and D_p^T denote the dissipation at the root hinge and at the travelling hinge, respectively; then the energy balance leads to

$$D_p = K_o - K - U_e, \quad D_p^R = M_p\int_0^t \left|\dot{\psi}_p\right| \mathrm{d}t, \quad D_p^T = D_p - D_p^R \tag{5.203}$$

Using the expressions above, Eq. (5.203a) may be recast into a nondimensional form,

$$\frac{D_p}{M_p} = \frac{1}{2}\gamma(v_o^2 - v^2) - \frac{1}{6}\left[\dot{\psi}^2 + \dot{\psi} v\lambda(3-\lambda) + v^2\lambda\right] - \frac{R\psi_e^2}{2\gamma v_o^2} \tag{5.204}$$

To elucidate the energy dissipation pattern for different ranges of input and structural parameters, some typical examples are given in Fig. 5.45, in which

$$d_p \equiv \frac{D_p}{K_o}, \quad d_p^R \equiv \frac{D_p^R}{K_o}, \quad d_p^T \equiv \frac{D_p^T}{K_o} =$$

Figure 5.45a shows the source of energy dissipation as a function of time for the rigid–perfectly plastic solution; while Fig. 5.45b shows the dissipation in pattern I response. These figures are obviously similar since in both cases the hinge transits the entire cantilever. Figure 5.45c, d shows the energy dissipation for pattern IIb and Pattern IIa, respectively. There is no plastic dissipation in Pattern IIb during the interval from the disappearance of the travelling hinge τ_{TD} to the formation of the root hinge τ_{RP}. If τ_{TD} tends to zero, the energy dissipation pattern

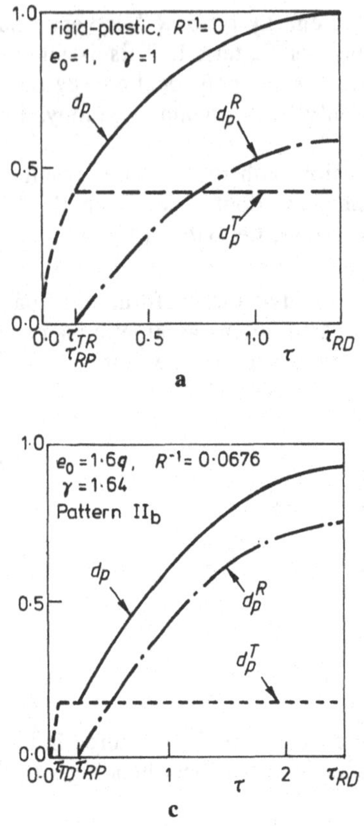

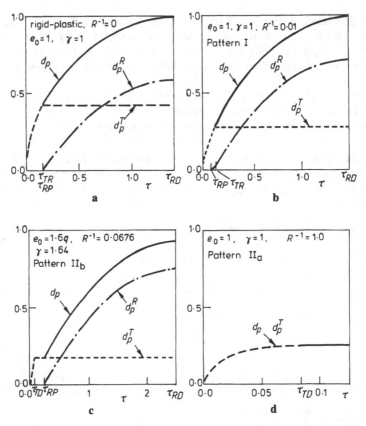

Fig. 5.45 Accumulation of energy dissipation with time: ——— total plastic dissipation (d_p), - — - — - dissipation at root hinge (d_p^R), and - - - dissipation at travelling hinge (d_p^T). τ_{RP} is the instant when the root spring enters into plastic state, τ_{TR} is the instant when the travelling hinge reaches the root, τ_{RD} is the instant when the root hinge disappears, τ_{TD} is the instant when the travelling hinge disappears; (a) rigid–perfectly plastic solution ($R^{-1}=0$); (b) pattern I; (c) pattern IIb; (d) pattern IIa.

approaches that of a modal solution. For Pattern IIa the root never dissipates any energy.

It may be noted that the parameters used for calculating the energy dissipation shown in Fig. 5.45c are exactly the same as Example 1 in Table 5.4. The finite element results indicate that in this example the interior plastic deformation develops first and this is nearly complete before appreciable rotation at the root hinge occurs. This is quite similar to the sequence of events given here for pattern IIb. Reid and Gui [1987] pointed out that the final modal response is delayed by an interaction between a reflected elastic flexural wave and the travelling hinge. For the current composite model, this feature is clearly present in pattern IIb. This characteristic feature of elastic–plastic response to impact is captured by this composite model that incorporates elastic deformation only at the root.

In most of the results shown here, we consider $e_0=1$; i.e. $K_0=M_p$. Wang and Yu [1991] reported that the effect of e_0 on the pattern of response is similar to that of R^{-1}. For specified values of γ and R^{-1}, the response tends to be pattern

I or pattern II if e_0 is very small or very large, respectively. These authors found that e_0 has little influence on the total energy dissipated d_p.

5.5.6 Remarks

1. In some respects the impact response of an elastic–plastic cantilever differs from that calculated for a rigid–perfectly plastic idealization of this structure; the dynamic response differs in distribution of bending moment, position and development of plastic regions, the deformation process, and the final distribution of plastic deformation. In particular, the elastic–plastic response of a cantilever struck at the tip exhibits the following features:

(a) If $R \approx 1$ or larger, a plastic region always propagates away from the impact point in an early period of the response. This small region of plastic deformation is similar to the travelling hinge predicted by the rigid–perfectly plastic theory.
(b) When the small plastically deforming region approaches the middle of the cantilever it slows and can oscillate back and forth; this causes more plastic dissipation and curvature in this central segment. Meanwhile, the root section usually undergoes reverse flexure.
(c) After the plastic region oscillates near midlength, it can resume travelling towards the root (if R is relatively large), or it can disappear (if R is relatively small). If the travelling hinge disappears near midlength, thereafter the root becomes a plastic hinge .
(d) The spatial distribution of energy dissipation in elastic–plastic response is usually rather different from that calculated with the rigid–perfectly plastic theory. While the perfectly plastic theory gives large final curvature near the root when the mass ratio γ is large, it also gives large final curvature near the tip when mass ratio γ is small. Only the elastic–plastic response gives large curvature midway between the ends of the cantilever.

2. From various approaches we conclude that although the input energy ratio R is important in judging the validity of the rigid–plastic theory, it is *not a unique* parameter since it does not characterize all of the approximations of the rigid–plastic theory. For a cantilever struck at the tip by a heavy rigid body, comparisons of the finite difference, finite element and the simplified model with an elastic–plastic root indicate that the energy ratio R, the mass ratio $\gamma = G/\rho L$ and the impact energy ratio e_0 all play significant roles in determining the validity of the rigid–plastic theory. The combined effect of input energy ratio R and mass ratio γ on dynamic behavior of impulsively loaded elastic–plastic cantilevers may be summarized in Table 5.5, which clearly states the essential influence of mass ratio γ on the deformation process.

3. The basic feature of dynamic response of elastic–plastic cantilevers is *not* significantly altered by some minor differences in modeling; e.g. the difference between a Timoshenko beam of rectangular cross-section and an ideal sandwich Euler–Bernoulli beam. Indeed the finite element, finite difference and rigid–plastic beam with elastic–plastic root all provide similar patterns for the deformation process.

Table 5.5 Dynamic response of impulsively loaded cantilevers: effects of energy ratio R and mass ratio γ

R	Large γ	Small γ
$R > 100$	Hinge passes through the entire length of the cantilever: similar to rigid–plastic solution (Sect. 4.5), or Pattern I (Sect. 5.5.5)	
$10 < R < 100$	Root dissipates most energy after travelling hinge terminates at midlength; Pattern IIb in Sect. 5.5.5 or Ex. 1 in Table 5.4	Root dissipates little energy after travelling hinge terminates at midlength; Pattern IIa in Sect. 5.5.5 or Ex. 2 in Table 5.4
$2 < R < 5$	Mode-like response, root dissipates almost all energy	Hinge travel delayed at midlength, Ex. 2 in Table 5.4
$0.5 < R < 2$	Almost entirely elastic response with small plasticity at root	Very localized plastic deformation near tip

5.6 Accuracy of Rigid–Plastic Analyses

Studies in Sects 5.1–5.5 about second–order effects in the dynamic plastic response of impulsively loaded cantilevers have enlightened us about the accuracy of the easy to use rigid–perfectly plastic idealization. In the present section, particular attention will be paid to the following questions: When is the rigid–plastic model valid? What conditions are necessary for elastic effects to be insignificant? Which factors control the 'error' of the rigid–plastic prediction in comparison with elastic–plastic solution for the final deflection?

Before we answer these questions, notice that (1) dynamic behavior of elastic–plastic structural members (e.g. cantilevers) is rather complex, as shown in Sect. 5.5; and (2) dynamic behavior of rigid–plastic structural members is complicated by the transient phase of deformation, see Sect. 4.5. It is difficult, therefore, to answer these questions by directly comparing elastic–plastic and rigid–plastic solutions. It is possible nevertheless to obtain some preliminary answers from comparisons with mode–type solutions. The concept of modal approximation has been explained in Chap. 2.

In the following, a single degree-of-freedom (DoF) mass spring system, which represents the mode–type motion of structures, is first studied in Sect. 5.6.1 in order to show what factors, apart from energy ratio R, significantly affect accuracy of the rigid–plastic prediction. Then the convergence of the response of a two DoF model of an elastic–plastic cantilever to the primary plastic mode is examined in Sect. 5.6.2; there the influence of initial momentum distribution is discussed.

5.6.1 Accuracy of Rigid–Plastic Analysis Estimated by Single DoF System

Previous investigations in Sect. 5.5 have confirmed that rigid–plastic analyses provide good approximations for dynamic plastic deformation of a structure subjected to intense short pulse loading if the energy ratio R, defined by Eq. (5.168), is large. In the limiting case of impulsive loading (i.e. an initial velocity

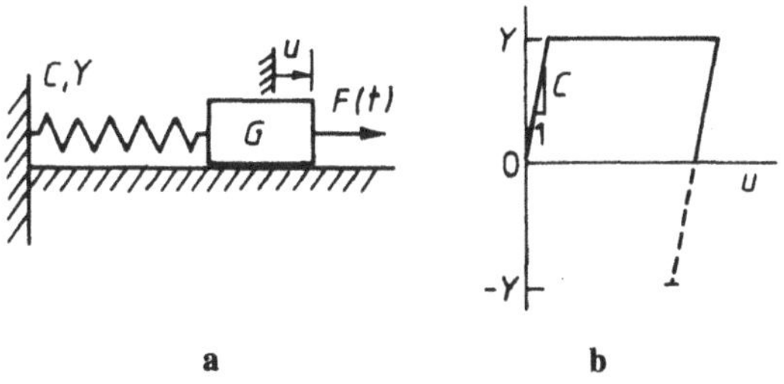

Fig. 5.46 Single DoF system loaded by pulse $F(t)$; (a) single DoF mass–spring system; (b) force–displacement relation for spring.

field), the 'error' in the final deflection caused by the rigid–plastic idealization is estimated to be of the order of $(1/R)$, see Symonds [1985]. In this case the error is positive since the initial kinetic energy must all be dissipated by plastic deformation rather than being divided between plastic work and elastic strain energy.

However, $R >> 1$ is a necessary but not sufficient condition for validity of the rigid–plastic idealization. When the loading pulse is relatively long, the error of a rigid–plastic solution can increase in magnitude and may become negative; i.e. the rigid–plastic solution can underestimate the final deformation of structures. Symonds [1981] has pointed out that if the duration of a loading pulse t_d, is not brief in comparison with the fundamental period for elastic vibration of the structure T_1, the errors caused by the rigid–plastic idealization may be large (e.g. 30–60%). This occurs although the actual plastic deformations are as large as 10–20 times the elastic limit strain. In general, therefore, rigid–plastic predictions are applicable only if both of the following conditions are satisfied:

$$R >> 1, \quad t_d < T_1 \tag{5.205}$$

Noting that a modal approximation for dynamic rigid–plastic deformation of a structure is a single DoF motion, Symonds and Frye [1988] examined the dynamic behavior of a single DoF mass–spring model with a spring made from either elastic–perfectly plastic or rigid–perfectly plastic materials. (A sketch of this single DoF system is shown in Fig. 5.46.) They investigated the response of this simple model to six different shaped pulse loads (rectangular, half-sine and triangular with four rise times) in order to assess the influence of rise time and pulse duration on the accuracy of the rigid–plastic predictions. Their calculations confirm the earlier observation that a large energy ratio R is necessary, but not sufficient to ensure that the rigid–plastic idealization gives a small error in comparison with the elastic–plastic solution. For example, Fig. 5.47 is obtained from this single DoF model when it is subjected to a linear decreasing pulse; i.e. a triangular pulse with zero rise time. Figure 5.47a shows the ratio u_f^{ep}/u_Y, where u_f^{ep} is the final displacement given by the elastic–plastic solution and $u_Y = N_Y/C$ is the displacement at which the spring in the single DoF system enters the plastic state. Here N_Y is the yield force of the spring, F_{max} denotes the peak force applied on the mass and a load parameter is defined as $\mu = F_{max}/N_Y$. Figure 5.47b shows the error of the rigid–plastic prediction relative to the elastic–plastic solution for the final displacement. Error is defined as $(u_f^{rp} - u_f^{ep})/u_f^{ep}$ where u_f^{rp} is the permanent displacement given by the rigid–plastic solution, as a function of the pulse duration (t_d/T_1) and R is the energy ratio defined by (5.168). For the present model, $R = 2u_f^{ep}/u_Y$. It is seen that while the error of the rigid–plastic prediction reduces with increasing R, the error increases with pulse duration t_d.

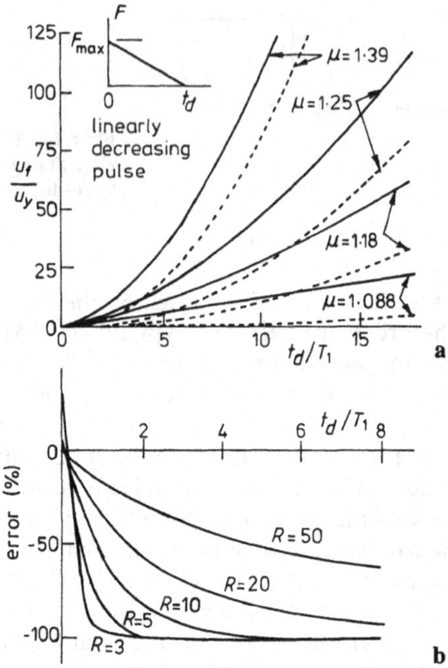

Fig. 5.47 Behavior of single DoF system subjected to linear decreasing pulse; (a) dependence of displacement ratio on loading parameters; (b) error of rigid–plastic prediction relative to elastic–plastic solution on final displacement.

A notable fact explored by Symonds and Frye [1988] is that when the loading pulse has a nonzero rise time, the error of the rigid–plastic prediction displays a wavy character over a large range of the ratio (t_d/T_1), and the discrepancies between the rigid–plastic and the elastic–plastic analyses may become quite significant. For example, Fig. 5.48 is obtained from the single DoF model when it is subjected to a half-sine pulse with duration t_d and peak force $F_{max} = \mu N_Y$. The error of the rigid–plastic prediction is shown in Fig. 5.48b, which indicates that the error is smaller for larger values of R, and the peaks in the error curves decrease as the pulse duration increases. Triangular pulses with nonzero rise time result in similar error curves. In cases where R is not very large, no simple formula based only on energy ratio R could provide an estimate of the error of the rigid–perfectly plastic approximation.

5.6.2 Convergence to Dynamic Plastic Mode Studied by Two DoF System

Simple lumped mass models of impulsively loaded beams can greatly simplify the beam dynamics by limiting the number of DoF; likewise, they limit the number of dynamic modes. Stronge and Hua [1990] adopted this kind of two DoF model as a tool to study the transient response of elastic–plastic beams and cantilevers, in comparsion with rigid–plastic solutions. Results from the two DoF model of a cantilever are briefly extracted below.

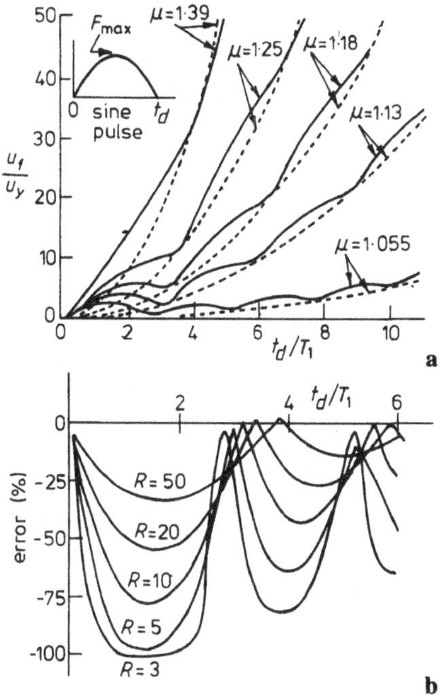

Fig. 5.48 Behavior of single DoF system subjected to half–sine pulse; (a) dependence of displacement ratio on loading parameters; (b) error of rigid–plastic prediction relative to elastic-plastic solution on final displacement.

To model a structure subjected to arbitrarily distributed impulsive loading, consider a cantilever of length L with two equally spaced masses $\overline{G}$, as shown in Fig. 5.49a. These masses are located at the midpoint and the tip; and they have initial velocities V_0 and ζV_0, respectively. Assume that deformations occur only at the discrete hinges located at the root and the midpoint. The cantilever consists of two rigid segments connected by these hinges. The hinges are assumed to have elastic–perfectly plastic behavior; that is, if the relative rotation at hinge i $(i = 1,2)$ is ψ_i, then the moment M_i during loading is given by

$$\begin{aligned} &M_i = C\psi_i, \qquad |\psi_i| < \psi_Y \\ & \qquad\qquad\qquad\qquad\qquad\qquad\qquad i=1,2 \\ &|M_i| = M_p, \qquad |\psi_i| \geq \psi_Y, \qquad \dot{\psi}_i > 0 \end{aligned} \tag{5.206}$$

where parameter ψ_Y is the yield rotation angle of each hinge. The deformation of this model is always initially elastic while both hinges have rotations $|\psi_i| < \psi_Y$; later, it can be elastoplastic after one of the hinges has a rotation magnitude $|\psi_i| \geq \psi_Y$; and finally it can become fully plastic if both hinges have rotation magnitudes $|\psi_i| \geq \psi_Y$. The analysis given by Stronge and Hua [1990] neglects elastic unloading so that the final deformed configuration is unambiguous.

By introducing nondimensional variables

$$\begin{aligned} m_i &\equiv M_i / M_p, \qquad \overline{\psi}_i \equiv \psi_i / \psi_Y \\ w_i &\equiv 2W_i / L\psi_Y, \qquad \overline{\tau} \equiv 2t(C/\overline{G}L^2)^{1/2} \end{aligned} \tag{5.207}$$

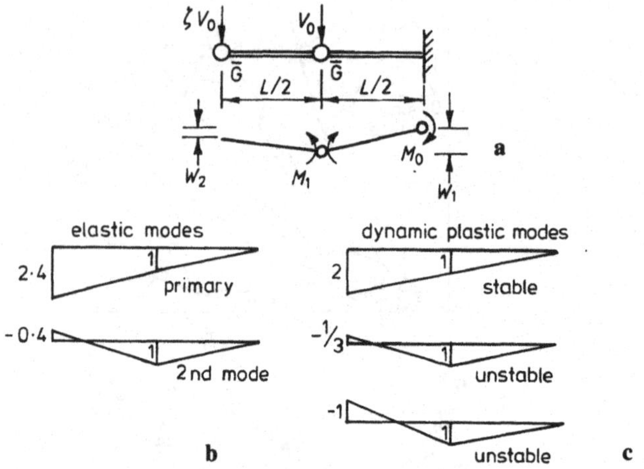

Fig. 5.49 Two DoF system; (a) two DoF model of impulsively loaded cantilever; (b) elastic modes; (c) dynamic plastic modes.

the equation of motion of this two DoF model is

$$\begin{Bmatrix} \ddot{\overline{w}}_1 \\ \ddot{\overline{w}}_2 \end{Bmatrix} + \begin{bmatrix} 1 & -2 \\ 0 & 1 \end{bmatrix} \begin{Bmatrix} m_1 \\ m_2 \end{Bmatrix} = \begin{Bmatrix} 0 \\ 0 \end{Bmatrix} \tag{5.208}$$

where $\overline{w}_i$ is the nondimensional transverse nodal displacement, m_i is the moment at node $(i-1)$, and $(\dot{\ }) \equiv d(\ )/d\overline{\tau}$. For impulsive loading the initial conditions are

$$\overline{w}_1(0) = 0, \quad \overline{w}_2(0) = 0, \quad \dot{\overline{w}}_1(0) = \overline{v}_o, \quad \dot{\overline{w}}_2(0) = \zeta \overline{v}_o \tag{5.209}$$

where $\overline{v}_o \equiv V_o (C/\overline{G})^{1/2} \psi_Y^{-1}$.

The cantilever has primary and secondary modes for fully plastic deformations as shown in Fig. 5.49c. The primary dynamic plastic mode is a rigid body rotation about the root. This mode is stable while the secondary plastic modes are unstable. The last secondary mode shown in Fig. 5.49c has moments of the same sense at both hinges. This curious mode only exists for $\dot{\overline{w}}_1 \dot{\overline{w}}_2 > 0$ and $\left|\dot{\overline{w}}_2 - 2\dot{\overline{w}}_1\right| > 0$ when it appears in conjunction with the primary mode — it does not exist independently. Stable modes are minima for the dissipation rate at any specified kinetic energy. For impulsively loaded rigid–plastic structures, deformation finally converges to a stable modal solution unless the initial momentum distribution is identical to an unstable mode shape.

Stronge and Hua [1990] gave formulations for the *elastic*, *elastoplastic* and *fully plastic phases* of motion of a structural model with two discrete masses; this model has only two locations where elastoplastic hinges can form. For the fully plastic phase, both of these hinges are plastic, and deformation occurs in a combination of dynamic plastic modes before coalescence with a stable plastic mode. After coalescence the final period of motion has a velocity distribution that is identical with a stable plastic mode; the deformation finally terminates when motion ceases.

The energy ratio for the model is $R = (\overline{v}_o^2/2)(1+\zeta^2)$ where ζ is the ratio of initial transverse speeds for the two masses. For $\overline{v}_o^2/2 = 5$ and $0 \le \zeta \le 3$, the velocity trajectories for elastoplastic deformations of the impulsively loaded

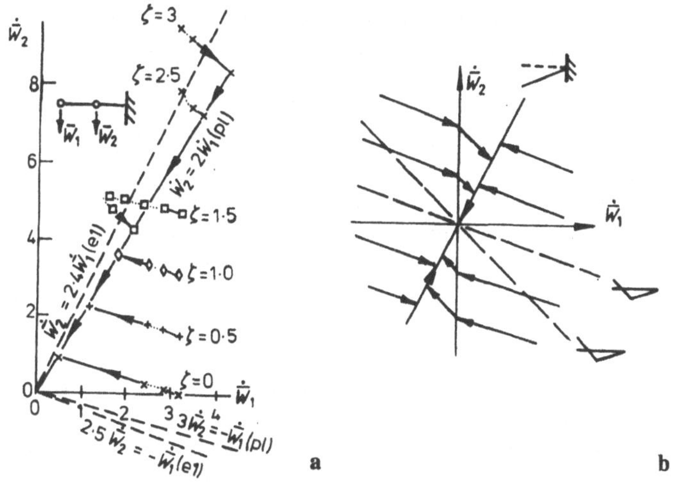

Fig. 5.50 Velocity trajectories for two DoF model; (a) elastic–plastic case; (b) rigid–perfectly plastic case.

cantilever are illustrated in Fig. 5.50a. The stable dynamic plastic mode corresponds to an initial momentum distribution, $\zeta = 0$. For initial conditions in the range $1.1 < \zeta < 2.0$, the deformation at the central hinge is still elastic when the velocity distribution is identical to that of a stable plastic mode so the elastoplastic trajectory crosses this mode line. For an initial momentum distribution near the stable plastic mode configuration, yield occurs first at the root hinge.

Velocity trajectories corresponding to the rigid–perfectly plastic approximation for the same model are shown in Fig. 5.50b. These solutions converge to the stable plastic mode along paths parallel to the unstable plastic modes; the discontinuities in these trajectories at $\dot{\overline{w}}_1 = 0$ are caused by a change in the sense of the moment at the root. For a specified energy ratio *R*, in comparison with the elastoplastic response, the error of the rigid–plastic approximation is largest for an initial momentum distribution that is not very different from the stable plastic mode shape.

Stronge and Hua [1990] also studied a two DoF model of a simply supported elastoplastic beam, which has two stable dynamic plastic modes. Their analysis shows that elastic effects during an intermediate phase of elastoplastic deformation can significantly affect subsequent widespread plastic deformation when multiple stable dynamic plastic modes exist. As a result, the accuracy of the rigid–plastic approximation is very sensitive to the initial momentum distribution, especially when the latter is close to the stable plastic mode. In these cases, *R* will not serve as the unique criterion in determining the error caused by the rigid–plastic idealization.

5.6.3 Remarks

1. From various approaches we may conclude that although energy ratio *R* is of importance in judging the validity of the rigid–plastic theory, it is *not the only factor* determining the 'error' of the rigid–plastic theory. A study of a pulse-

loaded one DoF model (Symonds and Frye [1988]) indicated that pulse duration has substantial influence, especially if rise time of the pulse is significant and the duration is long in comparison with the natural period. An analysis of a two DoF model (Stronge and Hua [1990]) shows that this 'error' strongly depends on whether the distribution of initial momentum is close to a mode shape.

2. The value of R is related to the elastic capacity $U_e^{\max}$ of the structure by (5.168), while $U_e^{\max}$ is usually calculated by assuming that all parts of the structure enter the yield state *simultaneously* (e.g. see (5.169)). This leads to an overestimate of $U_e^{\max}$ and a underestimate of R, especially when the structure involved is large. In fact, for a very large structure subjected to local loading, the rigid–plastic solution is a good approximation near the impact point and not accurate further away even for a small value of R.

3. According to the convergence theorem (refer to Chap. 2 or Martin and Symonds [1966]), the rigid–plastic approximation always converges to a stable plastic mode configuration; therefore, a plastic modal solution gives a good approximation provided the correct mode is selected and this is given the proper initial velocity. However, when elastic deformation is involved, the accuracy of a plastic mode approximation significantly depends on the initial momentum distribution, especially if the latter differs greatly from a stable plastic mode-shape.

References

Atkins, A.G. [1990]. Note on scaling in rigid–plastic fracture mechanics. *Int. J. Mech. Sci.* **32**, 547–548.

Bodner, S.R. and Symonds, P.S. [1960]. Plastic deformation in impact and impulsive loading of beams. *Plasticity*, Proceedings of the Second Symposium on Naval Structural Mechanics, (eds E.H. Lee and P.S. Symonds) Pergamon Press, 488–500.

Bodner, S.R. and Symonds, P.S. [1962]. Experimental and theoretical investigation of the plastic deformation of cantilever beams subjected to impulsive loading. *ASME J. Appl. Mech.* **29**, 719–728.

Conroy, M.F. [1952]. Plastic–rigid analysis of long beams under transverse impact loading. *ASME J. Appl. Mech.* **19**, 465–470.

Florence, A.L. and Firth, R.D. [1965]. Rigid–plastic beams under uniformly distributed impulses. *ASME J. Appl. Mech.* **32**, 481–488.

Forrestal, M.J. and Sagartz, M.J. [1978]. Elastic–plastic response of 304 stainless steel beams to impulse loads. *ASME J. Appl. Mech.* **45**, 685–687.

Hashmi, S.J., Al-Hassani, S.T.S. and Johnson, W. [1972]. Large deflexion elastic–plastic response of certain structures to impulsive load: Numerical solutions and experimental results. *Int. J. Mech. Sci.* **14**, 843–860.

Hodge, P. G., Jr. [1959]. *Plastic Analysis of Structures*. McGraw-Hill.

Hou, W.J., Yu, T.X. and Su, X.Y. [1995]. Elastic effect in dynamic response of plastic cantilever beam to impact, *Acta Mechanica Solida Sinica*, **16**, 13–21.

Johnson, W. and Travis, F.W. [1968]. High-speed blanking of steel. *Engineering Plasticity* (eds J. Heyman and F.A. Leckie) Cambridge University Press, 385–400.

Jones, N. [1967]. Influence of strain–hardening and strain-rate sensitivity on the permanent deformation of impulsively loaded rigid–plastic beams. *Int. J. Mech. Sci.* **9**, 777–796.

Jones, N. [1971]. A theoretical study of the dynamic plastic behaviour of beams and plates with finite-deflections. *Int. J. Solids Struct.* **7**, 1007–1029.

Jones, N. [1989]. On the dynamic inelastic failure of beams. *Structural Failure*, (eds T. Wierzbicki and N. Jones) John Wiley, 133–159.

Jones, N. and de Oliveira, J.G. [1979]. The influence of rotatory inertia and transverse shear on the dynamic plastic behaviour of beams. *ASME J. Appl. Mech.* **46**, 303–310.

Jones, N. and Song, B.Q. [1986]. Shear and bending response of a rigid–plastic beam to partly distributed blast-type loading. *J. Struct. Mech.* **14**(3), 275–320.

Jouri, W.S. and Jones, N. [1988]. The impact behaviour of aluminium alloy and mild steel double-shear specimens. *Int. J. Mech. Sci.* **30**, 153–172.

Liu, J. and Jones, N. [1987]. Experimental investigation of clamped beams struck transversely by a mass. *Int. J. Impact Engng.* **6**, 303–335.

Martin, J.B. [1989]. Dynamic bending collapse of strain-softening cantilever beams. *Structural Failure* (eds T. Wierzbicki and N. Jones) John Wiley, 365–388.

Martin, J.B. and Symonds, P.S. [1966]. Mode approximation for impulsively loaded rigid–plastic structures. *Proc. ASCE, J. Engng. Mech. Div.* **92**(EM5), 43–66.

Nonaka, T. [1967]. Some interaction effects in a problem of plastic beam dynamics, Parts 1–3. *ASME J. Appl. Mech.* **34**, 623–643.

Nonaka, T. [1977]. Shear and bending response of a rigid–plastic beam to blast-type loading. *Ingenieur-Archiv*, **46**, 35–52.

de Oliveira, J.G. [1982]. Beams under lateral projectile impact. *Proc. ASCE, J. Eng. Mech. Div.* **108**(EM1), 51–71.

de Oliveira, J.G. and Jones, N. [1979]. A numerical procedure for the dynamic plastic response of beams with rotatory inertia and transverse shear effects. *J. Struct. Mech.* **7**(2), 193–230.

Parkes, E.W. [1955]. The permanent deformation of a cantilever struck transversely at its tip. *Proc. Roy. Soc. Lond.* **A228**, 462–476.

Perrone, N. [1966]. A mathematically tractable model of strain–hardening, rate–sensitive plastic flow. *ASME J. Appl. Mech.* **33**, 210–211.

Reid, S.R. and Gui, X.G. [1987]. On the elastic–plastic deformation of cantilever beams subjected to tip impact. *Int. J. Impact Engng.* **6**, 109–127.

Reid, S.R., Yu, T.X. Yang, J.L. [1995]. Response of an elastic, plastic tubular cantilever beam subjected to force pulse at its tip, – small deflection analysis, *Int. J. Solids Struct.* **32** (in press).

Shioya, T. and Stronge, W.J. [1984]. Impact on rotating fan blades. *Inst. Phys. Conf. Ser.* No.70, 511–518.

Shu, D. [1990]. Structural arrangement and geometric effects on plastic deformations in collisions. Ph.D. thesis, Chap. 4, Cambridge University, U.K.

Shu, D., Stronge, W.J. and Yu, T.X. [1992]. Oblique impact at tip of cantilever. *Int. J. Impact Engng.* **12**, 37–47.

Stronge, W.J. and Hua Yunlong [1990]. Elastic effects on deformation of impulsively loaded elastoplastic beams. *Int. J. Impact Engng.* **9**, 253–262.

Stronge, W.J. and Shioya, T. [1984]. Impact and bending of a rigid–plastic fan blade. *ASME J. Appl. Mech.* **51**, 501–504.

Stronge, W.J. and Yu, T.X. [1989]. Dynamic plastic deformation in strain–hardening and strain–softening cantilevers. *Int. J. Solids Struct.* **25**, 769–782.

Symonds, P.S. [1965]. Viscoplastic behavior in response of structures to dynamic loading. *Behavior of Materials Under Dynamic Loading* (ed. N.J. Huffington) ASME, 106–124.

Symonds, P.S. [1968]. Plastic shear deformations in dynamic load problems. *Engineering Plasticity* (eds J. Heyman and F.A. Leckie) Cambridge University Press, 647–664. .

Symonds, P.S. [1981]. Elastic–plastic deflections due to pulse loading. *Dynamic Response of Structures* (ed. G. Hart) ASCE, New York, 887–901.

Symonds, P.S. [1985]. A review of elementary approximation techniques for plastic deformation of pulse-loaded structures. *Metal Forming and Impact Mechanics* (ed. S.R. Reid) Pergamon Press, Oxford, 175–194.

Symonds, P.S. and Fleming, W.T. Jr [1984]. Parkes revisited: on rigid–plastic and elastic–plastic dynamic structural analysis. *Int. J. Impact Engng.* **2**, 1–36.

Symonds, P.S. and Frye, C.W.G. [1988]. On the relation between rigid–plastic and elastic–plastic predictions of response to pulse loading. *Int. J. Impact Engng.* **7**, 139–149.

Symonds, P.S. and Mentel, T.J. [1958]. Impulsive loading of plastic beams with axial restraints. *J. Mech. Phys. Solids* **6**, 186–202.

Ting, T.C.T. [1964]. Plastic deformation of a cantilever beam with strain-rate sensitivity under impulsive loading. *ASME J. Appl. Mech.* **31**, 38–42.

Ting, T.C.T. [1965]. Large deformation of a rigid, ideally plastic cantilever beam. *ASME J. Appl. Mech.* **32**, 295–302.

Wang, X.D. and Yu, T.X. [1991]. Parkes revisited: effect of elastic deformation at the root of a cantilever beam. *Int. J. Impact Engng.* **11**, 197–209.

Yu, T.X. and Stronge, W.J. [1990]. Large deflections of a rigid–plastic beam-on-foundation from impact. *Int. J. Impact Engng.* **9**, 115–126.

[illegible]

Chapter 6

More Complex Configurations

This chapter considers the dynamic effects resulting from more complex loads and structural shapes. In some cases the more complicated configurations require additional degrees-of-freedom or other types of plastic hinges which undergo either stretching or twisting in addition to bending. For plastic deformations these combined states of generalized stress result in deformations and deformation rates that are related by the associated flow rule. Hence, this chapter begins with general considerations for the extremal properties and differentiability of the yield function at generalized hinges. Here we consider only rigid–perfectly plastic materials and hence plastic deformation that is localized at a finite number of discrete generalized plastic hinges. While continuous plastic regions are possible, they require a special distribution of applied force if the deformations are related to generalized stress through an associated flow rule.

Depending on the generalized stresses that are significant for dissipation, we may separate dynamic plastic problems into two groups: those involving only flexural deformation and those involving an interaction yield condition. Transversely loaded straight cantilevers with varying cross-section (Sect. 6.2) belong to the first group, although more than one flexural plastic hinge may be required if there is a crack or step in stiffness at a cross-section of a straight member (Sects 6.7 and 6.8). For a slender cantilever of curved or bent plan form, the effects of the shear and axial force on yielding can be negligible for in-plane loading; if so, these examples also have ordinary flexural plastic hinges. Thus they also fall into the first group.

It is important to consider an interaction yield criterion if the total work done on the structure by external forces is ultimately dissipated by more than one component of deformation. This occurs for example, in the problem of oblique impact on a structure (Sect. 6.3), where both normal and tangential components of initial momentum are brought to rest by the plastic dissipation at a generalized hinge (or hinges) that combines bending and stretching. The problems of curved or bent cantilevers subjected to out-of-plane loading also fall into this second group since here torque and bending moment are both significant in the plastically deforming section. Modal solutions that involve interactions between bending and twisting of curved members are dealt with in Sects 6.6 and 6.7.

6.1 General Considerations

6.1.1 Extremal Properties of Yield Function at Plastic Hinge

The generalized stresses $Q_\alpha = Q_\alpha(S)$ $(\alpha = 1,2,\cdots,\eta')$ that are functions of a spatial coordinate S are related by the yield function $\Psi(Q_\alpha)$. The yield condition $\Psi(Q_\alpha)$ is satisfied on at least one side of each plastic hinge $S = \Lambda_j$ $(j = 1,2,\cdots,\eta_{\mathrm{H}})$; however, it is not necessary that the yield condition is continuous at hinges. With regard to the differentiability of the yield function at a generalized plastic hinge, we can distinguish the following three cases.

1. If the yield function $\Psi(Q_\alpha)$ is a continuously differentiable function of S, in a neighborhood of an interior generalized plastic hinge, the function has a maximum at the hinge location $S = \Lambda_j$ $(0 < \Lambda_j < L)$, (Fig. 6.1a); consequently, the following conditions are satisfied:

$$\Psi(Q_\alpha(S))\Big|_{S=\Lambda_j} = 0 \tag{6.1a}$$

$$\left.\frac{\partial \Psi(Q_\alpha(S))}{\partial S}\right|_{S=\Lambda_j} = 0 \tag{6.1b}$$

$$\left.\frac{\partial^2 \Psi(Q_\alpha(S))}{\partial S^2}\right|_{S=\Lambda_j} \le 0 \tag{6.1c}$$

where (6.1c) requires $Q_\alpha(S) \in C^\mu$ $(\mu \ge 2)$; i.e. the generalized forces Q_α must have continuous second derivatives with respect to the spatial variable S.

2. If the yield function $\Psi(Q_\alpha)$ is continuous but *not* differentiable at an interior generalized plastic hinge located at $S = \Lambda_j$ $(0 < \Lambda_j < L)$, condition (6.1a) is maintained while the extremum conditions (6.1b) and (6.1c) are *no longer* valid (Fig. 6.1b); in this case one has

$$\Psi(Q_\alpha(S))\Big|_{S=\Lambda_j} = 0 \tag{6.2a}$$

$$\left.\frac{\partial \Psi(Q_\alpha(S))}{\partial S}\right|_{S=\Lambda_j-0} \ge 0, \qquad \left.\frac{\partial \Psi(Q_\alpha(S))}{\partial S}\right|_{S=\Lambda_j+0} \le 0 \tag{6.2b}$$

Here the derivatives ahead of and behind the hinge may not be equal to each other, since $Q_\alpha(S) \notin C^1$.

3. For a generalized plastic hinge at a boundary $(S = L)$ where displacements are constrained, the yield function Ψ is a nonanalytic extremum; hence conditions (6.1b) and (6.1c) are not necessary. As shown in Fig. 6.1c, at a displacement constraint,

$$\Psi(Q_\alpha(S))\Big|_{S=L} = 0 \tag{6.3a}$$

$$\left.\frac{\partial(Q_\alpha(S))}{\partial S}\right|_{S=L-0} \ge 0 \tag{6.3b}$$

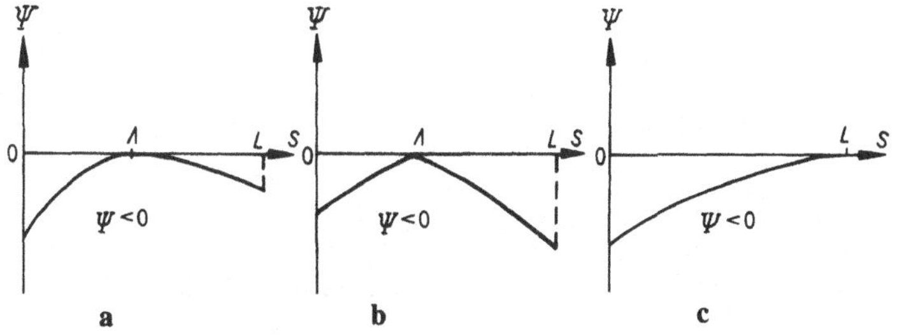

Fig. 6.1. Variation of yield function $\Psi(S)$ along cantilever: (a) at an interior generalized plastic hinge $\Psi(S)$ is a local extremum so $\Psi'(\Lambda)=0$; (b) $\Psi(\Lambda)=0$ but not differentiable; (c) hinge located at point of displacement constraint has $\Psi(L)=0$ but not an analytic extreme.

It must be noted, however, that each set of expressions, (6.1), (6.2) and (6.3), provides a different set of *local properties* for the yield function in a neighborhood of a plastic hinge located at $S=\Lambda_j$ $(0<\Lambda_j<L,\ j=1,\cdots,\eta_H)$; they do not guarantee that the yield criterion is not violated at other sections of the cantilever. Thus each of the above sets of equations form a set of *necessary but not sufficient conditions* for a deformation mechanism at an *interior* plastic hinge $S=\Lambda_j$ $(0<\Lambda_j<L)$ of a rigid–perfectly plastic cantilever. If this mechanism is a complete solution it satisfies the laws of motion, the yield condition and the flow law throughout the entire body. In cases 2 and 3, the spatial derivative of Ψ does not necessarily vanish at a plastic hinge, so the requirements for a complete solution stated in Chap. 2 must be applied directly to examine whether or not a mechanism is a complete solution. Conditions (6.2) or (6.3) can be employed as a check.

6.1.2 Differentiability of Arbitrary Function at Plastic Hinge

In order to construct a complete solution for a specified dynamic problem, attention must be paid to continuity conditions that are associated with the differentiability of the yield function at generalized plastic hinges. First let us consider arbitrary dependent variables that are functions of spatial coordinate S and time t, and examine their differentiability at plastic hinges.

If the μth order partial derivative of a function $F(S,t)$ with respect to S, $\partial^\mu F(S,t)/\partial S^\mu$, is continuous but not differentiable with respect to S in the range $0\le S\le L$ for $t\ge 0$, then the value of μ is defined as the *index* of differentiability of function $F(S,t)$ in $0\le S\le L$. This is indicated by the symbol $\langle F\rangle$; that is,

$$\mu \equiv \langle F(S,t)\rangle \tag{6.4}$$

The index of the differentiability of function $F(S,t)$ at any point in a lineal structure $0\le S\le L$ has the following properties:

1. If $F(S,t)\in C^\mu$ while $F(S,t)\notin C^{\mu+1}$, then $F'(S,t)\in C^{\mu-1}$ and $F'(S,t)\notin C^\mu$, where $F'(S,t)\equiv \partial F(S,t)/\partial S$. Hence, using the symbol $\langle\ \rangle$ for index of differentiability

$$\langle F'(S,t)\rangle = \langle F(S,t)\rangle - 1 \tag{6.5}$$

2. The second property concerns differentiability of a function $F(S,t)$ in a neighborhood of a generalized plastic hinge at $S=\Lambda$. If the jump of a function

vanishes at any time t, $[F(\Lambda,t)] \equiv F(\Lambda+,t) - F(\Lambda-,t) = 0$, then (refer to Sect. 4.3)

$$[\dot{F}(\Lambda,t)] + \dot{\Lambda}[F'(\Lambda,t)] = 0 \tag{6.6}$$

where $\dot{(\ )} \equiv \partial(\ )/\partial t$, $(\ )' \equiv \partial(\ )/\partial S$ and the speed of the travelling plastic hinge is $\dot{\Lambda}$. Thus, the following deductions can be made:

(a) For a stationary hinge $\dot{\Lambda} = 0$; thus Eq. (6.6) gives $[\dot{F}(\Lambda,t)] = 0$. Consequently

$$\langle \dot{F}(\Lambda,t) \rangle = \langle F(\Lambda,t) \rangle, \qquad \text{if} \quad \dot{\Lambda} = 0 \tag{6.7a}$$

(b) For a travelling hinge $\dot{\Lambda} \neq 0$, so $[\dot{F}(\Lambda,t)]$ and $[F'(\Lambda,t)]$ both are either zero or nonzero. It follows that

$$\langle \dot{F}(\Lambda,t) \rangle = \langle F'(\Lambda,t) \rangle = \langle F(\Lambda,t) \rangle - 1, \qquad \text{if } \dot{\Lambda} \neq 0 \tag{6.7b}$$

From Eqs (6.5) and (6.7b), it is seen that the index of differentiability of an arbitrary function could take a *negative* value. A negative value of μ, however, does not relate to the conventional meaning of the differentiability of a function; it is simply a result of the mathematical relationship (6.5) or (6.7b). Examples are given in Sect. 6.1.3.

3. Concerning the differentiability of the sum and the product of two functions $F_1(S,t)$ and $F_2(S,t)$, it can be proved that in general,

$$\langle F_1(S,t) + F_2(S,t) \rangle = \min\{\langle F_1(S,t) \rangle, \langle F_2(S,t) \rangle\} \tag{6.8a}$$

$$\langle F_1(S,t) \cdot F_2(S,t) \rangle = \min\{\langle F_1(S,t) \rangle, \langle F_2(S,t) \rangle\} \tag{6.8b}$$

6.1.3 Differentiability of Kinematic Variables at Plastic Hinge

In this section the differentiability of the kinematic variables that participate in a kinematically admissible velocity field is examined. This field represents a deformation mechanism with discrete generalized plastic hinges.

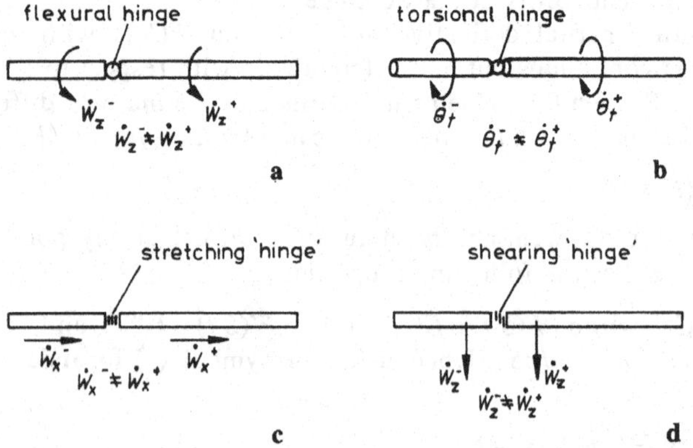

Fig. 6.2. Discontinuity at generalized plastic hinge: (a) flexural hinge — discontinuity of angular velocity related to transverse motion; (b) torsional hinge — discontinuity of angular velocity about the centroidal axis; (c) extensional hinge — discontinuity of axial velocity; (d) shear hinge — discontinuity of transverse velocity.

For ideal flexural hinges, as already discussed in Sect. 4.3, the discontinuity in angular velocity is a common and essential feature at a plastic hinge, no matter whether they are stationary or travelling; e.g. see Fig. 6.2a, where subscripts *plus* and *minus* pertain to the values at the right and left side of the hinge, respectively. Furthermore, a discontinuity in the axial component of angular velocity develops if a generalized plastic hinge involves torsional plastic deformation; see Fig. 6.2b.

However, if a generalized plastic hinge involves stretching or shear deformation, then there is a difference in translational velocity between the two sides of the hinge. Consequently, a discontinuity in axial or transverse (shearing) velocity develops at a stationary hinge if the length of the plastically deforming segment is negligible.

Let W_z and W_x denote the transverse and axial components of displacement, respectively. The following deductions can be made if the displacement field is continuous at an ideal plastic hinge:

1. For flexural plastic deformation, the transverse velocity must be continuous if transverse displacement is continuous so $\left[\dot{W}_z\right]=0$ but $\left[\dot{W}_z'\right]\neq 0$, thus

$$\left\langle \dot{W}_z \right\rangle = 0, \qquad \left\langle \dot{W}_z' \right\rangle = -1, \tag{6.9a}$$

$$\left\langle W_z \right\rangle = \begin{cases} 0, & \text{for stationary hinges (s.h.)} \\ 1, & \text{for travelling hinges (t.h.)} \end{cases} \tag{6.9b}$$

$$\left\langle \ddot{W}_z \right\rangle = \begin{cases} 0, & \text{(s.h.)} \\ -1, & \text{(t.h.)} \end{cases} \tag{6.9c}$$

2. For torsional plastic deformation, the differentiability of rotational deformation quantities is at least of the same order as those of flexure.

3. For axial plastic deformation, suppose $\left[\dot{W}_x\right]\neq 0$ so that

$$\langle \dot{W}_x \rangle = -1 \tag{6.10a}$$

$$\left\langle \ddot{W}_x \right\rangle = \begin{cases} -1, & \text{(s.h.)} \\ -2, & \text{(t.h.)} \end{cases} \tag{6.10b}$$

$$\left\langle W_x \right\rangle = \begin{cases} -1, & \text{(s.h.)} \\ 0, & \text{(t.h.)} \end{cases} \tag{6.10c}$$

From a physical point of view the axial displacement W_x has to be continuous at plastic hinges, so that (6.10c) excludes any deformation mechanism with a *stationary* plastic hinge that involves axial plastic deformation; however, a deformation mechanism with both flexural and axial plastic deformation at a travelling hinge remains acceptable.

4. For continuous displacement the plastic shear deformation $\left[W_z\right]=0$ but $\left[\dot{W}_z\right]\neq 0$, so that

$$\left\langle \dot{W}_z \right\rangle = -1 \tag{6.11a}$$

$$\left\langle \ddot{W}_z \right\rangle = \begin{cases} -1, & \text{(s.h.)} \\ -2, & \text{(t.h.)} \end{cases} \tag{6.11b}$$

Similar to the case of axial plastic deformation, plastic shear at a stationary hinge also results in discontinuous transverse displacement.

6.1.4 Differentiability of Generalized Stresses at Plastic Hinge

The differentiability of generalized stresses is related to the differentiability of kinematic variables that participate in the dynamic equations. In the case of impulsive loading, typical dynamic equations in the transverse and axial directions are

$$Q' = \rho \ddot{W}_z \tag{6.12a}$$

$$M' = -Q \tag{6.12b}$$

$$N' = \rho \ddot{W}_x \tag{6.12c}$$

where $\rho = \rho(S)$ is the density per unit length of the cantilever, which may vary with S. Note that a strong or weak discontinuity of ρ is caused by a discontinuity of cross-section (e.g. a stepped beam) or by a discontinuity in material density per unit length (e.g. a composite beam consisting of two beams of different materials that are butted together).

According to the property (6.8b) the differentiabilities of Q and N depend on those of $\ddot{W}_z$ and $\ddot{W}_x$, as well as the differentiability of ρ. For instance, from (6.12a),

$$\langle Q' \rangle = \min\{\langle \rho(S) \rangle, \langle \ddot{W}_z \rangle\} \tag{6.13}$$

where $\langle \ddot{W}_z \rangle$ is found from (6.9c).

In the case of a homogeneous beam of uniform cross-section, (6.12) leads to

$$\langle Q \rangle = \langle \ddot{W}_z \rangle + 1 \tag{6.14a}$$

$$\langle M \rangle = \langle Q \rangle + 1 = \langle \ddot{W}_z \rangle + 2 \tag{6.14b}$$

$$\langle N \rangle = \langle \ddot{W}_x \rangle + 1 \tag{6.14c}$$

It should be noted that the rotary inertia of beam differential elements is disregarded in dynamic equation (6.12b). If this effect is considered, that equation becomes

$$M' = -Q + I_{ox} \ddot{\theta}_m \tag{6.12b'}$$

where I_{ox} is the second moment of mass per unit length about a transverse axis through the centroid of the cross-section (in general, I_{ox} may vary with S if the cantilever is not uniform); θ_m is the usual rotation angle of the beam center line due to bending alone. The total inclination of the center line W_z' and the shear angle (i.e. transverse shear strain) $\tilde{\gamma}$ are related by

$$W_z' = \tilde{\gamma} + \theta_m \tag{6.15}$$

Hence, when the rotary inertia of beam differential elements is incorporated in the dynamic equations, the differentiabilities of M and Q will be mainly determined by the angular acceleration term $I_{ox} \ddot{\theta}_m$:

$$\langle M \rangle = \langle \ddot{\theta}_m \rangle + 1 = \langle \ddot{W}_z' \rangle + 1 = \langle \ddot{W}_z \rangle \tag{6.16a}$$

$$\langle Q \rangle = \langle M \rangle - 1 = \langle \ddot{W}_z \rangle - 1 \tag{6.16b}$$

Comparison of Eq. (6.16) with (6.14) shows that for the same degree of smoothness in the displacement field, bending moment and shear force are two orders less differentiable when rotary inertia is considered.

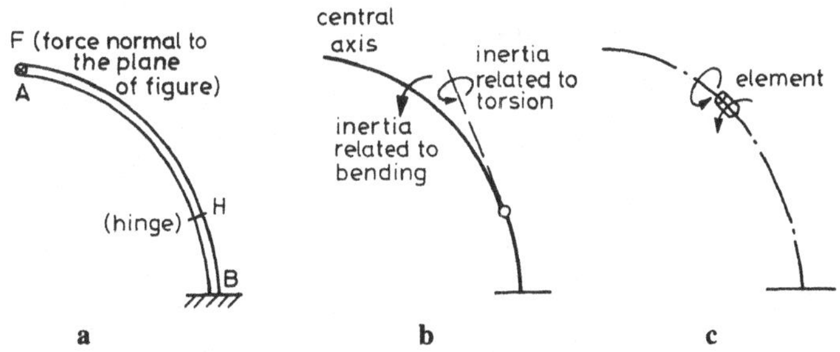

Fig. 6.3. (a) Curved cantilever subjected to transverse force at tip. The force is normal to the plane of curvature; (b) rotary inertia of curved cantilever about tangent line through hinge H if all mass is concentrated at center line; (c) rotary inertia of beam element with respect to longitudinal axis.

In general, if a crooked or curved thin beam is subjected to bending and/or torsion, as sketched in Fig. 6.3, and a plastic hinge forms at section H, then the rotary inertia of the beam segment between the tip and the hinge, AH, consists of two parts. The first part comes from the second moment of mass of segment AH about the centroid of the hinge section H, as illustrated in (Fig. 6.3b); this is due to the out-of-straightness of the member. The second part comes from the polar second moment of mass in a cross-section about the centroid (Fig. 6.3c). The latter is called the *rotary inertia of the beam element about the longitudinal axis of the beam*. If a crooked or curved beam is slender, the second part of the rotary inertia will be much smaller than the first part. In most problems this second part is disregarded in the same way that transverse rotary inertia is usually disregarded.

Bending–torsion problems of curved and bent cantilevers have dynamic equations that are obtained in Sects 6.6 and 6.7. In these sections it is confirmed that negligible polar second moment of inertia for the cross-section about the central axis results in better differentiability for torque rather than for bending moment,

$$\langle T \rangle = \langle M \rangle + 1 \tag{6.17}$$

If the rotary inertia of a beam element about the central axis of the beam is taken into account, however, the differentiability of torque T will be similar to that of bending moment M as given in (6.16a). This is because the torque is related to the polar second moment of mass about the central axis of the element in a way similar to that of transverse rotary inertia (see Eq. (6.12b′)).

6.1.5 Differentiability of Yield Function at Plastic Hinge

If the yield function is an analytic function of the generalized stresses $Q_\alpha(S,t)$ $(\alpha = 1,2,\cdots,\eta')$, i.e. $\Psi(Q_\alpha) \in C^\mu$ where μ is an arbitrary positive number, then at generalized plastic hinges the yield function has the same differentiability as that of the generalized stresses appearing in Ψ:

$$\langle \Psi(Q_\alpha) \rangle = \min_{\alpha=1,\cdots,\eta'} \langle Q_\alpha(S,t) \rangle \tag{6.18}$$

Some yield surfaces $(\Psi = 0)$ have singular points where the yield function $\Psi(Q_\alpha)$ is continuous but not continuously differentiable; that is, $\Psi(Q_\alpha) \in C^\mu$ with $\mu < 1$. If the stress state at a hinge is at one of these singular points, then at the hinge the differentiability of the yield function is

Table 6.1 Index of differentiability of various quantities for continuous displacements at a generalized plastic hinge

		Stationary hinge $\Lambda = 0$	Travelling hinge $\Lambda \neq 0$
Bending only			
Deflection	W_z	0	1
Velocity	$\dot{W}_z$	0	0
Acceleration	$\ddot{W}_z$	0	–1
Slope or inclination	$W_z' = \theta_m$	–1	0
Angular velocity	$\dot{W}_z' = \dot{\theta}_m$	–1	–1
Angular acceleration	$\ddot{\theta}_m$	–1	–2
Curvature	$W_z'' = \kappa$	–2	–1
Shear force	Q	1	0
Bending moment	M^a	2	1
Bending moment	M^b	0	–1
Bending + twist			
Rotational angle	θ_t	–1	0
Rotational velocity	$\dot{\theta}_t$	–1	–1
Rotational acceleration	$\ddot{\theta}_t$	–1	–2
Torque	T^a	3	2
Yield function	$\Psi(M,T)^a$	2	1
Torque	T^b	0	–1
Yield function	$\Psi(M,T)^b$	0	–1
Bending + stretching			
Axial displacement	W_x	–	0
Axial velocity	$\dot{W}_x$	–	–1
Axial acceleration	$\ddot{W}_x$	–	–2
Axial force	N	–	–1
Yield function	$\Psi(M, N)$	–	–1
Bending + shear			
Shear displacement	W_z	–1	0
Shear velocity	$\dot{W}_z$	–1	–1
Shear acceleration	$\ddot{W}_z$	–1	–2
Shear force	Q	0	–1
Yield function	$\Psi(M, Q)^a$	0	–1
Shear force	Q^b	–1	–2
Yield function	$\Psi(M, Q)^b$	–1	–2

[a] Effect of rotary inertia of beam elements is disregarded.
[b] Effect of rotary inertia of beam elements is considered.

$$\langle \Psi(Q_\alpha) \rangle = 0 \tag{6.19}$$

and the extremal conditions (6.1b) and (6.1c) for an interior plastic hinge are not valid. For example, if a maximum normal stress criterion is employed for a beam of rectangular cross-section that is simultaneously subjected to bending and stretching, then the yield function is

$$\Psi(M,N) = \frac{|M|}{M_p} + \left(\frac{N}{N_p}\right)^2 - 1 \tag{6.20}$$

Consequently, at points where the yield function is analytic (i.e. smooth),

$$\langle \Psi(M,N) \rangle = \min\langle |M(S,t)|, N(S,t) \rangle \tag{6.21a}$$

while at singular points on the yield surface (e.g. $M = 0$, $N = \pm N_p$ of (6.20)) we obtain

$$\langle \Psi(M,N) \rangle = 0 \tag{6.21b}$$

An example of a beam with an interaction yield function $\Psi(M,Q)$ subjected to simultaneous bending and shear was examined by Symonds [1968]. His conclusions regarding differentiability of the yield function were similar to (6.21).

In general, at any nonsingular point of the yield surface, the differentiability of the yield function $\Psi(Q_\alpha)$ is determined by Eq. (6.18); that is, it depends on differentiability of the participating generalized stresses Q_α $(\alpha = 1, 2, \cdots, \eta')$. For instance, in the case of bending–stretching interaction, the yield condition has the index given in (6.21a). Equation (6.10c) excludes any deformation mechanism involving axial deformation at a stationary plastic hinge; hence combined flexural and axial plastic deformations can occur only at a travelling hinge. At a travelling hinge, the combination of (6.14b) and (6.9c) gives $\langle M \rangle = 1$, while the combination of (6.14c) and (6.10b) leads to $\langle N \rangle = -1$. Thus, $\langle \Psi(M,N) \rangle = \min\{\langle M \rangle, \langle N \rangle\} = -1$.

Table 6.1 is obtained by this process of deduction; the table summarizes the differentiability conditions which components of deformation must satisfy in order to give continuous displacements. The table lists the minimum requirements for indices of differentiability based on the assumptions: (1) the density per unit length $\rho = \rho(S)$ is a smooth function of S; and (2) only nonsingular points on the yield surface are considered. These requirements are necessary to obtain continuous displacements if the flow law is associated with the yield condition.

From Table 6.1 it can be seen that the yield function is only an analytic extreme at a plastic hinge for the cases of pure bending and bending–torsion and then only if there is negligible rotary inertia of the beam element about a transverse axis. There is no necessity, however, for the yield function to be an analytic extreme at a plastic hinge if the yield function Ψ is not continuous at this location.

Discontinuities in stress resultants (and thus the yield function) at a travelling hinge in a rigid-perfectly plastic beam are considered in Sect. 6.3. Prager [1954] has termed such discontinuities as 'artificial' rather than natural since acceleration of mass passing through the travelling hinge is caused by stress jumps that are not along the yield surface – rather they are from an interior point to the yield surface. The analysis in Sect. 6.3 assumes that at a travelling hinge, axial strain develops from these jumps in axial stress and particle velocity.

6.2 Straight Cantilevers With Smoothly Varying Cross-Sections

6.2.1 Yield Function and Conditions at Plastic Hinge

An analysis of an initially straight rigid–perfectly plastic beam with smoothly varying cross-sectional properties of mass per unit length $\rho(X)$, second moment of area $I(X)$ and fully plastic bending moment $M_p(X)$, is a first step towards generalizing the previous results for application to more complex structural configurations. In this section we consider members with cross-sectional properties that vary smoothly; i.e. the properties are parameters that can be differentiated with respect to the spatial coordinate X at least twice. Later in this chapter (Sect. 6.7), examples of members with discontinuous cross-sectional properties are considered. In this section our aim is to identify consequences of the rigid–perfectly plastic constitutive approximation in more complex configurations.

Only flexural deformations are considered so the yield function is

$$\Psi(X,t) = \frac{|M(X,t)|}{M_p(X)} - 1 \tag{6.22}$$

where the fully plastic moment $M_p(X)$ is a prescribed positive function of the spatial coordinate X, and the index of differentiability (defined in Sect. 6.1) is $\langle M_p(X)\rangle \geq 2$. According to the discussion in Sect. 6.1, if the cross-section properties have continuous derivatives at every point, then continuity of displacements requires the dynamic moment distribution $M(X,t)$ to have an index of differentiability that is no less than $\langle M(X,t)\rangle = 2$ or 1 at stationary or travelling hinges, respectively. At a plastic hinge, therefore, the yield function must satisfy certain limiting conditions given by Eqs (6.1) or (6.3).

1. At a plastic hinge located at $X = \Lambda$ in the interior of the beam ($0 < \Lambda < L$) the bending moment satisfies

$$M(\Lambda) = M_p(\Lambda)\ \operatorname{sgn}(M) \tag{6.23a}$$

$$\frac{\partial M(\Lambda)}{\partial X} = \frac{\mathrm{d}M_p(\Lambda)}{\mathrm{d}X}\ \operatorname{sgn}(M) \tag{6.23b}$$

2. At a plastic hinge located at the root $\Lambda = L$ where the cross-section properties are continuous, the bending moment $M(L) = \lim_{\varepsilon \to 0} M(L-\varepsilon)$ satisfies

$$M(L) = M_p(L)\ \operatorname{sgn}(M) \tag{6.24a}$$

$$\frac{\partial M(L)}{\partial X} = \frac{\mathrm{d}M_p(L)}{\mathrm{d}X}\ \operatorname{sgn}(M) \tag{6.24b}$$

In particular, note that at a travelling hinge in the interior of the beam, the shear force $Q(\Lambda) = \partial M(\Lambda)/\partial X \neq 0$ if the fully plastic moment $M_p(X)$ is not uniform along the beam. This is in contradiction to an assumption made in some previous analyses (e.g. Al-Hassani et al. [1973]). The conditions above were pointed out first by Zhou and Yu [1987] who provided analyses of dynamic response of tapered cantilevers subjected to either a suddenly applied force or impact by a rigid body.

6.2.2 Suddenly Applied Steady Force at Tip of Tapered Cantilever: An Example

Deformation Mechanism and Equations of Motion Consider a cantilever of variable cross-section subjected to step loading of magnitude F suddenly applied at the tip, (see Fig. 6.4a). Recalling the deformation mechanism for a cantilever of uniform cross-section found in Sect. 4.1, we begin by assuming that a *stationary*

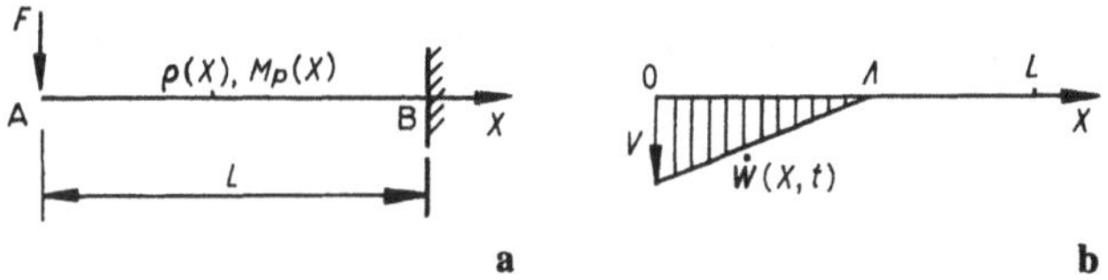

Fig. 6.4. (a) Cantilever of variable cross-section subjected to suddenly applied transverse step force F at the tip; (b) velocity field of a single hinge mechanism.

plastic hinge forms at an interior cross-section $X = L$. The kinematically admissible velocity field related to this mechanism is shown in Fig. 6.4b, i.e.

$$\dot{W} = \begin{cases} V(1 - X/\Lambda), & 0 \le X \le \Lambda \\ 0, & \Lambda \le X \le L \end{cases} \tag{6.25}$$

where V is the velocity at the tip of the cantilever. Accordingly, the acceleration field is

$$\ddot{W} = \begin{cases} \dot{V}(1 - X/\Lambda), & 0 \le X \le \Lambda \\ 0, & \Lambda \le X \le L \end{cases} \tag{6.26}$$

The distributions of bending moment and shear force can be found from the equations of motion for a segment of length X that begins at the tip,

$$M(X) = \begin{cases} FX - \int_0^X \rho(\xi)\dot{V}(1 - \xi/\Lambda)(X - \xi)\,d\xi, & 0 \le X \le \Lambda \\ FX - \int_0^\Lambda \rho(\xi)\dot{V}(1 - \xi/\Lambda)(X - \xi)\,d\xi, & \Lambda \le X \le L \end{cases} \tag{6.27}$$

$$Q(X) = -\frac{dM}{dX} = \begin{cases} -F + \int_0^X \rho(\xi)\dot{V}(1 - \xi/\Lambda)\,d\xi, & 0 \le X \le \Lambda \\ -F + \int_0^\Lambda \rho(\xi)\dot{V}(1 - \xi/\Lambda)\,d\xi, & \Lambda \le X \le L \end{cases} \tag{6.28}$$

It is seen that both bending moment $M(X)$ and shear force $Q(X)$ are independent of time t, since the plastic hinge is stationary.

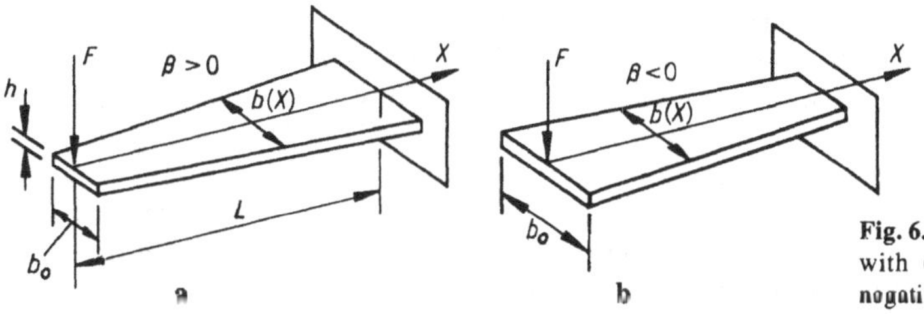

Fig. 6.5. Tapered cantilever with (a) positive β ; (b) negative β.

Solution for a Tapered Cantilever Now let us suppose that the cantilever has uniform density per unit volume, constant thickness h and a width that varies linearly from b_o at the tip to $b_o(1+\beta)$ at the root,

$$b(X) = b_o\left(1+\beta\frac{X}{L}\right) \tag{6.29a}$$

A positive width $b(X)>0$ at sections X in the range $0 \le X \le L$ requires $\beta > -1$. Tapered cantilevers with either positive or negative β are sketched in Fig. 6.5a, b, respectively. For this uniformly varying section, the density per unit length and the fully plastic bending moment $M_p(X)$ can be found in relation to the respective values at the tip as

$$\rho(X) = \rho_o\left(1+\beta\frac{X}{L}\right) \tag{6.29b}$$

$$M_p(X) = M_o\left(1+\beta\frac{X}{L}\right) \tag{6.29c}$$

From Table 6.1 and Eq. (6.29c) it is seen that the present example has a yield function (6.22) with an index of differentiability $\langle\Psi\rangle \ge 2$, so the bending moment at a hinge is governed by Eq. (6.23). Substituting (6.29b) into (6.27) and (6.28) and using (6.23) results in the following expressions for moment of momentum about the tip and transverse momentum,

$$F\Lambda - \frac{1}{12}\rho_o\dot{V}\Lambda^2\left(4+\frac{\beta\Lambda}{L}\right) = M_p(\Lambda) = M_o\left(1+\frac{\beta\Lambda}{L}\right) \tag{6.30a}$$

$$F - \frac{1}{6}\rho_o\dot{V}\Lambda\left(3+\frac{\beta\Lambda}{L}\right) = \frac{\partial M_p(\Lambda)}{\partial X} = \frac{\beta M_o}{L} \tag{6.30b}$$

By eliminating $\rho_o\dot{V}$ from these equations and defining

$$f \equiv \frac{FL}{M_o}, \qquad \lambda \equiv \frac{\Lambda}{L} \tag{6.31}$$

one finds that the hinge location λ satisfies a quadratic equation

$$\beta(f-\beta)\lambda^2 + 2(f-2\beta)\lambda - 6 = 0 \tag{6.32}$$

Consequently, the nondimensional hinge position is

$$\lambda \equiv \frac{\Lambda}{L} = \frac{1}{\beta(f-\beta)}\left(-f+2\beta+\sqrt{f^2+2f\beta-2\beta^2}\right) \tag{6.33}$$

If a plastic hinge is located at an interior cross-section of a tapered cantilever, $0<\lambda<1$, the diagrams of shear force and bending moment along the cantilever are similar to those sketched in Fig. 6.6, where (a) and (b) illustrate cases of $\beta>0$ and $\beta<0$, respectively. The case of $\beta=0$ represents the usual cantilever of uniform cross-section. Figure 6.6 shows for cantilevers of nonuniform cross-section ($\beta \ne 0$), shear force does not vanish at the plastic hinge; correspondingly, the bending moment curve $M(X)$ is not a maximum at the hinge point $X=\Lambda$, but it is tangential to the fully plastic moment curve $M_p(X)$. Thus 'zero shear force' is not a general condition for determining the location of a plastic hinge; this condition applies only to uniform beams.

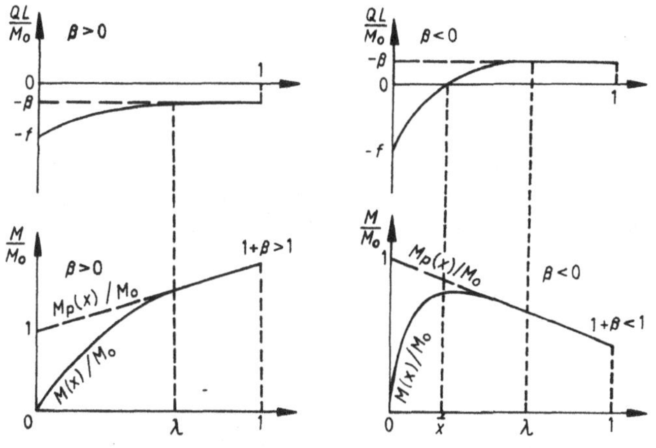

Fig. 6.6. Distributions of shear force and bending moment along cantilever for (a) positive β; (b) negative β.

In a tapered cantilever, a fully plastic segment of the beam $X > \Lambda$ supports a limited shear force without accelerating. In this respect, this kinetic solution is similar to the solution for a structure subjected to a suddenly applied force equal in magnitude to the static collapse force F_c. In both cases the state of stress yields fully plastic sections and a mechanism of collapse but the structure does not accelerate unless $F > F_c$.

Location of Plastic Hinge as Function of Force Magnitude Expression (6.30a) indicates that positive tip acceleration $\dot{V} > 0$ appears if and only if the force is larger than a collapse force,

$$\frac{FL}{M_0} \equiv f \geq f_1(\beta) \equiv 1 + \beta \tag{6.34}$$

If $f < f_1(\beta)$, the tapered cantilever remains static; if f slightly exceeds $f_1(\beta)$, a plastic hinge appears at the root of the cantilever; i.e. at $\lambda = 1$.

If the applied force is only slightly larger than the plastic collapse force the hinge remains at the root. The plastic hinge moves inboard of the root if the value of λ given by Eq. (6.33) is less than unity. Thus, the minimum force for an interior hinge is

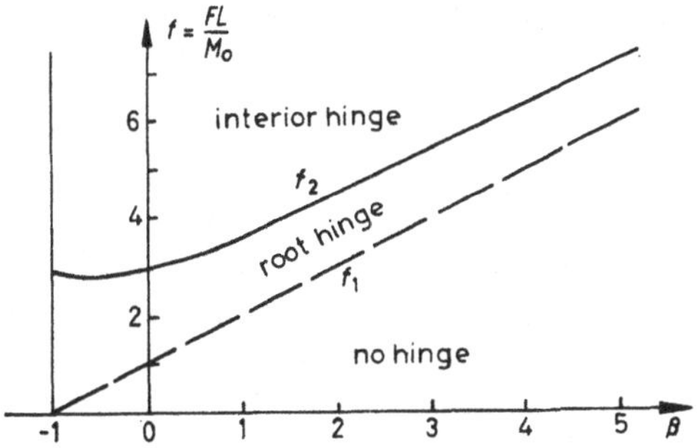

Fig. 6.7. Occurrence of three possible deformation mechanisms depends on the parameters β and f.

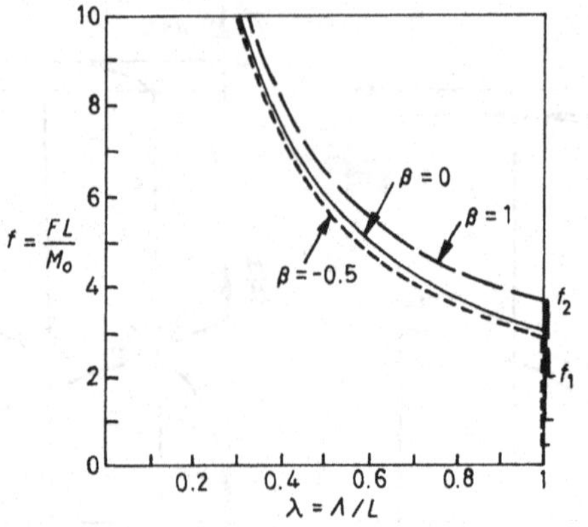

Fig. 6.8. Dependence of plastic hinge location on step force f, for various values of the taper parameter β.

$$\frac{FL}{M_0} \equiv f > f_2(\beta) \equiv 3 + \frac{\beta(1+\beta)}{2+\beta} \tag{6.35}$$

Figure 6.7 gives a map of the (β, f) plane divided into three regions by curves $f_1(\beta)$ and $f_2(\beta)$: (1) if $f < f_1(\beta) \equiv 1+\beta$ no hinge forms; (2) if $f_1(\beta) \leq f \leq f_2(\beta)$ a hinge forms at the root; and (3) if $f > f_2(\beta)$ a hinge forms at an interior cross-section.

The dependence of the plastic hinge location on the magnitude of the step force is shown in Fig. 6.8. It is seen that for tapered cantilevers, the taper ratio β does not significantly influence the plastic hinge location $\lambda(\tau)$.

6.3 Oblique Impact on Straight Cantilever

6.3.1 Problem and Assumptions

In many practical cases, impact loading on a structure is not entirely transverse, i.e. it involves both bending and axial stretching. For a perfectly plastic structure with a yield condition that couples bending moment and axial tension, a model with discrete travelling hinges which combine bending and stretching deformations cannot satisfy both the flow rule and yield condition throughout the deforming region. A similar problem arises in transversely loaded beams with clamped end supports due to the axial constraints at the ends of these beams. Axial stretching from deflection between the fixed ends, however, results in negligibly small axial accelerations. Hence Symonds and Mentel [1958] assumed that in these beams, the simultaneous stretching and bending that occur at plastic hinges are only kinematically related. For oblique impact on a cantilever where axial motion is not constrained, Shu et al. [1992] reported on experiments which showed that the axial disturbance decays quickly in an early part of the transient response period. Subsequently the response is almost entirely flexural and similar to that caused by the transverse component of the initial momentum. To represent this behaviour,

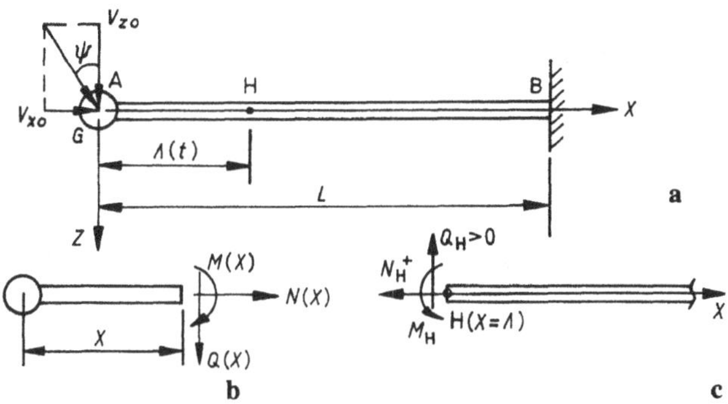

Fig. 6.9. (a) A single 'hinge' mechanism for oblique impact on a cantilever; (b) positive senses of axial force N, shear force Q and bending moment M; (c) free-body diagram of segment HB ($X > \Lambda$) when $M_H > 0$, $Q_H > 0$.

we assume that deformation occurs at a discrete travelling plastic hinge but require that yield is satisfied on one side only of the (infinitesimal) deforming region.

Let a straight cantilever of uniform rectangular cross-section with length L and density ρ per unit length be subjected to oblique impact on a heavy particle attached to the tip, as shown in Fig. 6.9a. The particle has mass G. After impact the initial velocity of mass G has a transverse component V_{zo} and an axial component V_{xo}. Apart from the heavy particle at the tip, the cantilever is initially at rest. In the present analysis, the following assumptions are made.

1. The structure is composed of a rigid–perfectly plastic, rate-independent material.
2. The effect of shear force on yield is neglected, so the yield criterion involves only two generalized stresses, i.e. bending moment M and axial force N. The member has a rectangular cross-section with width b and thickness h, so the yield function at any cross-section X is

$$\Psi = \Psi(M,N) \equiv \left|\frac{M}{M_p}\right| + \left(\frac{N}{N_p}\right)^2 - 1 \tag{6.36}$$

where $M_p = Ybh^2/4$ and $N_p = Ybh$ are the fully plastic bending moment and fully plastic axial force, respectively.
3. Deflections are small so that all geometrical relations are measured relative to the initial configuration. Furthermore, the additional bending moment (beam-column effect) caused by the axial force has been disregarded.

6.3.2 Formulation Based on Single–Hinge Mechanism

Deformation Mechanism and Kinematically Admissible Field Let X be an axial coordinate measured from the tip and Z be a transverse coordinate that is orthogonal to X and in-plane with the initial velocity (V_{xo}, V_{zo}), as shown in Fig. 6.9a. Suppose the deformation mechanism has a single hinge. At time t after impact, the hinge is located a distance $X = \Lambda$ from the impact point at the tip; at this instant the rigid segment AH is rotating about H while segment HB remains at rest. In addition to rigid body rotation about hinge H, segment AH also has axial motion in response to the axial momentum imparted initially by the axial velocity

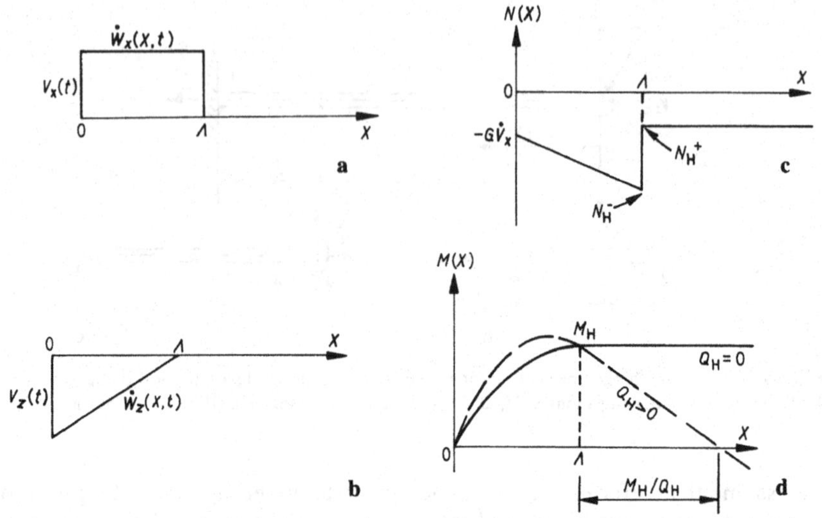

Fig. 6.10. Distributions of (a) axial velocity; (b) transverse velocity; (c) axial force; (d) bending moment.

V_{xo} of G. This axial motion requires axial plastic deformation and this occurs only at the generalized hinge H; hence, the rotating segment AH has uniform axial speed $\dot{W}_x$. Therefore, the axial and transverse velocity fields related to this single–hinge mechanism are

$$\dot{W}_x(X,t) = \begin{cases} V_x(t), & 0 \le X \le \Lambda \\ 0, & \Lambda \le X \le L \end{cases} \tag{6.37}$$

$$\dot{W}_z(X,t) = \begin{cases} -\dot{\theta}_m(\Lambda - X) = V_z(t)\left(1-\dfrac{X}{\Lambda}\right), & 0 \le X \le \Lambda \\ 0, & \Lambda \le X \le L \end{cases} \tag{6.38}$$

Here, $W_x(X,t)$ and $W_z(X,t)$ denote the axial and transverse displacements, respectively; $\dot{\theta}_H(t) = V_z(t)/\Lambda$ is the angular velocity of segment AH rotating about hinge H; $V_x(t)$ and $V_z(t)$ are the axial and transverse velocity components, respectively, of the concentrated mass at the tip, as illustrated in Fig. 6.10a and b.

Equation (6.10c) shows that axial plastic deformation can occur only at a *travelling* hinge; in other words, if continuity of displacements requires that axial velocity is not uniform, axial stretching occurs only at concentrated travelling 'hinges' moving at speed $\dot{\Lambda}(t) \neq 0$. By differentiating (6.37) and (6.38), the axial and transverse accelerations are

$$\ddot{W}_x(X,t) = \begin{cases} \dot{V}_x(t), & 0 \le X \le \Lambda \\ 0, & \Lambda \le X \le L \end{cases} \tag{6.39}$$

$$\ddot{W}_z(X,t) = \begin{cases} \dot{V}_z(t)\left(1-\dfrac{X}{\Lambda}\right) + V_z(t)\dot{\Lambda}(t)\dfrac{X}{\Lambda^2}, & 0 \le X \le \Lambda \\ 0, & \Lambda \le X \le L \end{cases} \tag{6.40}$$

respectively. Note that both axial and transverse accelerations have discontinuities at hinge $X = \Lambda$; the transverse acceleration is similar to that shown in Fig. 4.23c.

Equations of Motion We assume that dissipation due to shear is negligible so that the generalized stresses which are active in the yield condition are only bending moment $M(X)$ and axial force $N(X)$. The positive senses of these stress resultants is defined in Fig. 6.9b. Then by making use of the transverse acceleration given by (6.40), the equations of motion of segment AH in the transverse direction can be written as

$$\frac{\mathrm{d}}{\mathrm{d}t}\left(\frac{1}{2}\rho\Lambda V_z\right) = -G\frac{\mathrm{d}V_z}{\mathrm{d}t} + Q_{\mathrm{H}} \tag{6.41}$$

$$\frac{\mathrm{d}}{\mathrm{d}t}\left(\frac{1}{6}\rho\Lambda^2 V_z\right) = M_{\mathrm{H}} + Q_{\mathrm{H}}\Lambda \tag{6.42}$$

Where Q_{H} is the shear force at hinge H. Note that both shear force $Q(X)$ and bending moment $M(X)$ are continuous at the hinge since shear deformation and rotary inertia of beam elements are disregarded, see Table 6.1.

Similarly, by making use of the axial acceleration given by (6.39), the equations of motion of segment AH in the axial direction can be written as

$$(G+\rho\Lambda)\frac{\mathrm{d}V_x}{\mathrm{d}t} = N_{\mathrm{H}}^- \tag{6.43}$$

$$\frac{\mathrm{d}}{\mathrm{d}t}\{(G+\rho\Lambda)V_x\} = N_{\mathrm{H}}^+ \tag{6.44}$$

where $N_{\mathrm{H}}^- = N(\Lambda - 0, t)$ and $N_{\mathrm{H}}^+ = N(\Lambda + 0, t)$ denote the axial forces on left and right sides, respectively, of the plastic hinge H. Consequently, at the moving hinge the axial force is discontinuous,

$$N_{\mathrm{H}}^+ = N_{\mathrm{H}}^- + \rho V_x \frac{\mathrm{d}\Lambda}{\mathrm{d}t} \tag{6.45}$$

At the travelling hinge this difference in axial force accelerates added mass to the axial speed of the tip segment.

For initial conditions shown in Fig. 6.9a, during the first phase of motion the axial velocity V_x satisfies

$$V_x(t) \geq 0, \qquad \frac{\mathrm{d}V_x(t)}{\mathrm{d}t} \leq 0, \qquad \frac{\mathrm{d}\Lambda}{\mathrm{d}t} > 0 \tag{6.46}$$

Consequently, (6.43) and (6.45) provide

$$N_{\mathrm{H}}^- < N_{\mathrm{H}}^+, \qquad N_{\mathrm{H}}^- < 0 \tag{6.47}$$

The distribution of axial force along the cantilever for the case of $N_{\mathrm{H}}^+ < 0$ is as sketched in Fig. 6.10c. It will become clear that in the early stage of motion while $\Lambda \ll L$, the hinge speed $\mathrm{d}\Lambda/\mathrm{d}t$ is large enough that the axial force N_{H}^+ just ahead of the hinge which is given by (6.44) could be positive (i.e. a tensile force); it can be so large that it exceeds the fully plastic axial force N_p. This is an unacceptable consequence of rigid–perfectly plastic modeling; it can be removed by taking the initial position of the travelling hinge a short distance away from the tip. This small distance is chosen such that the following relation is always satisfied:

$$\left|N_{\text{H}}^{-}\right| \geq \left|N_{\text{H}}^{+}\right| \tag{6.48}$$

Yield Criterion and Flow Rule By substituting M_{H}, N_{H}^{-} and N_{H}^{+} into the yield function Ψ given by (6.36), one finds that function Ψ takes different values on the left and right sides of the hinge. For segment AH with length that satisfies (6.48), however, the value of Ψ is algebraically larger on the side of the hinge closer to the impact point, H^{-}. Thus, the yield criterion is satisfied *behind the travelling plastic hinge*; this can be written as

$$\Psi(\Lambda) = \left|\frac{M_{\text{H}}}{M_p}\right| + \left(\frac{N_{\text{H}}^{-}}{N_p}\right)^2 - 1 = 0 \tag{6.49}$$

The discontinuities in rate-of-rotation and axial velocity at the travelling hinge are related by the flow rule associated with yield criterion (6.49)

$$\frac{V_x}{-\dot{\theta}_m} = \frac{h\left|N_{\text{H}}^{-}\right|}{2N_p} = -\frac{hN_{\text{H}}^{-}}{2N_p}, \qquad \text{if } N_{\text{H}}^{-} > -N_p \tag{6.50a}$$

$$\frac{V_x}{-\dot{\theta}_m} \geq \frac{h}{2}, \qquad \text{if } N_{\text{H}}^{-} = -N_p \tag{6.50b}$$

so that

$$V_x = -\frac{hN_{\text{H}}^{-}V_z}{2N_p\Lambda}, \qquad \text{if } N_{\text{H}}^{-} > -N_p \tag{6.51a}$$

$$V_x \geq -\frac{hV_z}{2\Lambda}, \qquad \text{if } N_{\text{H}}^{-} = -N_p \tag{6.51b}$$

The equations of motion and kinematic conditions (6.41)–(6.44), (6.49) and (6.51) provide *six* equations for *seven* unknown functions of time, i.e. $V_x(t)$, $V_z(t)$, $\Lambda(t)$, $N_{\text{H}}^{-}(t)$, $N_{\text{H}}^{+}(t)$, $M_{\text{H}}(t)$ and $Q_{\text{H}}(t)$. One more condition is needed to completely determine the dynamic response of the cantilever after oblique impact.

Limitations on Shear Force at Hinge As shown above, the yield function Ψ given by (6.36) takes its largest value ($\Psi = 0$) at the generalized hinge. However, this does not ensure that Ψ is an analytic extreme at $X = \Lambda$, because according to Table 6.1 for (M, N) interaction the yield function $\Psi(X)$ is not continuous at the hinge. Nevertheless, the yield condition requires either $\lim_{\varepsilon \to 0} \mathrm{d}\Psi(\Lambda - \varepsilon)/\mathrm{d}X \geq 0$ or $0 \geq \lim_{\varepsilon \to 0} \mathrm{d}\Psi(\Lambda + \varepsilon)/\mathrm{d}X$. For the yield function (6.36) these inequalities lead to

$$\frac{\partial}{\partial X}\Psi(X) = \text{sgn}(M)\frac{1}{M_p}\frac{\partial M(X)}{\partial X} + \frac{2N(X)}{N_p^2}\frac{\partial N(X)}{\partial X} \geq 0, \text{ at } X = \Lambda - \tag{6.52a}$$

$$\frac{\partial}{\partial X}\Psi(X) = \text{sgn}(M)\frac{1}{M_p}\frac{\partial M(X)}{\partial X} + \frac{2N(X)}{N_p^2}\frac{\partial N(X)}{\partial X} \leq 0, \text{ at } X = \Lambda + \tag{6.52b}$$

The choice between these two conditions depends on whether the yield condition is satisfied ahead of $(\Lambda +)$ or behind $(\Lambda -)$ the plastic hinge. It will be confirmed that $M(X) > 0$ holds for most of the cantilever in the present problem so $\text{sgn}(M) = 1$. From equilibrium of the beam element and continuity of shear force at the hinge,

$$\left.\frac{\partial M(X)}{\partial X}\right|_{X=\Lambda-} = \left.\frac{\partial M(X)}{\partial X}\right|_{X=\Lambda+} = -Q(\Lambda) \equiv -Q_{\mathrm{H}} \tag{6.53}$$

Combining (6.52), (6.53) and $M_p/N_p = h/4$ gives

$$\left.\frac{h}{2}\frac{N_{\mathrm{H}}^-}{N_p}\frac{\partial N(X)}{\partial X}\right|_{X=\Lambda-} \geq Q_{\mathrm{H}} \geq \left.\frac{h}{2}\frac{N_{\mathrm{H}}^+}{N_p}\frac{\partial N(X)}{\partial X}\right|_{X=\Lambda+} \tag{6.54}$$

Ahead of the travelling hinge, sections of the beam in segment HB ($\Lambda+ < X \leq L$) remain stationary so the axial force $N(X)$ is constant in this segment; hence

$$\left.\frac{\partial N(X)}{\partial X}\right|_{X=\Lambda+} = 0$$

Consequently, the limitation on shear force at a travelling hinge (6.54) is recast as

$$\left.\frac{h}{2}\frac{N_{\mathrm{H}}^-}{N_p}\frac{\partial N(X)}{\partial X}\right|_{X=\Lambda-} \geq Q_{\mathrm{H}} \geq 0 \tag{6.55}$$

Choice of Shear Force at Hinge According to inequality (6.55), there is only a limited range of admissible shear force at a travelling hinge. The range falls between two extreme cases:

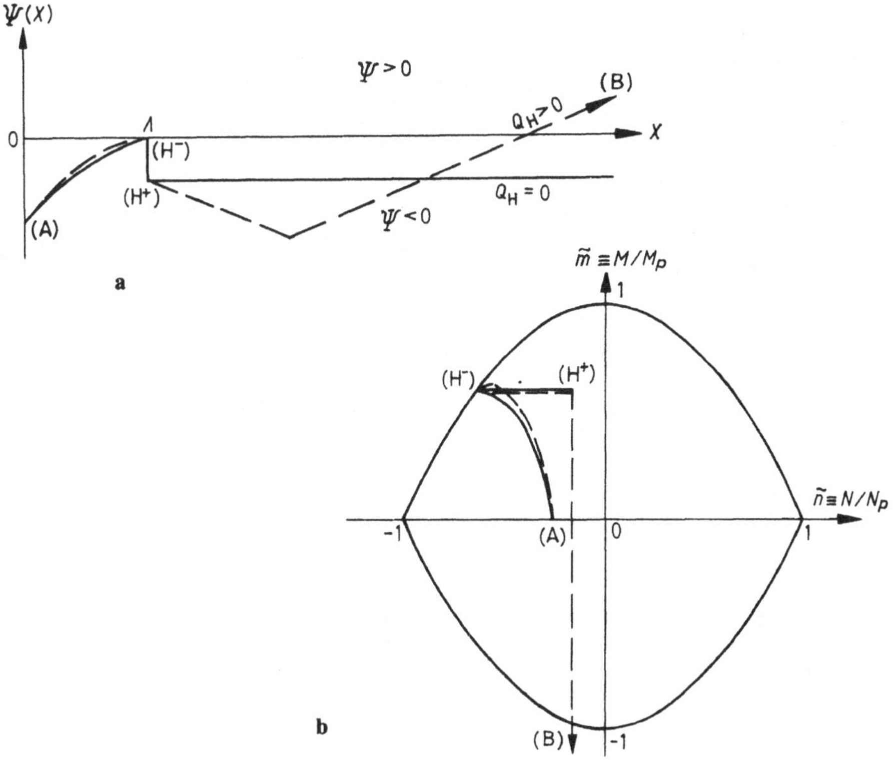

Fig. 6.11. (a) Variation of yield function $\Psi(X)$ along the cantilever, (b) stress profile in the ($\tilde{m}$, $\tilde{n}$) plane.

1. $Q_H = \frac{h}{2}\frac{N_H^-}{N_p}\left.\frac{\partial N(X)}{\partial X}\right|_{X=\Lambda-}$ if $\partial\Psi(\Lambda)/\partial X = 0$

2. $Q_H = 0$ if $\partial M/\partial X = 0$

If Q_H takes a positive value as in case (1), then the bending moment in segment HB ($\Lambda \le X \le L$) is determined by $M(X) = M_H - Q_H(X - \Lambda)$. As a result, the bending moment distribution is sketched as the broken line in Fig. 6.10d; the moment changes sign when $X > \Lambda + M_H/Q_H$. On the other hand, if $Q_H = 0$ as in case (2), then the bending moment in segment HB remains constant, i.e. $M(X) = M_H$ in $\Lambda \le X \le L$. This bending moment distribution is sketched as the solid line in Fig. 6.10d.

The corresponding distributions of the yield function $\Psi(X)$ along the cantilever are sketched in Fig. 6.11a, where the broken line and the solid line refer to cases of (1) nonvanishing shear force at the generalized hinge $Q_H > 0$ and (2) vanishing shear force at $Q_H = 0$, respectively.

In the plane of generalized stresses (N/N_p, M/M_p), the stress distributions at an instant during the transient phase are represented by the broken and solid line, respectively, in Fig. 6.11b. In the moving segment AH, as distance from the hinge decreases, the stress state approaches the yield surface at H^-. At this point case (1) requires the stress variation in AH to be tangent to the yield (limit) locus if $\partial\Psi/\partial X = 0$; while case (2) requires the stress variation to have a horizontal slope.

Both Fig. 6.11a and b indicate that when $Q_H > 0$, as in case (2), there can be other sections ahead of the travelling hinge that also satisfy yield if X is sufficiently large. In other words, this condition implies that in the early part of the transient phase, there is more than one travelling hinge and only some of these emanate from the impact point. On the other hand, $Q_H = 0$ as in case (2) ensures satisfaction of the yield criterion ahead of a single travelling hinge. While it is simpler to analyze only a single travelling hinge, there are no additional theoretical conditions that determine the shear force at the travelling hinge other than the limits specified in (6.54); i.e. *there is not a unique solution when stretching and flexural motions are coupled through an interaction yield surface.*

Equations of Motion with $Q_H = 0$ A complete transient solution with only a single generalized hinge can be imposed by assuming that shear force vanishes at a travelling hinge $Q_H = 0$; at this hinge the flow rule relates flexure and stretching. With $Q_H = 0$ the previous equations of motion (6.41) and (6.42) are recast as

$$\frac{\mathrm{d}}{\mathrm{d}t}\left(\frac{1}{2}\rho\Lambda V_z\right) = -G\frac{\mathrm{d}V_z}{\mathrm{d}t} \tag{6.41'}$$

$$\frac{\mathrm{d}}{\mathrm{d}t}\left(\frac{1}{6}\rho\Lambda^2 V_z\right) = M_H \tag{6.42'}$$

These equations, together with (6.43), (6.44), (6.49) and (6.51), constitute six governing equations for six unknown functions $V_x(t)$, $V_z(t)$, $\Lambda(t)$, $N_H^-(t)$, $N_H^+(t)$ and $M_H(t)$.

A direct integration of Eq. (6.41') results in

$$V_z = \frac{V_{z0}}{1 + \rho\Lambda/2G} \tag{6.56}$$

Then, from (6.51),

$$\frac{N_{\mathrm{H}}^{-}}{N_p} = -\frac{2V_x\Lambda(1+\rho\Lambda/2G)}{hV_{zo}} \tag{6.57}$$

where the initial value of Λ is small enough that the axial force at the hinge satisfies the yield condition $N_{\mathrm{H}}^{-}/N_p \geq -1$. In accord with (6.43) this axial force gives an axial acceleration for the tip segment; using (6.43) and (6.49), one finds

$$\frac{\mathrm{d}V_x}{\mathrm{d}t} = -\frac{2N_p V_x \Lambda(1+\rho\Lambda/2G)}{hV_{zo}(G+\rho\Lambda)} \tag{6.58}$$

while Eqs (6.43) – (6.45) together with the yield condition (6.49) give

$$\frac{\mathrm{d}\Lambda}{\mathrm{d}t} = \frac{6}{\rho V_{zo}}\frac{(1+\rho\Lambda/2G)^2 M_p}{\left\{2\Lambda(1+\rho\Lambda/2G)-\rho\Lambda^2/2G\right\}}\left\{1-\left(\frac{2V_x\Lambda(1+\rho\Lambda/2G)}{hV_{zo}}\right)^2\right\} \tag{6.59}$$

Equations (6.58) and (6.59) require $N_{\mathrm{H}}^{-} > -N_p$. Note that they are first order o.d.e. for two unknown functions $V_x(t)$ and $\Lambda(t)$; the appropriate initial conditions for the equations are

$$V_x(0) = V_{xo}, \qquad \Lambda(0) = 0 \tag{6.60}$$

Nondimensionalization To develop general understanding of the system behavior, it is helpful to obtain solutions in terms of the following nondimensional variables:

$$w_x \equiv \frac{W_x}{L}, \quad w_z \equiv \frac{W_z}{L}, \quad x \equiv \frac{X}{L}, \quad \lambda \equiv \frac{\Lambda}{L}, \quad T_{\mathrm{o}} \equiv L\sqrt{\frac{\rho L}{M_p}}$$

$$\tau \equiv \frac{t}{T_{\mathrm{o}}}, \quad v_x \equiv \frac{V_x}{L/T_{\mathrm{o}}}, \quad v_z \equiv \frac{V_z}{L/T_{\mathrm{o}}}, \quad v_{xo} \equiv \frac{V_{xo}}{L/T_{\mathrm{o}}}, \quad v_{zo} \equiv \frac{V_{zo}}{L/T_{\mathrm{o}}} \tag{6.61}$$

$$\gamma \equiv \frac{G}{\rho L}, \quad \tilde{m} \equiv \frac{M}{M_p}, \quad n \equiv \frac{NL}{M_p} \equiv \frac{\zeta N}{N_p}$$

where the slenderness ratio for beams of rectangular cross-section is defined as $\zeta \equiv 4L/h$. Then, Eqs (6.56), (6.57), (6.45), (6.49), (6.58) and (6.59) can be recast in nondimensional form

$$v_z = \frac{v_{zo}}{1+\lambda/2\gamma} \tag{6.56'}$$

$$n_{\mathrm{H}}^{-} = -\frac{\zeta^2 v_x \lambda(1+\lambda/2\gamma)}{2v_{zo}} \qquad (\geq -\zeta) \tag{6.57'}$$

$$n_{\mathrm{H}}^{+} = n_{\mathrm{H}}^{-} + v_x\frac{\mathrm{d}\lambda}{\mathrm{d}\tau} \qquad (\leq \zeta) \tag{6.45'}$$

$$\tilde{m}_{\mathrm{H}} = 1-\left(\frac{\zeta v_x \lambda(1+\lambda/2\gamma)}{2v_{zo}}\right)^2 \tag{6.49'}$$

$$\frac{\mathrm{d}v_x}{\mathrm{d}\tau} = -\frac{\zeta^2 v_x \lambda(1+\lambda/2\gamma)}{2v_{zo}(\gamma+\lambda)} \tag{6.58'}$$

$$\frac{d\lambda}{d\tau}=\frac{6}{v_{zo}}\frac{(1+\lambda/2\gamma)^2}{\{2\lambda(1+\lambda/2\gamma)-\lambda^2/2\gamma\}}\left\{1-\left(\frac{\zeta v_x\lambda(1+\lambda/2\gamma)}{2v_{zo}}\right)^2\right\} \tag{6.59'}$$

The initial conditions related to (6.58') and (6.59') are

$$v_x=v_{xo},\quad \lambda=0,\quad \text{at}\quad \tau=0 \tag{6.60'}$$

6.3.3 Solution Based on Single-Hinge Mechanism

Removing Singularity from the Initial Conditions As shown in Eq. (6.60'), the travelling hinge is initially located at the impact point so $\lambda=0$ at $\tau=0$; consequently (6.58') and (6.59') indicate that

$$\frac{dv_x}{d\tau}=0,\quad \frac{d\lambda}{d\tau}\to\infty,\quad \text{at}\quad \tau=0 \tag{6.62}$$

But if both transverse and axial momentum are imparted to the tip by impulsive loading, Eq. (6.45') leads to $n_H^+\to\infty$ at $\tau=0$, and this violates the yield criterion. Whenever axial forces are coupled to the bending moment through an interaction yield criterion, this analytical difficulty arises immediately from discontinuous impulsive loading. It occurs because of the infinitely fast moving plastic hinge that emanates from points where momentum is discontinuous.

To avoid the singularity at $\tau=0$, the initial conditions (6.60') can be altered by assuming that the hinge is initially a short distance away from the impact point. Let this short distance from the tip be $\lambda_o \ll 1$; with this numerical approximation, the initial values of hinge speed $d\lambda/d\tau$ and axial force n_H^+ can be estimated from Eqs (6.59'), (6.57') and (6.45'),

$$\frac{d\lambda}{d\tau}\approx\frac{3}{v_{zo}\lambda_o},\quad n_H^-\approx-\frac{\zeta^2 v_{xo}\lambda_o}{2v_{zo}}<0,\quad n_H^+\approx-\frac{\zeta^2 v_{xo}\lambda_o}{2v_{zo}}+\frac{3v_{xo}}{v_{zo}\lambda_o}>0 \tag{6.63}$$

Hence, the requirement of condition (6.48), i.e. $|n_H^-|\ge|n_H^+|$ can be satisfied if the hinge is initially located a small distance λ_o from the tip,

$$\lambda_o\ge\sqrt{3}/\zeta \tag{6.64}$$

or in dimensional form,

$$\Lambda_o\ge\frac{h\sqrt{3}}{4} \tag{6.64'}$$

This shows that as long as V_{xo} is of the same order as V_{zo}, the assumed starting distance Λ_o should be the same order as the depth h of the beam and therefore very small in comparison with the length L of the cantilever. Based on this discussion, the initial conditions (6.60') are replaced by

$$v_x=v_{xo},\quad \lambda=\lambda_o=\sqrt{3}/\zeta,\quad \text{at}\quad \tau=0 \tag{6.65}$$

The present modeling is based on predominately flexural motion and thus is not appropriate for primarily axial impact where $V_{xo}/V_{zo}\gg 1$.

Iterative Procedure for Numerical Solution The following procedure is employed to obtain numerical solutions for this problem.

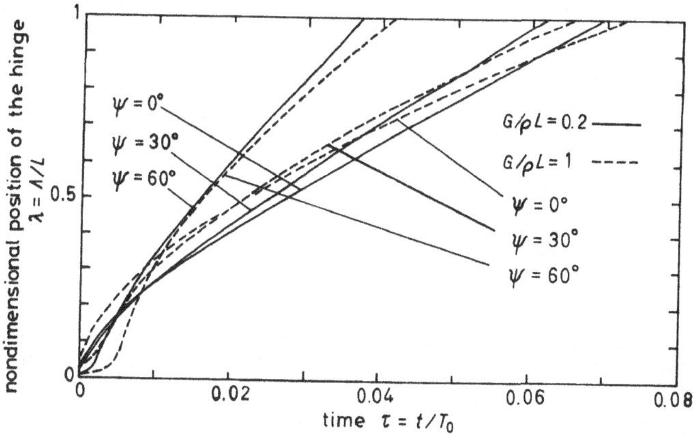

Fig. 6.12. Movement of plastic hinge along cantilever subjected to oblique impact. Energy ratio $e_0 \equiv K_0 / M_P = 1.0$; mass ratio $\gamma \equiv G/\rho L = 0.2$ and 1.0; angle of obliquity $\psi = 0°$, $30°$ and $60°$.

1. Specify the four nondimensional parameters that appear in the governing equations and initial conditions, i.e. γ, ζ, v_{xo} and v_{zo}; or better still, γ, ζ, (v_{xo}/v_{zo}) and the nondimensional initial kinetic energy $e_0 \equiv \gamma(v_{xo}^2 + v_{zo}^2)/2$.
2. Numerically integrate Eqs (6.58') and (6.59') with initial conditions (6.65) to find $\lambda(\tau)$ and $v_x(\tau)$ during period before time $\tau = \tau_u$ when axial motion ceases, $v_x(\tau_u) = 0$.
3. Obtain $v(\tau)$, $n_H^-(\tau)$, $n_H^+(\tau)$ and $\tilde{m}_H(\tau)$, for the coupled phase of motion $\tau_0 \le \tau \le \tau_u$ with the help of (6.56'), (6.57'), (6.45') and (6.49'), respectively.
4. After axial motion ceases (i.e. for $\tau \ge \tau_u$), transverse motion and flexural deformation continue. The governing equations during $\tau_u \le \tau \le \tau_f$, are the same as those in Sect. 4.5; the solution can be obtained in the previous manner. The transverse motion finally ceases at $\tau = \tau_f$.

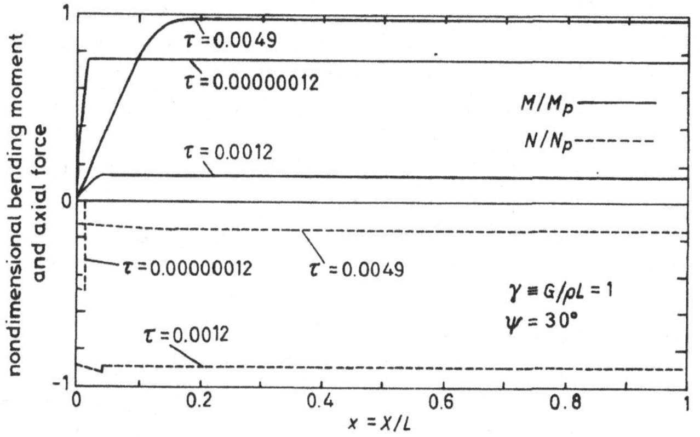

Fig. 6.13. Distributions of bending moment and axial force at various instants after oblique impact, $e_0 = K_0 / M_P = 1.0$, $\gamma = G/\rho L = 1.0$ and $\psi = 30°$.

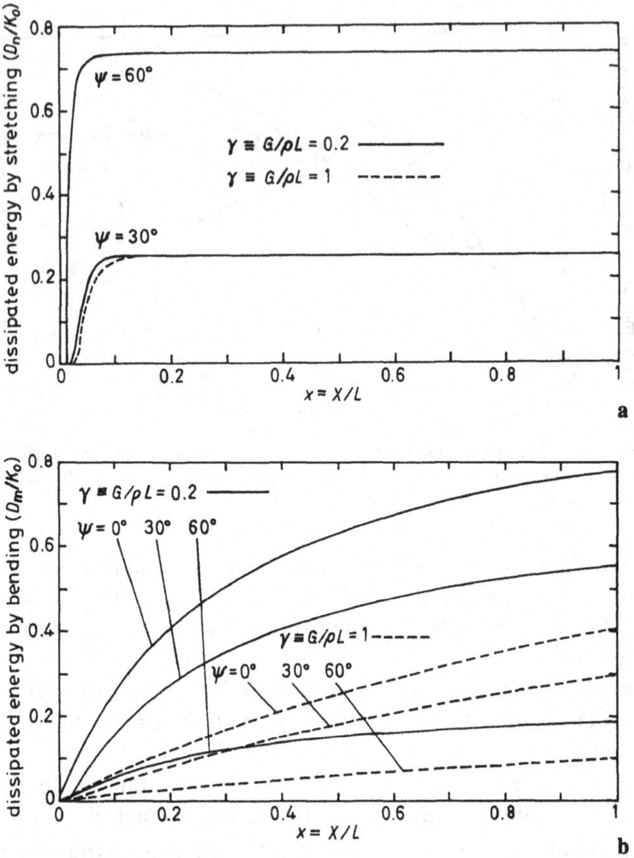

Fig. 6.14. Energy dissipation accumulated from tip to cross-section at $x = X/L$ from (a) axial deformation and (b) bending.

Numerical Examples For mass ratios $\gamma = 0.2$ and 1.0, and angles of obliquity for impact $\psi = \tan^{-1}(V_{xo}/V_{zo}) = 0°, 30°$ and $60°$, numerical calculations have been carried out; some results are depicted in Figs 6.12–6.15. In all these examples, the nondimensional input energy $e_o \equiv K_o/M_p = 1.0$ remains the same, and the slenderness ratio $L/h = 25$ (i.e. $\zeta = 100$).

Figure 6.12 shows the variation of the nondimensional position of the travelling hinge with time. The figure shows that for larger angles of obliquity the average speed of the hinge during the transient phase is reduced while axial motion exists. The bending moment and axial force throughout the beam at various instants of time are shown in Fig. 6.13 for a mass ratio $\gamma = 1$ and impact angle, $\psi = 30°$. The axial force increases rapidly in the initial period while bending moment decreases; subsequently, the axial force decreases before axial motion ceases. After axial motion ceases, the hinge is still near the tip. Thereafter the bending moment equals the fully plastic moment. This indicates that an axial component of impact only affects the early part of the transient phase of dynamic response. The axial momentum is completely eliminated while the hinge is still close to the

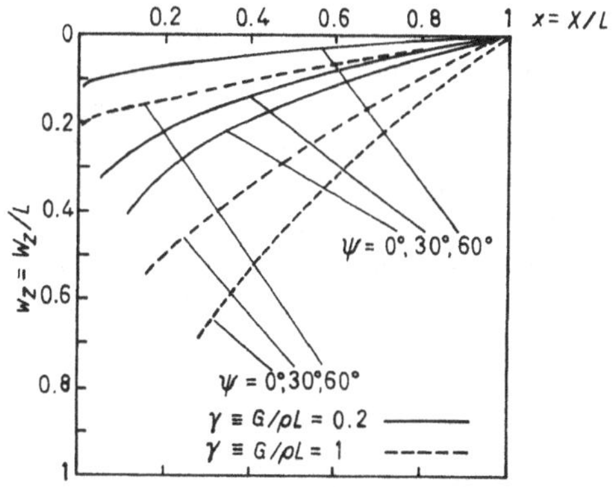

Fig. 6.15. Final deformed shapes of cantilevers subjected to impact. Energy ratio $e_0 \equiv K_0 / M_p = 1.0$; mass ratio $\gamma \equiv G / \rho L = 0.2$ and 1.0; angle of obliquity $\psi = 0°$, $30°$ and $60°$.

impact point; thereafter, the response is essentially the same as that of the cantilever subjected to a transverse impact (see Sect. 4.5).

Figure 6.14a and b shows the energy dissipated by axial and flexural deformations before the hinge transits any location x. As shown in Fig. 6.14a, if the axial deformation ceases before the travelling hinge reaches the root of the cantilever, then the total energy dissipated by axial deformation (tension or compression) is independent of the mass ratio γ and depends only on the obliquity ψ of impact. Figure 6.14b shows, however, that the energy dissipated by bending varies significantly with both the mass ratio γ and the obliquity ψ of impact. The sum of energies dissipated at the interior cross-sections $(0 \le X < L)$ by flexural and axial deformations is usually less than the input energy K_0; the remaining energy is dissipated by rotation about the stationary hinge at the root.

The final deformed shape resulting from oblique impact at a nondimensional initial energy $e_0 \equiv K_0 / M_p = 1.0$ is depicted in Fig. 6.15. This illustrates that the axial component of impact has almost no influence on the final shape of the cantilever.

6.4 Circular Arc Cantilever Subjected to In-plane Step Force

6.4.1 Engineering Background and Assumptions

Structural members that are not straight are frequently employed in machine parts, buildings and pipe networks; common examples are curved beams, bent beams, arches, circular rings and simple frames. In particular, the dynamic behavior of curved beams is useful for risk analyses of pipework systems.

A sudden or guillotine break in pressurized piping gives a suddenly applied follower force acting on the broken end of the pipe and this causes pipe-whip.

The jet reaction forces exerted on a fractured pipe by escaping fluid can be large enough so that the broken pipe undergoes large dynamic deflection and rotation. This is a hazard to other pipes or equipment in the vicinity. Since plastic deformation in a pipe-whip process is usually much larger than the elastic deformation, a rigid–plastic model for pipe deformation has been employed in studies of pipe-whip. According to measurements from experiments, the jet forces exerted on broken pipes are somewhat similar to rectangular pulses so analyses of straight cantilevers subjected to step or rectangular pulse loading may be used directly as first-order approximations for pipe-whip prediction. This type of analysis was given in Sects 4.1 and 4.2.

When the postulated break is near an elbow, it is necessary to analyze the dynamic behavior of a curved cantilever. Since the jet reaction force can be in any direction, the pipe may be subjected not only to a bending moment, but also to torque and axial force. If the dynamic force or impulse acts in the plane of the curved beam, then the response is primarily bending. If the load is out-of-plane, torsional deformations can be equally large.

With pipe-whip of a curved section of tubing in mind, Yu and Johnson [1981] first examined a quadrantal cantilever beam of circular plan form that is subjected to in-plane step loading at the tip. They employed an interactive yield criterion between bending moment and axial force and found that a stationary plastic hinge at an internal cross-section requires a step force within a certain range of magnitude and direction. For some directions of the force, any deformation mechanism with an internal hinge violates the yield criterion.

In Sects 6.4 – 6.6 we examine the dynamic response of circular curved cantilevers under various dynamic loading conditions. In these analyses, in addition to the rigid–perfectly plastic idealization, it is assumed that

1. the curved cantilever has uniform cross-section and constant density ρ per unit length;
2. the cross-section dimensions are much smaller than the radius of the center line R, so the influence of shear and axial forces on yielding can be neglected (however, the interaction between bending and torsion will be considered for the case of out-of-plane loading, in Sect. 6.6);

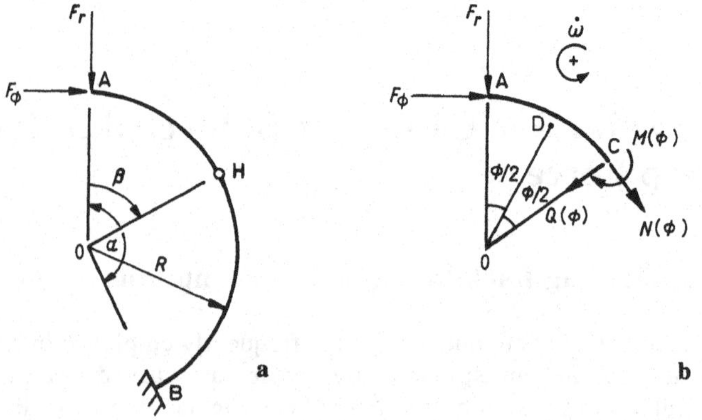

Fig. 6.16. Circular cantilever subjected to step force at the tip: (a) configuration with plastic hinge H; (b) free-body diagram of arc element AC within segment AH, showing the sign convention of bending moment M, shear force Q and axial force N.

3. the deflections are so small in comparison with R, that the dynamic equations and geometric relations can all be written on the basis of the initial configuration.

In the present section, we first consider a curved cantilever with a center line at radius R; the center line subtends an angle α ($\pi/2 \le \alpha \le \pi$) of a circular arc. End B of the circular cantilever is clamped while the free end A, is subjected to a suddenly applied radial force F_r or tangential force F_ϕ that acts in the plane of the arc, as shown in Fig. 6.16.

If the magnitude of the force is constant, the dynamic deformation mechanism does not change with time; i.e. the response takes a stationary pattern of deformation. In other words, the dynamic response to a steady force is a modal solution, provided the material is assumed to be rigid–perfectly plastic and only small deflections are considered. Also, the effects of shear and axial forces on yield are neglected. Under these assumptions, the circular cantilever starts to deform if F_i (radial force F_r or tangential force F_ϕ) is as large as the static collapse load F_c. The collapse mechanism of the circular cantilever is characterized by the location of one or more plastic hinges. If $F_i > F_c$, the cantilever deforms in a dynamic mode which is different from that of static collapse because of inertia. This dynamic mode has a constant acceleration field; the number of hinges and their locations depend on both the applied and inertia forces, i.e. they are functions of F_i and the mass per unit length ρ, the radius of the curved beam R and the fully plastic bending moment M_p of the cross-sections.

6.4.2 Radial Force at Tip

For a circular arc cantilever the spatial coordinate is chosen as the arc length along the center line $S = R\phi$, where ϕ is the angular coordinate measured from tip A. By taking the sign conventions for bending moment M and shear force Q as shown in Fig. 6.16, M and Q are related by $\mathrm{d}M/\mathrm{d}S = -Q$ or $\mathrm{d}M/\mathrm{d}\phi = -RQ$. The shear distribution $Q(S)$ can be calculated from the equations of motion of a generic segment such as AC. Integration of the shear resultant along the cantilever then gives the bending moment distribution $M(S)$. Using this formulation, both global variables (i.e. the rotational accelerations and hinge positions) and the local distribution of moment can be obtained simultaneously. For a cantilever subjected to a radial force F_r at the tip (Fig. 6.16), there is no deformation if $F_r < F_{rc} = M_p/R$; if $F_r = F_{rc}$, a plastic hinge H forms at $\beta = \pi/2$.

Force $F_{rc} < F_r < F_{r2}$ For $F_r > F_{rc}$, the position angle β of the hinge H moves from $\beta = \pi/2$ towards the loading point A as the force increases. This single-hinge mode is valid for $F_{rc} < F_r < F_{r2}$. If $F_r \ge F_{r2}$, a second hinge occurs at the root B. The complete analysis is carried out as follows.

If $F_{rc} < F_r < F_{r2}$, the cantilever deforms with a single hinge H located at angle β from tip A. The equation of motion for acceleration of segment AC in the direction CO is used to derive the shear force Q at a generic point C on segment AH (Fig. 6.16); this relation does not involve the axial force at C. The center of mass of segment AC is designated by D in Fig. 6.16; it is located a distance $\overline{\mathrm{OD}}$ from the center of the circular arc,

$$\overline{\mathrm{OD}} = \frac{R\sin(\phi/2)}{\phi/2}$$

where $\phi = \angle AOC$. The radial line OD bisects $\angle AOC$.

The acceleration of point D has a component in direction CO,

$$\dot{\omega}\left[\overline{OD}\sin(\phi/2) + R\sin(\beta-\phi)\right] = \dot{\omega}R\left[(1-\cos\phi)/\phi + \sin(\beta-\phi)\right]$$

where $\dot{\omega}$ is the anticlockwise angular acceleration of segment AH. Using the above expression, the translational equation of motion for segment AC in direction CO can be written as

$$\begin{aligned} F_r\cos\phi - Q &= (\rho R\phi)\dot{\omega}R\left[(1-\cos\phi)/\phi + \sin(\beta-\phi)\right] \\ &= \rho\dot{\omega}R^2\left[1-\cos\phi + \phi\sin(\beta-\phi)\right] \end{aligned} \qquad (6.66)$$

As the effects of shear and axial forces on yield are neglected, the bending moment M is a local extremum at hinge H where $\phi = \beta$, so at the hinge the shear $Q = 0$. Hence, from Eq. (6.66),

$$F_r\cos\beta = \rho\dot{\omega}R^2(1-\cos\beta) \qquad (6.67)$$

In the rotating segment, the moment $M(\theta)$ is found by integrating the expression for the shear force with the boundary condition $M(0) = 0$; this results in

$$M = F_rR\sin\phi - \rho R^3\dot{\omega}\left[\phi - \sin\phi + \phi\cos(\beta-\phi) + \sin(\beta-\phi) - \sin\beta\right] \qquad (6.68)$$

The bending moment at the hinge $M(\beta) = M_p$, which gives

$$M_p = F_rR\sin\beta - 2\rho R^3\dot{\omega}(\beta - \sin\beta) \qquad (6.69)$$

From Eqs (6.67) and (6.69), F_r and $\dot{\omega}$ are determined as functions of β,

$$F_r = \frac{M_p}{R}\frac{1-\cos\beta}{\sin\beta(1+\cos\beta) - 2\beta\cos\beta} \qquad (6.70)$$

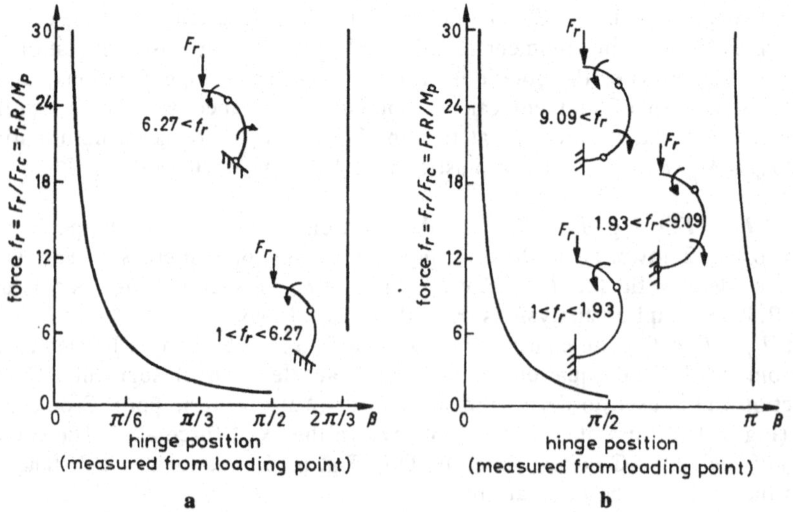

Fig. 6.17. Dependence of hinge position β on the force magnitude $f_r \equiv F_r/F_{rc}$ for a circular cantilever subjected to a transverse step force F_r at the tip, in case of (a) $\alpha = 2\pi/3$ and (b) $\alpha = \pi$.

$$\dot{\omega} = \frac{M_p}{\rho R^3}\frac{\cos\beta}{\sin\beta(1+\cos\beta)-2\beta\cos\beta} \tag{6.71}$$

The hinge location β is shown in Fig. 6.17a and b for cantilevers with included angles $\alpha = 120°$ and $180°$, respectively. The corresponding moment distributions are shown in Fig. 6.18a and b.

Force $F_r > F_{r2}$ The single-hinge mode of deformation will only be valid for a limited range of force. If $F_r \geq F_{r2}$, two hinges are required to satisfy the yield condition; the location of the second hinge can be determined by considering the moment distribution in segment HB. This moment depends on the axial force N_H at hinge H. The axial force N_H can be obtained by formulating the translational equation of motion for segment AH in the direction of the tangent at H,

$$F_r \sin\beta + N_H = \rho\beta R\dot{\omega}\left[R - \overline{OD}\cos(\beta/2)\right] \tag{6.72}$$

The axial force N_H at H is due to both a component of the applied force and a component of the acceleration of the tip segment; thus the preceding equation can be rearranged as

$$N_H = \rho R^2\dot{\omega}(\beta - \sin\beta) - F_r \sin\beta = -\frac{M_p}{R}\frac{\sin\beta - \beta\cos\beta}{\sin\beta(1+\cos\beta)-2\beta\cos\beta} \tag{6.73}$$

Using this expression, the moment at the root M_B can be written as

$$M_B = N_H R[1-\cos(\alpha-\beta)] + M_p \tag{6.74}$$

Substituting Eq. (6.73) into Eq. (6.74), the limiting condition $M_B = -M_p$ yields

$$(\sin\beta - \beta\cos\beta)[1-\cos(\alpha-\beta)] = 2\sin\beta(1+\cos\beta) - 4\beta\cos\beta \tag{6.75}$$

The angle β for formation of a second plastic hinge can be calculated for a given value of α from this expression; the minimum force for development of this second mode is denoted as F_{r2}, which is a function of the included angle α. For a radial force of arbitrarily large magnitude at the tip and a statically admissible single-hinge mode of deformation, the hinge can be located no further from the tip than the quarter circle $\beta = \pi/2$.

If $F_r > F_{r2}$, the cantilever deforms in a double-hinge mode with the newly formed hinge located at the root B. The location of the first hinge H moves towards the loading point as F_r increases (Fig. 6.17). The equations of motion for this mode are formulated as follows.

1. From the rate of change of moment of momentum of HB about the additional hinge at B (see Fig. 6.16) we obtain

$$-N_H R[1-\cos(\alpha-\beta)] - 2M_p = 2\rho R^3\dot{\Omega}[\alpha - \beta - \sin(\alpha-\beta)] \tag{6.76}$$

2. The equation of translational motion of segment AH in the radial direction HO gives

$$F_r\cos\beta = \rho\beta R\left[\dot{\omega}R\frac{\sin^2(\beta/2)}{\beta/2} - 2\dot{\Omega}R\sin\left(\frac{\alpha-\beta}{2}\right)\cos\left(\frac{\alpha-\beta}{2}\right)\right] \tag{6.77}$$

where the term in the square bracket is the component of acceleration of the center of AH in direction HO. Note that a positive angular acceleration $\dot{\Omega}$ of segment HB is clockwise while positive $\dot{\omega}$ of segment AH is anticlockwise.

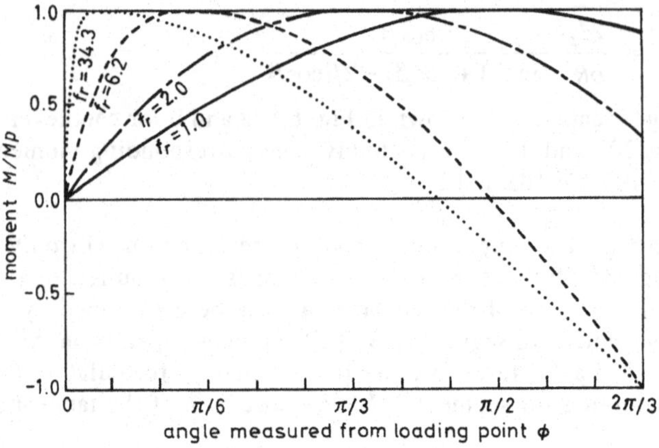

a

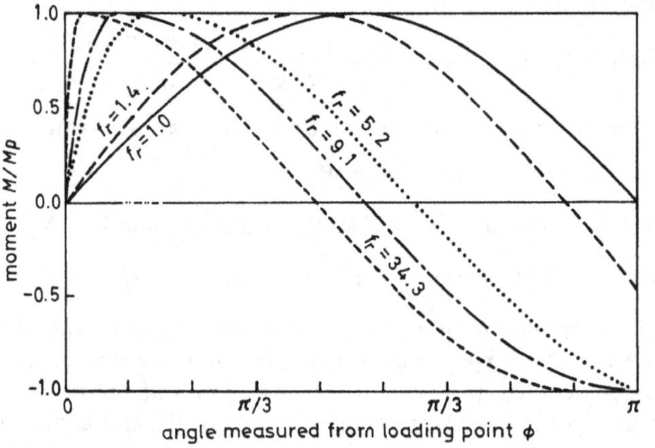

b

Fig. 6.18. Bending moment distribution in cantilever of circular plan form subjected to transverse step force F_r at the tip, in case of (a) $\alpha = 2\pi/3$ and (b) $\alpha = \pi$.

3. The equation of translational motion of segment AH in the tangential direction at H gives

$$N_{\mathrm{H}} + F_r \sin\beta = \rho\beta R\left[\dot{\omega}R\left(1 - \frac{\sin(\beta/2)\cos(\beta/2)}{\beta/2}\right) + 2\dot{\Omega}R\sin^2\left(\frac{\alpha-\beta}{2}\right)\right] \tag{6.78}$$

4. For segment AH, the rate of change in moment of momentum about the centroid results in

$$F_r R\frac{\sin^2(\beta/2)}{\beta/2} - N_{\mathrm{H}}R\left(1 - \frac{\sin(\beta/2)}{\beta/2}\right) = \rho\beta R^3\dot{\omega}\left[1 - \left(\frac{\sin(\beta/2)}{\beta/2}\right)^2\right] + M_p \tag{6.79}$$

When a force F_r is given, the corresponding hinge location β, hinge force N_{H}, and angular accelerations of the segments $\dot{\omega}$ and $\dot{\Omega}$ can be determined from

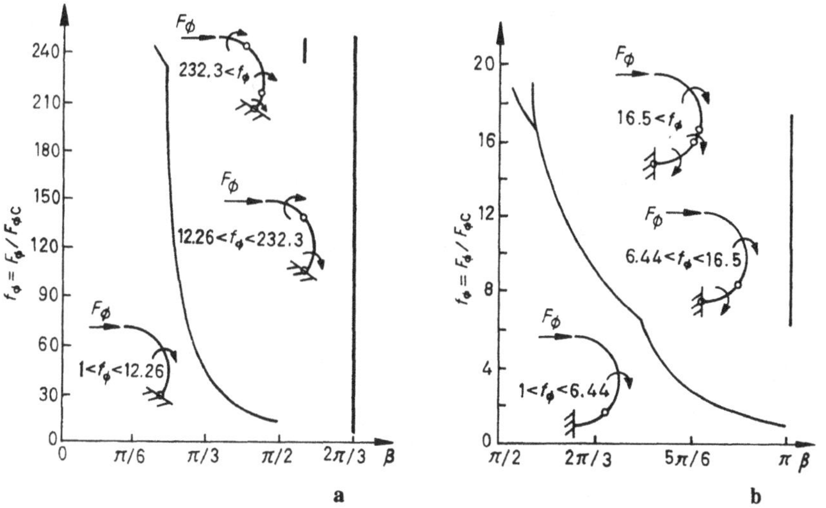

Fig. 6.19. Dependence of hinge position β on the force magnitude $f_\phi \equiv F_\phi / F_{\phi c}$ for a circular cantilever subjected to a tangent step force F_ϕ at the tip, in case of (a) $\alpha = 2\pi/3$ and (b) $\alpha = \pi$.

Eqs (6.76) – (6.79). The f_r vs. b relation is shown in Fig. 6.17. Using the same procedure as before, the moment distributions are computed and shown in Fig. 6.18 for various ranges of F_r. It is found that when $\alpha < 158.0°$, this double-hinge mode is valid for all subsequent values of F_r, while if $\alpha > 158.0°$, the hinge at root B moves away from the root as F_r increases.

To summarize, under a radial force F_r, the static collapse force $F_{rc} = M_p/R$ results in deformation with a single hinge located at $\beta = \pi/2$. If $F_{rc} < F_r < F_{r2}$, the location of the single hinge moves towards the loading point as F_r increases. If $F_r = F_{r2}$, a second hinge develops; this is located at the root. Radial applied forces $F_r > F_{r2}$ require a second hinge located at the root for included angles $90° \le \alpha \le 158°$, while larger angles have a second hinge located inboard of the root if the force is sufficiently large.

6.4.3 Tangential Force at Tip

Consider the same cantilever as in Sect. 6.4.2 but with a tangential force F_ϕ instead of the radial force of the previous case. All other conditions are identical. Using the same procedure we obtain somewhat different results.

The fully plastic collapse force for a tangential force F_ϕ applied at the tip of a curved cantilever with constant radius R that subtends an angle α is $F_{\phi c} = M_p/[R(1-\cos\alpha)]$. If the applied force $F_\phi = F_{\phi c}$, plastic collapse occurs with a hinge at the root B; in this mechanism the entire cantilever rotates about B. The equation of motion is written as follows:

$$F_\phi R(1-\cos\alpha) - M_p = 2(\alpha - \sin\alpha)\dot{\omega} R^3 \qquad (6.80)$$

In calculating hinge positions for larger values of F_ϕ, it is found that a second hinge forms at locations that depend on the included angle α. When α < 133.6°, the second hinge H forms between B and A at a characteristic force $F_{\phi 2}(\alpha)$

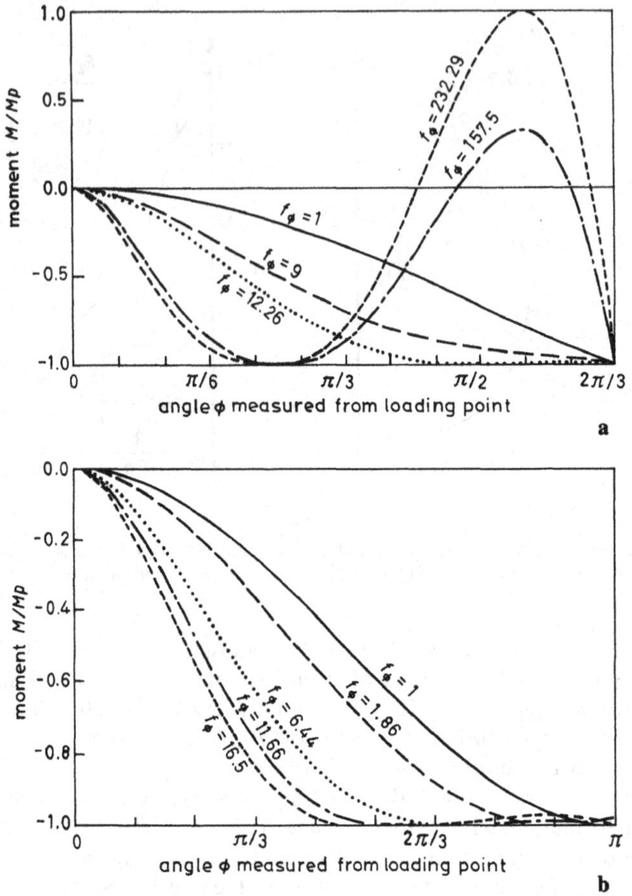

Fig. 6.20. Bending moment distribution in cantilever of circular plan form subjected to tangent step force F_ϕ at the tip, in case of (a) $\alpha = 2\pi/3$ and (b) $\alpha = \pi$.

(Fig. 6.19a). As F_ϕ increases, the cantilever deforms with two hinges for $F_{\phi 2} \le F_\phi \le F_{\phi 3}(\alpha)$. With an applied force $F_\phi > F_{\phi 3}(\alpha)$ a third hinge forms between H and the root B; modes for $F_\phi > F_{\phi 3}$ have three or more hinges.

If $\alpha > 133.6°$, the location of the hinge originally at B moves towards the tip as F_ϕ increases (Fig. 6.19b). If the force $F_\phi = F_{\phi 2}(\alpha)$ the first hinge is located at H away from the root and a second hinge forms at the root B. If $F_\phi > F_{\phi 2}$, the mode has two hinges; the first hinge H continues to approach the loading point as F_ϕ increases while the other hinge stays at the root. The hinge at H bifurcates at a characteristic value of applied force, $F_{\phi 3}$.

The modal hinge locations as a function of applied load are shown in Fig. 6.19a and b for $\alpha = 120°$ and $180°$ respectively. The corresponding moment distributions are shown in Fig. 6.20a and b.

In summary, the circular cantilever has a static plastic collapse force $F_\phi = M_p/[R(1-\cos\alpha)]$. For somewhat larger forces, $F_\phi > F_{\phi 2}$, the beam deforms with one hinge either at or near the root B. If $\alpha < 133.6°$, one hinge remains at

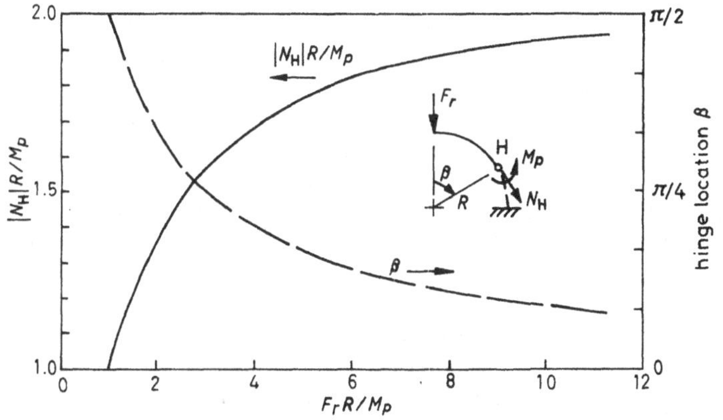

Fig. 6.21. Hinge position and magnitude of axial force at hinge as functions of the magnitude of a transverse step force F_r at tip of circular cantilever, in case of $\alpha = \pi/2$.

the root and when $F_\phi > F_{\phi 2}(\alpha) > F_{\phi c}$ a second hinge forms within the beam. If $\alpha > 133.6°$, the first hinge forms near rather than at the root. The location of this hinge moves away from the root as F_ϕ increases; a second hinge forms at the root when the applied force exceeds a limit value $F_{\phi 2}$. For still larger force there is another limiting magnitude $F_{\phi 3}$ where the hinge closest to the loading point bifurcates. If $F_\phi > F_{\phi 3}$ a three hinge mode appears.

6.4.4 Discussion

With a radial force F_r, there are two different patterns of response depending on whether α is larger or smaller than 158.0°; similarly, for a tangential force F_ϕ two distinctive patterns of modes are separated by an arc of 133.6°. With either a radial or tangential force, the curved segment immediately adjacent to the force has an arc length that monotonically decreases towards zero as the magnitude of the force increases.

There are some differences between the modes for radial and tangential forces. For the same curved structure, the number of hinges increases more rapidly for a tangential force because there is a large axial component of force at hinges near the tip. For a curved cantilever, the number of plastic hinges in a mode increases without bound as the magnitude of a tangential applied force increases. For the same structure with a radial force, dynamic modes have only a small number of plastic hinges although the force may be very large.

This effect in curved beams can be understood by considering a quadrantal cantilever ($\alpha = \pi/2$). With a radial force at the tip the axial resultant force at the hinge H is $N_H < 2M_p/R$ for all F_r, as shown in Fig. 6.21. The axial force N_H at hinge H is not large enough to form another hinge at the root B and hence the modes for this direction of force have only a single hinge irrespective of the magnitude of F_r. In contrast, with a tangential force F_ϕ at the hinge closest to the loading point, the axial force increases without bound as F_ϕ increases; correspondingly, the number of hinges for a mode also increases in order to satisfy yield.

Stronge et al. [1990] presented modal analyses that brought out these characteristics for several different arch, ring and curved beam structural elements.[1] They found that dynamic modes for slender curved structures are sensitive to displacement constraints that induce axial forces. When a step force acts at an interior point of a structure, displacement constraints at the loaded point result in an axial force at the point where the load is applied that increases with the load. This axial resultant force depends on both the radial and tangential components of the applied force. In curved structures where axial resultant forces increase without limit as the load increases, the number of rigid segments in a dynamic mode can also increase without limit.

These conclusions arise when interactions between axial and bending stress resultants are neglected; i.e. when the yield condition depends solely on the bending moment. With a constant yield moment at plastic hinges and an impulsive load, a closed curved structure such as a ring has no perfectly plastic solution that is both kinematically and dynamically admissible. For large forces (or localized impulsive loads) the initial stage of motion is dominated by plastic axial compression at hinges close to the load and this cannot be neglected. Applied forces that induce large axial stress resultants require use of a yield condition that couples axial and bending stress resultants in deforming segments; i.e. an analysis similar to that in Sect. 6.3. The solutions presented by Stronge et al. [1990] demonstrate that the effect of the axial stress resultant on yield is not negligible when large axial resultant forces are induced by load and geometry. The example of a rigid–plastic ring with a yield polygon that couples axial and bending stress resultants was analyzed by Cline and Jahsman [1967] for a smoothly distributed impulsive load. The early time solution exhibited a plastic compression segment of changing length within the loaded section; this segment finally dissipated a substantial part of the initial energy. For a force applied at a point on a curved structure, however, the corresponding plastic compression segment degenerates to a hinge at the loading point. Although the analysis developed in the present section neglects the effect of axial stress resultants on yield, it is useful for developing general understanding of mechanisms for deformation of suddenly loaded, curved structures.

6.5 Circular Arc Cantilever Subjected to In-plane Impact

In this section, we consider a quadrantal circular cantilever beam subjected to radial impact at the tip by a rigid particle mass travelling in the plane of the curved beam; this problem was first investigated by Yu et al. [1985b, 1986].

The basic assumptions have been declared at the beginning of Sect. 6.4. As depicted by Fig. 6.22a, suppose that the collision imparts an initial inward velocity V_0 in the radial direction to a particle of mass G at the tip of the curved cantilever. Here only a small deflection analysis is presented; a further analysis of this problem based on a large deflection formulation is given by Zhang and Yu [1986].

[1] Dynamic response of a rigid–plastic ring was previously studied by Owens and Symonds [1955], Cline and Jahsman [1967] and Zhang and Yu [1989]; the large deflection dynamic response of arches was analyzed by Palomby and Stronge [1988].

Section 6.4 showed the existence of multi-hinge mechanisms, but these are related to various subtending angles of curved cantilevers that are larger than $\alpha=\pi/2$. In the present problem where $\alpha=\pi/2$, the quadrantal circular configuration and impulsive loading allow the complete solution to be constructed using a single hinge mechanism which travels from the tip to the root, provided the effects of shear and axial force on yielding are neglected.

6.5.1 Rigid–Plastic Formulation

Velocity Field In a single-hinge mechanism, the dynamic deformation can be expected to have two distinct phases: a transient phase (*phase I*) and a modal phase (*phase II*). In phase I, a plastic hinge travels from the impact point (i.e. the tip of the curved cantilever) towards the root. Suppose that the collision occurs at time $t=0$; and at time $t>0$ a plastic hinge forms at an interior cross-section H, located at angle β, as shown in Fig. 6.22b. Arc segment AH rotates about the instantaneous center H as a rigid body, with angular velocity ω, while arc segment HB remains at rest. Hence, magnitudes of radial velocities related to this mechanism are given by

$$\dot{W}(\phi) = \begin{cases} 2\omega R\sin\dfrac{(\beta-\phi)}{2}, & 0\le\phi\le\beta \\ 0, & \beta\le\phi\le\pi/2 \end{cases} \tag{6.81}$$

where ϕ denotes an angle measured from the vertical Z-axis to an arbitrary cross-section C in the curved cantilever, see Fig. 6.22b. Here angle ϕ plays the same role as arc length measured from the tip. In particular, the velocity at the tip is given by taking $\phi=0$ in (6.81); that is,

$$V = \dot{W}(0) = 2\omega R\sin\frac{\beta}{2} \tag{6.82}$$

A comparison of (6.81) and (6.82) leads to

$$\dot{W}(\phi) = \begin{cases} V\sin\dfrac{(\beta-\phi)}{2}\Big/\sin\dfrac{\beta}{2}, & 0\le\phi\le\beta \\ 0, & \beta\le\phi\le\pi/2 \end{cases} \tag{6.83}$$

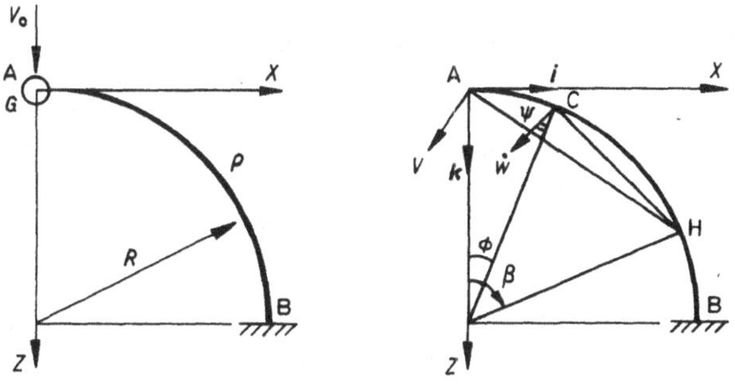

a b

Fig. 6.22. (a) A quadrantal circular cantilever subjected to in-plane radial impact by heavy particle at the tip; (b) deformation mechanism with single hinge at interior section H.

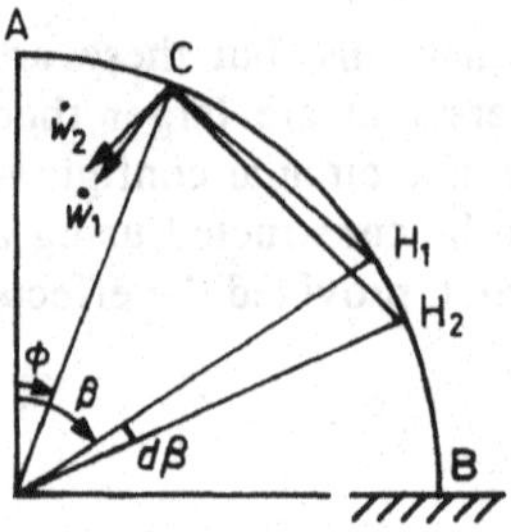

Fig. 6.23. Change in direction of velocity at section C in arc AH when plastic hinge moves from H_1 to H_2.

which relates to the distribution of the velocity magnitude to the tip velocity V and the hinge location β.

Let **i** and **k** denote unit vectors in the directions of coordinates X and Z, respectively. Note that the direction of velocity $\dot{W}$ is perpendicular to chord $\overline{HC}$, see Fig. 6.22b, so that the velocity vector at cross-section C within arc segment AH can be written as

$$\begin{aligned}\dot{\mathbf{W}} &= -\dot{W}\sin\psi\,\mathbf{i}+\dot{W}\cos\psi\,\mathbf{k}\\ &= -\dot{W}\sin\left(\frac{\beta+\phi}{2}\right)\mathbf{i}+\dot{W}\cos\left(\frac{\beta+\phi}{2}\right)\mathbf{k}\end{aligned} \tag{6.84}$$

where a geometric relation is defined as $\psi=(\beta+\phi)/2$.

Acceleration Field To obtain the acceleration field, the differentiation of (6.84) with respect to time t gives

$$\begin{aligned}\ddot{\mathbf{W}} = \frac{\mathrm{d}\dot{\mathbf{W}}}{\mathrm{d}t} &= -\left\{\ddot{W}\sin\left(\frac{\beta+\phi}{2}\right)+\dot{W}\frac{\dot{\beta}}{2}\cos\left(\frac{\beta+\phi}{2}\right)\right\}\mathbf{i}\\ &+\left\{\ddot{W}\cos\left(\frac{\beta+\phi}{2}\right)-\dot{W}\frac{\dot{\beta}}{2}\sin\left(\frac{\beta+\phi}{2}\right)\right\}\mathbf{k}\end{aligned} \tag{6.85}$$

Accordingly, the magnitude of the acceleration is found to be

$$(\ddot{\mathbf{W}})^2 = \ddot{\mathbf{W}}\cdot\ddot{\mathbf{W}} = \ddot{W}^2+\dot{W}^2\dot{\beta}^2/4 \tag{6.86}$$

for a cross-section C within arc segment AH ($0\le\phi\le\beta$); and in particular,

$$(\dot{\mathbf{V}})^2 = \dot{\mathbf{V}}\cdot\dot{\mathbf{V}} = \dot{V}^2+V^2\dot{\beta}^2/4 \tag{6.87}$$

for tip A ($\phi=0$). The first term on the right-hand side of (6.86) and (6.87) indicates that the accelerations at C and A are partly caused by the variation of velocity magnitude related to the change in the angular velocity about hinge H; while the second term shows that the accelerations are also partly caused by the change in the direction of velocities at C and A due to the movement of hinge H. As shown in Fig. 6.23, when a hinge moves from H_1 to H_2 the velocity vector at point C changes from $\dot{\mathbf{W}}_1$ to $\dot{\mathbf{W}}_2$, and this change in direction causes an axial component of acceleration.

The magnitude of $\ddot{W}$, which is only the variation of $\dot{W}$ in the current direction, can be obtained directly from (6.83):

$$\ddot{W}(\phi) = \dot{V}\frac{\sin((\beta-\phi)/2)}{\sin(\beta/2)}+\frac{1}{2}V\dot{\beta}\frac{\sin(\phi/2)}{\sin^2(\beta/2)} \tag{6.88}$$

To apply Tamuzh's variational principle for rigid–plastic dynamics, Eq. (2.20), it should be noted that in the present case, the functional related to this principle is

$$J = J(\dot{V}, \dot{\beta}) = \frac{1}{2}\int_0^\beta \rho \ddot{\mathbf{W}} \cdot \ddot{\mathbf{W}} R\, \mathrm{d}\phi + \frac{1}{2} G \dot{\mathbf{V}} \cdot \dot{\mathbf{V}} + M_p \dot{\omega} \tag{6.89}$$

where $\omega = V/[2R\sin(\beta/2)]$ is the instantaneous angular velocity of segment AH about point H. Substituting Eqs (6.83), (6.86), (6.87) and (6.88) into (6.89), one finds

$$\begin{aligned} J = J(\dot{V}, \dot{\beta}) = \frac{1}{2}\rho R \int_0^\beta \Bigg\{ & \dot{V}^2 \frac{\sin^2((\beta-\phi)/2)}{\sin^2(\beta/2)} + V\dot{V}\dot{\beta} \frac{\sin((\beta-\phi)/2)\sin(\phi/2)}{\sin^3(\beta/2)} \\ & + \frac{1}{4} V^2 \dot{\beta}^2 \frac{\sin^2(\phi/2)}{\sin^2(\beta/2)} + \frac{1}{4} V^2 \dot{\beta}^2 \frac{\sin^2((\beta-\phi)/2)}{\sin^2(\beta/2)} \Bigg\} \mathrm{d}\phi \\ & + \frac{1}{2} G \left\{ \dot{V}^2 + \frac{V^2 \dot{\beta}^2}{4} \right\} + \frac{M_p}{2R} \frac{\dot{V}\sin(\beta/2) - 0.5 V\dot{\beta}\cos(\beta/2)}{\sin^2(\beta/2)} \end{aligned} \tag{6.90}$$

According to Tamuzh's principle the actual acceleration field makes J a minimum; that is,

$$\frac{\partial J}{\partial \dot{V}} = 0, \qquad \frac{\partial J}{\partial \dot{\beta}} = 0 \tag{6.91}$$

which gives the following pair of equations

$$\dot{V}\left\{\beta - \sin\beta + \frac{G}{\rho R}(1-\cos\beta)\right\} + V\dot{\beta}\left\{1 - \frac{\beta}{2}\frac{1+\cos\beta}{\sin\beta}\right\} = -\frac{M_p}{\rho R^2}\sin(\beta/2) \tag{6.92}$$

$$\dot{V}\left\{1 - \frac{\beta}{2}\frac{1+\cos\beta}{\sin\beta}\right\} + \frac{V\dot{\beta}}{4}\left\{\frac{\beta - \sin\beta}{\sin^2(\beta/2)} + \beta - \sin\beta + \frac{2G}{\rho R}\sin^2(\beta/2)\right\} = \frac{M_p}{2\rho R^2}\cos(\beta/2) \tag{6.93}$$

From the sum of twice (6.93) and the product of $\cot(\beta/2)$ and (6.92) we obtain

$$\dot{V}\left\{1 - \cos\beta + \frac{G}{\rho R}\sin\beta\right\} + \frac{V\dot{\beta}}{2}\left\{2\beta - \sin\beta + \frac{G}{\rho R}(1-\cos\beta)\right\} = 0 \tag{6.94}$$

By defining the following set of nondimensional variables

$$T_0 \equiv R\sqrt{\frac{\rho R}{M_p}}, \quad \tau \equiv \frac{t}{T_0}, \quad w \equiv \frac{W}{R}, \quad v \equiv \frac{V}{R/T_0}, \quad \gamma \equiv \frac{G}{\rho R} \tag{6.95}$$

the basic equations (6.94) and (6.92) are recast in nondimensional form as

$$\frac{\mathrm{d}v}{\mathrm{d}\tau}[1 - \cos\beta + \gamma\sin\beta] + \frac{v}{2}\frac{\mathrm{d}\beta}{\mathrm{d}\tau}[2\beta - \sin\beta + \gamma(1-\cos\beta)] = 0 \tag{6.96a}$$

$$\frac{\mathrm{d}v}{\mathrm{d}\tau}[\beta - \sin\beta + \gamma(1-\cos\beta)] + \frac{v}{2}\frac{\mathrm{d}\beta}{\mathrm{d}\tau}\left[2 - \beta\frac{1+\cos\beta}{\sin\beta}\right] = -\sin(\beta/2) \tag{6.96b}$$

Solving (6.96) for $(\mathrm{d}v/\mathrm{d}\tau)$ and $(\mathrm{d}\beta/\mathrm{d}\tau)$, one obtains

$$\frac{dv}{d\tau} = \sin(\beta/2)[2\beta - \sin\beta + \gamma(1-\cos\beta)]/2\Gamma$$
$$\frac{d\beta}{d\tau} = -\sin(\beta/2)[1-\cos\beta + \gamma\sin\beta]/v\Gamma \tag{6.97}$$

with

$$\Gamma = 1-\cos\beta + \beta\sin\beta - \sin^2(\beta/2) + \gamma[2\sin\beta - 2\beta + \beta\cos\beta + \sin\beta\cos\beta] - \gamma^2(1-\cos\beta)^2/2 \tag{6.98}$$

In view of the singularity at $\tau = 0$ the numerical integration of Eqs (6.97) can be started by choosing slightly modified initial conditions,

$$\beta = \sqrt{\frac{6\tau_0}{v_0}}, \qquad v = \frac{v_0}{1+\dfrac{1}{2\gamma}\sqrt{\dfrac{6\tau_0}{v_0}}} \tag{6.99}$$

where $\tau_0 << 1$ is an arbitrary small value and $v_0 \equiv V_0T_0/R$ is the nondimensional initial velocity of the colliding mass. The initial condition (6.99) is obtained by taking series expansions of Eqs (6.97) and (6.98) for small β and then integrating these expressions.

6.5.2 Discussion of Solution

Complete Rigid–Plastic Solution When nondimensional parameters for initial velocity v_o and mass ratio γ are specified, a numerical solution can be obtained by integrating Equation (6.97) using the Runge–Kutta procedure, starting from the initial condition (6.99).

Like the response of a straight cantilever to impact at the tip (Sect. 4.5), the dynamic response of the curved cantilever also has two phases — a transient phase with a travelling hinge and a subsequent modal phase where the cantilever rotates about a plastic hinge at the root. Figure 6.24 shows the variation of the hinge position β with time (t/t_1). Here t_1 denotes the time when the travelling hinge reaches the root of the curved cantilever; i.e. the time when phase I ends. This indicates that the smaller the mass ratio $\gamma = G/\rho R$, the more uniform is the

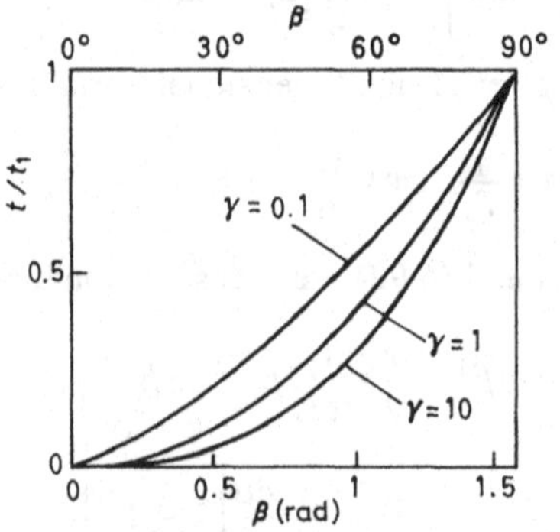

Fig. 6.24. Variation of hinge location β with time.

travelling speed of the plastic hinge along the curved cantilever. For phase II (the modal phase) the rotation angle about the root hinge θ_B, can be easily calculated from the remaining kinetic energy at the end of phase I.

To illustrate the admissibility of this rigid–plastic solution, we should examine whether the yield condition is satisfied in the entire cantilever. First an expression for the axial force at the hinge, N_H, can be deduced according to the acceleration field (6.85); then the bending moment at the root, M_B, is found to be

$$\begin{aligned}\tilde{m}_B &\equiv \frac{M_B}{M_p} = 1 + \frac{N_H R}{M_p}(1 - \sin\beta) \\ &= 1 + \frac{1-\sin\beta}{4\Gamma}\{[2\beta - \sin\beta + \gamma(1-\cos\beta)]\,[\beta - \sin\beta + \gamma(1-\cos\beta)] \\ &\qquad + \sin\beta[1-\cos\beta + \gamma\sin\beta](\beta - \sin\beta)/(1-\cos\beta)\}\end{aligned}$$

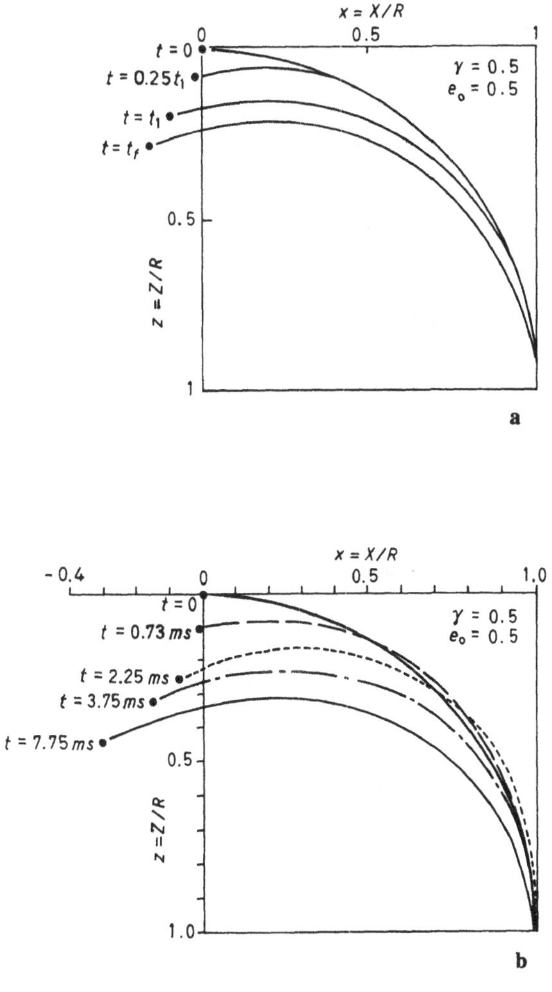

Fig. 6.25. Deformed shapes of circular cantilever in case of $e_0 = 0.5$, $\gamma = 0.5$: (a) complete rigid–plastic solution; (b) finite-element elastic–plastic solution.

where Γ is defined by (6.98). For an inward radial force F_r, calculation indicates that (1) the axial force at the travelling hinge H is always compressive; (2) by taking $\beta \to 0$ in the expression above, it is known that immediately after the impulsive loading (i.e. when $\tau \ll 1$ and $\beta \ll 1$) $N_H R/M_p \approx -5/4$ and $\tilde{m}_B \approx -1/4$; and (3) during the transient phase, the bending moment at the root increases monotonically from $\tilde{m}_B \approx -0.25$ at $\beta = 0$ to $\tilde{m}_B \approx 1.0$ at $\beta = \pi/2$. During the entire dynamic response, the yield condition is not violated anywhere in the curved cantilever; hence the single-hinge rigid–plastic solution is complete.

Deformed Shapes If the tip velocity $v_0(\tau)$ and the hinge location $\beta(\tau)$ are known for phase I and the hinge rotation θ_B is known in phase II, the deflection of a circular arc cantilever can be calculated at every instant. Figure 6.25a shows the deformed shapes based on a complete *rigid–plastic* solution for a typical example of a light mass, $\gamma = 1/2$ and input energy ratio $e_0 = 1/2$. In the figure the termination times of phases I and II are t_1 and t_f, respectively. For comparison, Fig. 6.25b gives deformed shapes obtained from a finite-element calculation using the same energy ratio; the finite element analysis, however, considered an elastic–perfectly plastic material, see Yu et al. [1986]. Note that in the first part of the elastic–plastic response period, the segment close to the root undergoes rotation in a *reverse direction* (clockwise in the figure) before it begins to rotate in an anti-clockwise direction. Also it should be noted that the time t = 7.75 ms is the termination of plastic deformation in the elastic–plastic cantilever, but the deformed shape is still changing due to elastic vibration.

Like the case of a straight rigid–plastic cantilever, the small deflection solution for a curved cantilever also gives a final inclination at the tip A that is *always* equal to the initial kinetic energy divided by the fully plastic bending moment of the cantilever,

$$|\theta_{Af}| = e_0 \equiv K_0/M_p \tag{6.100}$$

Energy Dissipation The part of the initial impact energy that is dissipated in phase I is independent of the initial velocity v_0; it is a function of only the mass ratio $\gamma = G/\rho R$. As shown in Fig. 6.26, the value D_1/K_0 decreases with increasing mass ratio γ .

For very large mass ratio, an asymptotic limit can be obtained by taking $\gamma \to \infty$ in expression (6.98) and Eq. (6.97); thus, one finds

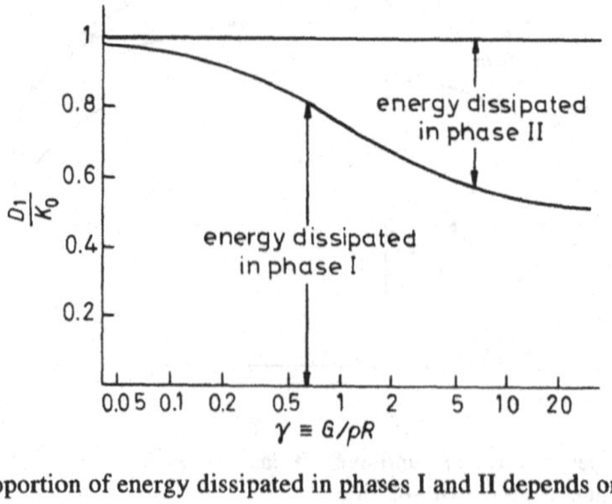

Fig. 6.26. Proportion of energy dissipated in phases I and II depends on mass ratio γ.

$$\Gamma \rightarrow -\gamma^2(1-\cos\beta)^2/2$$

$$\frac{dv}{d\tau} \rightarrow \sin(\beta/2)\gamma(1-\cos\beta)/2\Gamma = -\left[2\gamma\sin(\beta/2)\right]^{-1}$$

$$\frac{d\beta}{d\tau} \rightarrow -\sin(\beta/2)\gamma\sin\beta/v\Gamma = \cos(\beta/2)\left[v\gamma\sin^2(\beta/2)\right]^{-1}$$

It follows that

$$\frac{dv}{d\beta} = -\frac{v\sin(\beta/2)}{2\cos(\beta/2)} \tag{6.101}$$

The differential equation has a solution

$$v = v_0\cos(\beta/2) \tag{6.102}$$

which also satisfies the initial condition $v = v_0$ at $\tau = 0$ when the hinge is initially at the impact point $\beta = 0$. At the end of phase I, the hinge has moved to $\beta = \pi/2$, so (6.102) gives a tip speed $v = v_0\cos(\pi/4) = v_0/\sqrt{2}$. The part of the impact energy dissipated in phase I is found to be

$$D_1/K_0 \rightarrow 1/2 \tag{6.103}$$

This is illustrated by Fig. 6.26, where the curve approaches 1/2 as mass ratio $\gamma \rightarrow \infty$. Expression (6.103) indicates that for a curved cantilever of circular plan form subjected to transverse impact at the tip, the travelling plastic hinge dissipates at least half the entire energy, even if the colliding mass is very heavy. This conclusion is notably different from that for a straight rigid–plastic cantilever; in that case $D_1/K_0 \rightarrow 0$ if $\gamma \rightarrow \infty$. In other words, for a straight cantilever all the impact energy is dissipated in the modal phase when the mass ratio is very large, see Eq. (4.141) and Fig. 4.29.

6.5.3 Modal Approximation

A modal approximation consists of a kinematically admissible velocity field that is a separable function of spatial and temporal independent variables (see Chap. 2). For a curved cantilever, a modal velocity field has the form

$$\dot{\mathbf{W}}(\phi,t) = V^*(t)\mathbf{\Phi}^*(\phi) \tag{6.104}$$

where $V^*(t)$ is the tip speed of a modal approximation, and $\mathbf{\Phi}^*(\phi) = \Phi_1^*(\phi)\cdot\mathbf{i} + \Phi_2^*(\phi)\cdot\mathbf{k}$ is the shape vector, with $\mathbf{i}$ and $\mathbf{k}$ being unit vectors shown in Fig. 6.22b. In the present problem, the principal mode for the curved cantilever is a rigid-body rotation about the root B; thus at the tip where $\phi = 0$, the mode shape has components $\Phi_1^*(0) = -1$ and $\Phi_2^*(0) = 1$. Hence, the components of the shape function can be written as

$$\Phi_1^*(\phi) = -\cos\phi, \qquad \Phi_2^*(\phi) = 1-\sin\phi \tag{6.105}$$

The initial value of the tip velocity $V^*(t)$ in the modal approximation can be determined by the *min Δ_0 technique* (refer to Sect. 2.5). Thus one finds

$$V^*(0) \equiv \frac{V(0)}{2(1+0.5708/\gamma)} \tag{6.106}$$

where $\gamma = G/\rho R$ is the mass ratio. Accordingly, the initial kinetic energy of the modal field is

$$K_0^* = \frac{1}{2}[V^*(0)]^2 \int_0^{\pi/2} \rho(\Phi_1^{*2} + \Phi_2^{*2}) R \, d\phi + \frac{1}{2} G [\sqrt{2} V^*(0)]^2$$
$$\equiv \frac{K_0}{2(1+0.5708/\gamma)} \tag{6.107}$$

The final rotation angle at the root B is then

$$\left|\theta_{Bf}^*\right| = \frac{K_0}{M_p} = \frac{e_0}{2(1+0.5708/\gamma)} \tag{6.108}$$

from which the final displacement at the tip can be easily found.

For the limiting case $\gamma \to \infty$, expressions (6.106)–(6.108) give

$$V^*(0) \to V_0/2, \qquad K_0^* \to K_0/2, \qquad \left|\theta_{Bf}^*\right| \to e_0/2 \tag{6.109}$$

Note that the modal approximation corresponds to phase II in the complete rigid–plastic solution; hence (6.109) is consistent with (6.103), and both expressions indicate that phase I and phase II each dissipate one half of the initial kinetic energy when the mass ratio is very large.

The fact that only half of the impact energy is dissipated by flexure can be explained as follows. The initial velocity of the colliding mass at point A can be resolved into two components; these are perpendicular and parallel to the line BA, respectively. Only the component that is perpendicular to BA contributes to the bending moment at the root where the modal hinge is located. Therefore, only half of the initial kinetic energy is dissipated by the rotation at this hinge in either phase II or the modal approximation. The remainder of the impact energy is dissipated at impact by the reaction impulse at the hinge.

6.6 Circular Arc Cantilever Subjected to Out-of-Plane Step Force

In Sects 6.4 and 6.5, a circular curved cantilever was dynamically loaded in the plane of initial curvature, so yielding was dominated by the bending moment. In the problem considered below the curved cantilever is subjected to a step force suddenly applied in a direction normal to the plane of the bend, so the loading is a combination of bending moment and torque.

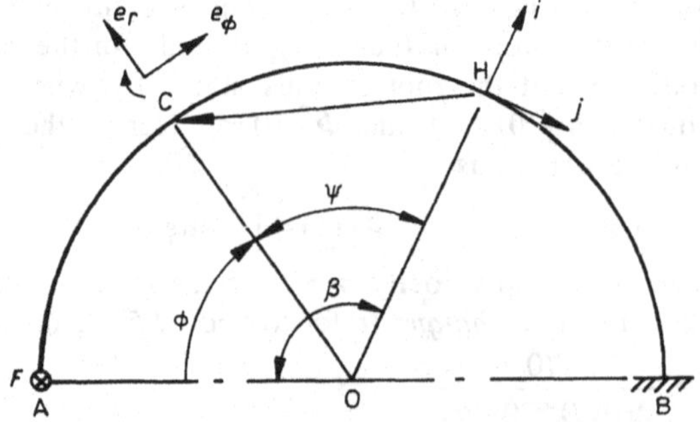

Fig. 6.27. Cantilever of circular plan form subjected to step force F acting normal to the plane; H denotes the position of a generalized plastic hinge.

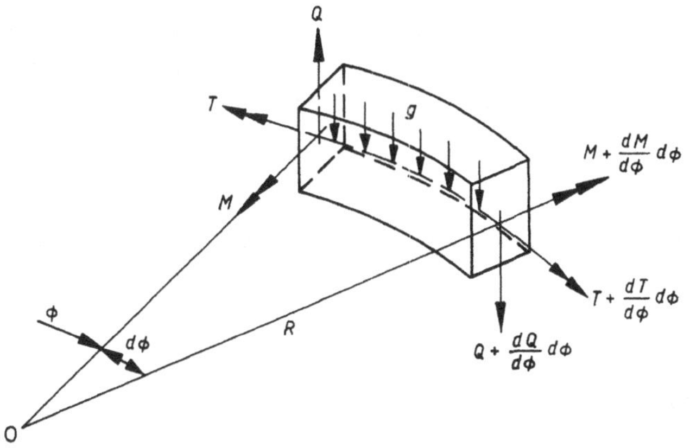

Fig. 6.28. Free–body diagram of small segment of curved cantilever, showing the positive sense of external force, shear force, bending moment and torque.

6.6.1 Equations of Motion

Problem and Assumptions Consider a uniform curved cantilever of constant radius R, that subtends a circular arc of 180°, as shown in Fig. 6.27. End B is built-in while the other end A is free. Suppose that an out-of-plane step force of magnitude F is applied at the tip A of the curved cantilever. The direction of the force is normal to the plane of the cantilever, (Fig. 6.27). An analysis of this problem with the assumptions stated in the beginning of Sect. 6.4 was originally given by Yu, Hua and Johnson [1985a].

Previous analysis of a straight (or curved) cantilever subjected to a step force at the tip (see Sects 4.1 and 6.4, respectively) has indicated that a steady pattern of deformation develops in response to this suddenly applied steady force; this pattern is a dynamic mode for the structure subjected to this force. For a steady force, the stationary pattern appears because the accelerations are independent of the velocities. In the present problem, therefore, it can be expected that the step force will create a stationary generalized plastic hinge at a cross-section H; this hinge can take any position along the circular arc. The aim of the present analysis is to find the modal solution for a curved cantilever. The mode shape (i.e. the location of the generalized plastic hinge) is a function of the force magnitude F.

Static Equilibrium Let an angular coordinate ϕ measured from the tip be used to locate any section; then *static* equilibrium of a differential element along the length of a curved cantilever such as shown in Fig. 6.28 leads to

$$\frac{dQ}{d\phi} = -gR \tag{6.110}$$

$$\frac{dT}{d\phi} = -M \tag{6.111}$$

$$\frac{dM}{d\phi} - T = -QR \tag{6.112}$$

where $g(\phi)$, $Q(\phi)$, $M(\phi)$ and $T(\phi)$ denote the distributed external force, shear force, bending moment and torque, respectively. The positive sense of forces and couples are indicated in Fig. 6.28. A combination of Eqs (6.110)–(6.112) results in

$$\frac{d^2M}{d\phi^2} + M = gR^2 \tag{6.113}$$

Equations (6.111)–(6.113) can be recast into nondimensional form

$$\frac{d\tilde{t}}{d\phi} = -\frac{1}{\eta}\tilde{m} \tag{6.111'}$$

$$\frac{d\tilde{m}}{d\phi} - \eta\tilde{t} = -q \tag{6.112'}$$

$$\frac{d^2\tilde{m}}{d\phi^2} + \tilde{m} = \frac{gR^2}{M_p} \tag{6.113'}$$

where

$$\tilde{m} \equiv \frac{M}{M_p}, \qquad \tilde{t} \equiv \frac{T}{T_p}, \qquad q \equiv \frac{QR}{M_p}, \qquad \eta \equiv \frac{T_p}{M_p} \tag{6.114}$$

Parameter η depends solely on the shape of the cross-section. For some typical cross-sections the values of η are listed in Table 6.2. The table indicates that $\eta < 1$ holds for most cross-sections, but $\eta \geq 1$ is possible if the cross-section is very thin and wide. For example, $\eta \geq 1$ for rectangular cross-sections if $b/h \geq 2.488$ with $\alpha = \sqrt{3}$; also $\eta \geq 1$ for thin-walled rectangular box sections if $b/h \geq 3.232$ with $\alpha = \sqrt{3}$. Here, the ratio of uniaxial yield stress Y to yield stress in pure shear τ_y is $\alpha \equiv Y/\tau_y$ — this ratio depends on the yield criterion. For example, $\alpha = 2$ for the Tresca criterion and $\alpha = \sqrt{3}$ for the von Mises criterion.

Table 6.2 Ratio of fully plastic torque to fully plastic bending moment η

Cross-section	M_p	T_p	$\eta \equiv T_p/M_p$	η (with $\alpha = \sqrt{3}$)
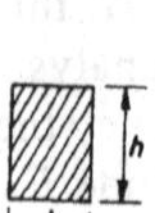	$\frac{1}{4}Ybh^2$	$\frac{1}{6\alpha}Yb^2(3h-b)$ $(b \leq h)$	$\frac{2b}{3\alpha h}\left(3-\frac{b}{h}\right)$ $(b \leq h)$	$\frac{2b}{3\sqrt{3}h}\left(3-\frac{b}{h}\right)$ $(b \leq h)$
		$\frac{1}{6\alpha}Yh^2(3b-h)$ $(b \geq h)$	$\frac{2}{3\alpha}\left(3-\frac{h}{b}\right)$ $(b \geq h)$	$\frac{2}{3\sqrt{3}}\left(3-\frac{h}{b}\right)$ $(b \geq h)$
	$\frac{1}{4}Yb^3$	$\frac{1}{3\alpha}Yb^3$	$\frac{4}{3\alpha}$	0.770
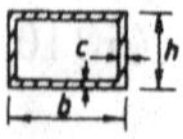	$\frac{1}{2}Y(2b+h)hc$	$\frac{2}{\alpha}Ybhc$	$\frac{4}{\alpha}\frac{b}{(2b+h)}$	$\frac{4}{\sqrt{3}}\frac{b}{(2b+h)}$
	$\frac{4}{3}Ya^3$	$\frac{2\pi}{3\alpha}Ya^3$	$\frac{\pi}{2\alpha}$	0.907
	$4Ya^2c$	$\frac{2\pi}{3\alpha}Ya^2c$	$\frac{\pi}{2\alpha}$	0.907

Yield Function and Flow Rule In Sect. 1.5, it was shown that the interactive yield function for bending moment and torque for beams of circular or rectangular cross-section can be approximated by

$$\Psi(\tilde{m},\tilde{t}) \equiv \tilde{m}^2 + \tilde{t}^2 - 1 \tag{6.115}$$

where $\Psi(\tilde{m},\tilde{t}) = 0$ represents a circular yield locus in the plane of generalized stresses $(\tilde{m},\tilde{t})$, see Fig. 1.11.

Suppose a generalized plastic hinge forms at a cross-section H. The location of the hinge subtends an angle β from the tip. At the hinge the bending moment, torque and yield function are as follows:

$$\tilde{m}_{\mathrm{H}} \equiv \frac{M_{\mathrm{H}}}{M_p}, \qquad \tilde{t}_{\mathrm{H}} \equiv \frac{T_{\mathrm{H}}}{T_p}, \qquad \Psi_{\mathrm{H}} \equiv \tilde{m}_{\mathrm{H}}^2 + \tilde{t}_{\mathrm{H}}^2 - 1 \tag{6.116}$$

Associated with the yield criterion $\Psi(\tilde{m},\tilde{t}) = 0$ that is satisfied at the hinge, the flow rule gives

$$\dot{\theta}_t = \mu\dot{\theta}_m/\eta \tag{6.117}$$

where $\mu \equiv \tilde{t}_{\mathrm{H}}/\tilde{m}_{\mathrm{H}}$, while $\dot{\theta}_m$ and $\dot{\theta}_t$ denote the rotation rates in the directions of the unit vectors $\mathbf{e}_r$ and $\mathbf{e}_\phi$, respectively, see Fig. 6.27. Thus, arc segment AH has angular acceleration $\dot{\boldsymbol{\omega}}$ about a stationary hinge where

$$\dot{\boldsymbol{\omega}} = \frac{\mathrm{d}}{\mathrm{d}t}(\dot{\theta}_m\mathbf{e}_r + \dot{\theta}_t\mathbf{e}_\phi) = \ddot{\theta}_m(\mathbf{e}_r + \frac{\mu}{\eta}\mathbf{e}_\phi) \tag{6.118}$$

Let the angle subtended by arc HC be $\psi = \beta - \phi$. Then the acceleration of section C is

$$\ddot{\mathbf{w}} = \dot{\omega} \times \overline{\mathrm{HC}} = -R\ddot{\theta}_m\left\{\sin\psi - \frac{\mu}{\eta}(1-\cos\psi)\right\}\mathbf{k} \tag{6.119}$$

where $\mathbf{k} = -\mathbf{e}_r \times \mathbf{e}_\phi$ is a unit vector normal to the plane of the figure and inward. Related to this acceleration field, the inertia force per unit length is given by

$$g(\phi) = -|\rho\ddot{\mathbf{w}}| = -\rho R\ddot{\theta}_m\left\{\sin\psi - \frac{\mu}{\eta}(1-\cos\psi)\right\} \tag{6.120}$$

Substituting (6.120) into equation of motion (6.113') leads to

$$\frac{\mathrm{d}^2\tilde{m}}{\mathrm{d}\phi^2} + \tilde{m} = -\frac{\rho R^3}{M_p}\ddot{\theta}_m\left\{\sin\psi - \frac{\mu}{\eta}(1-\cos\psi)\right\} \tag{6.121}$$

Equations (6.111'), (6.112') and (6.121) can be solved subject to the following boundary conditions at the tip A,

$$q = -f, \qquad \tilde{m} = 0, \qquad \tilde{t} = 0, \qquad \text{at} \quad \psi = \beta \quad (\text{i.e. } \phi = 0) \tag{6.122}$$

where $f \equiv FR/M_p$ is the nondimensional force.

6.6.2 Solution

Bending Moment and Torque Combining Eqs (6.122) and (6.112') gives

$$\tilde{m} = 0, \qquad \frac{\mathrm{d}\tilde{m}}{\mathrm{d}\phi} = f, \qquad \text{at} \quad \psi = \beta \quad (\text{i.e. } \phi = 0) \tag{6.123}$$

The solution of the second-order differential equation (6.121) that satisfies boundary conditions (6.123), leads to the following expression for bending moment $\tilde{m}$

$$\tilde{m} = C_1 \sin\psi + C_2 \cos\psi + \frac{\rho R^3}{2M_p}\ddot{\theta}_m \left\{ \left(-\frac{\mu}{\eta}\sin\psi + \cos\psi \right)\psi + \frac{2\mu}{\eta} \right\} \quad (6.124)$$

with

$$C_1 = -\frac{\rho R^3}{2M_p}\ddot{\theta}_m \left(A_1 \sin\beta + A_2 \cos\beta \right) - f\cos\beta$$

$$C_2 = -\frac{\rho R^3}{2M_p}\ddot{\theta}_m \left(A_1 \cos\beta - A_2 \sin\beta \right) + f\sin\beta$$

$$A_1 = \left(-\frac{\mu}{\eta}\sin\beta + \cos\beta \right)\beta + \frac{2\mu}{\eta}$$

$$A_2 = -\frac{\mu}{\eta}\sin\beta + \cos\beta + \left(-\frac{\mu}{\eta}\cos\beta - \sin\beta \right)\beta$$

Substituting (6.124) into Eq. (6.111') and integrating, an expression for torque $\tilde{t}$ is obtained that satisfies boundary condition (6.122),

$$\tilde{t} = -\frac{1}{\eta}\Bigg\{ C_1 \cos\psi - C_2 \sin\psi + f$$

$$-\frac{\rho R^3}{2M_p}\ddot{\theta}_m \left[-\frac{\mu}{\eta}(\sin\psi - \psi\cos\psi) + \cos\psi + \psi\sin\psi + \frac{2\mu}{\eta}\psi - A_2 - C_3 \right]\Bigg\} \quad (6.125)$$

where

$$C_3 = -\frac{\mu}{\eta}(\sin\beta - \beta\cos\beta) + \cos\beta + \beta\sin\beta + \frac{2\mu}{\eta}\beta$$

Conditions at Plastic Hinge For mode identification, only a deformation mode with stationary hinges needs to be considered if the applied force is steady. In the analysis concerned with differentiability of the yield function (see Table 6.1), we showed that at a *stationary hinge* combined bending–torsion gives an interaction yield function $\Psi \in C^2$. Hence, at the plastic hinge H, conditions (6.1) apply; that is at $\phi = \beta$,

$$\Psi = 0, \qquad \frac{d\Psi}{d\phi} = 0, \qquad \frac{d^2\Psi}{d\phi^2} \le 0, \quad (6.126)$$

Since

$$\frac{d\Psi}{d\phi} = \frac{d}{d\phi}(\tilde{m}^2 + \tilde{t}^2 - 1) = 2\tilde{m}\frac{d\tilde{m}}{d\phi} + 2\tilde{t}\frac{d\tilde{t}}{d\phi}$$

by using (6.111'), the second equation in (6.126) leads to

$$\left.\frac{d\Psi}{d\phi}\right|_{\phi=\beta} = 2\tilde{m}\left(\frac{d\tilde{m}}{d\phi} - \tilde{t}\frac{1}{\eta} \right)\Bigg|_{\phi=\beta} = 0 \quad (6.127)$$

This implies that a stationary plastic hinge at an interior section H must be one of two possible types. Either

Type I: section H is under pure torsion so $\tilde{m} = 0$ at $\phi = \beta$; or
Type II: section H is subjected to combined bending and torsion, where

$$\left.\left(\frac{d\tilde{m}}{d\phi} - \tilde{t}\frac{1}{\eta}\right)\right|_{\phi=\beta} = 0 \tag{6.128}$$

Analysis of Type I Hinge For a hinge of type I, there is pure torsion at H, so $\tilde{m}_H = 0$ and $|\mu| \to \infty$, while $\ddot{\theta}_t = \mu\ddot{\theta}_m/\eta$ still holds. By applying conditions (6.123) and $\tilde{m}_H = 0$, Eq. (6.124) gives

$$\tilde{m} = f\left\{C_4 \sin\psi - C_5\left(1 - \cos\psi - \frac{\psi}{2}\sin\psi\right)\right\} \tag{6.129}$$

with

$$C_4 = \frac{2\left(1 - \cos\beta - \frac{\beta}{2}\sin\beta\right)}{(1-\cos\beta)^2}, \qquad C_5 = \frac{2\sin\beta}{(1-\cos\beta)^2}$$

After substituting (6.129) into equation (6.111'), the torque $\tilde{t}$ can be found by integration and use of the boundary condition $\tilde{t} = 0$ at $\psi = \beta$,

$$\tilde{t} = -\frac{f}{\eta}\left\{C_4 \cos\psi + C_5\left(\psi - \frac{3}{2}\sin\psi + \frac{\psi}{2}\cos\psi\right) + C_6\right\} \tag{6.130}$$

with

$$C_6 = \frac{2(1-\cos\beta - \beta\sin\beta) + \sin^2\beta}{(1-\cos\beta)^2}$$

The plastic hinge conditions $\Psi_H \equiv \tilde{m}_H^2 + \tilde{t}_H^2 - 1 = 0$ and $\tilde{m}_H = 0$ require $\tilde{t}_H = \pm 1$. However, $\tilde{t}_H = +1$ leads to a negative f according to (6.130). Hence for $f > 0$, only $\tilde{t}_H = -1$ is possible, and this leads to a relationship between the step force f and the hinge position β

$$f = \eta C^* \tag{6.131}$$

with

$$C^* = \frac{(1-\cos\beta)^2}{4 - 4\cos\beta - 3\beta\sin\beta + \sin^2\beta}$$

This expression gives the location of a pure torsion hinge of type I if force f is held steady. Now we examine conditions required for this type of hinge to develop. Since

$$\frac{d^2\Psi}{d\phi^2} = 2\left(\frac{d\tilde{m}}{d\phi}\right)^2 + 2\tilde{m}\frac{d^2\tilde{m}}{d\phi^2} - \frac{2\tilde{m}}{\eta}\frac{d\tilde{t}}{d\phi} - \frac{2\tilde{t}}{\eta}\frac{d\tilde{m}}{d\phi}$$

for a type I hinge, the third condition in (6.126) and $\tilde{t}_H = -1$ result in

$$\left.\frac{d^2\Psi}{d\phi^2}\right|_{\phi=\beta} = 2\frac{d\tilde{m}_H}{d\phi}\left(\frac{d\tilde{m}_H}{d\phi} - \frac{1}{\eta}\tilde{t}_H\right) = 2\frac{d\tilde{m}_H}{d\phi}\left(\frac{d\tilde{m}_H}{d\phi} + \frac{1}{\eta}\right) \le 0 \tag{6.132}$$

It is known from (6.129) that $d\tilde{m}_H/d\phi = -fC_4 < 0$ since $f > 0$ and $C_4 > 0$ for hinge positions in the range $0° \le \beta \le 180°$. Hence, (6.132) requires

$$fC_4 \le 1/\eta \tag{6.133}$$

or, by using (6.131),

$$C_4 C^* = \frac{2 - 2\cos\beta - \beta\sin\beta}{4 - 4\cos\beta - 3\beta\sin\beta + \sin^2\beta} \le \frac{1}{\eta^2} = \left(\frac{M_p}{T_p}\right)^2 \tag{6.134}$$

Satisfying inequality (6.134) establishes a *necessary condition* for forming a generalized plastic hinge of type I at a point H which subtends angle β. In order to judge whether this hinge is unique, however, it is also necessary to examine whether or not the generalized yield condition is satisfied throughout the undeformed portion of the curved cantilever; i.e. in arc segment HB, where B is at the root of the cantilever. Uniqueness requires

$$\Psi \equiv \left(\frac{M}{M_p}\right)^2 + \left(\frac{T}{T_p}\right)^2 - 1 < 0 \tag{6.135}$$

for an arbitrary section C in arc segment HB. It can be shown (see Yu, Hua and Johnson [1985a]) that Eq. (6.135) results in an inequality

$$C_4C^* = \frac{2-2\cos\beta-\beta\sin\beta}{4-4\cos\beta-3\beta\sin\beta+\sin^2\beta} < \frac{2(1-\cos\gamma_c)}{(1-\cos\gamma_c)^2+\eta^2\sin^2\gamma_c} \equiv \Gamma \tag{6.136}$$

where γ_c is the angle subtended by the arc HC. When $\gamma_c \to 0$, $\Gamma \to 1/\eta^2$, so that (6.136) coincides exactly with Eq. (6.134) when point C approaches hinge H. Consequently, after some calculation the following conclusions can be drawn.

1. If $\eta < 1$, then $\Gamma_{\min} = 1$ at $\gamma_c = 180°$, so that $C_4C^* \le \Gamma_{\min}$ requires $\beta \ge \beta^* \equiv 113°$, where $\beta^* = 113°$ is a root of the equation

$$1-\cos\beta-\beta\sin\beta+\frac{1}{2}\sin^2\beta = 0 \tag{6.137}$$

   Hence, the hinge is unique if it is located at $\beta \ge 113°$.
2. If $\eta \ge 1$, then $\Gamma_{\min} = 1/\eta^2$ at $\gamma_c \to 0$, so uniqueness of the hinge H can be examined by applying inequality (6.134).

Analysis of Type II Hinge For a stationary type II hinge, combining expressions (6.124) and (6.125) with condition (6.128) leads to

$$\frac{\rho R^3}{2M_p}\ddot{\theta}_m = \frac{f}{G_0} \tag{6.138}$$

where

$$G_0 = \left[\left(1-\frac{1}{\eta^2}\right)G_2 - \frac{1}{\eta^2}(1-G_1) - 1\right]\left[\left(1-\frac{1}{\eta^2}\right)\cos\beta + \frac{1}{\eta^2}\right]^{-1}$$

$$G_1 = A_2 + C_3 = 2\left(-\frac{\mu}{\eta}(\sin\beta-\beta)+\cos\beta\right)$$

$$G_2 = A_1\sin\beta + A_2\cos\beta = -\frac{\mu}{\eta}(\beta - 2\sin\beta + \sin\beta\cos\beta) + \cos^2\beta$$

In these expressions the value of $\mu = \tilde{t}_H/\tilde{m}_H$ is the negative root of a quadratic equation,

$$B_2\mu^2 + B_1\mu + B_o = 0 \tag{6.139}$$

with

$$B_o = (1-\cos\beta)^2$$

$$B_1 = -\left(1-\frac{1}{\eta^2}\right)\eta\sin^2\beta\cos^2\beta + \frac{1}{\eta}(\beta - 2\sin\beta + \sin\beta\cos\beta) - \frac{2}{\eta}(2\sin\beta-\beta)\cos\beta$$

$$+\left(1+\frac{1}{\eta^2}\right)\eta\sin\beta-\left[\left(1-\frac{1}{\eta^2}\right)\cos\beta+\frac{1}{\eta^2}\right]\eta(\beta-\cos\beta\sin\beta)$$

$$B_2 = \left(1-\frac{1}{\eta^2}\right)(\beta-2\sin\beta+\sin\beta\cos\beta)\sin\beta+\frac{2}{\eta^2}(\sin\beta-\beta)\sin\beta$$
$$+\left[\left(1-\frac{1}{\eta^2}\right)\cos\beta+\frac{1}{\eta^2}\right]\left(2-2\cos\beta-\sin^2\beta\right)$$

The yield condition at a plastic hinge $\Psi_{\mathrm{H}} \equiv \tilde{m}_{\mathrm{H}}^2+\tilde{t}_{\mathrm{H}}^2-1=0$ gives a relationship for the hinge location β as a function of force f at a type II hinge as

$$f = \eta\left\{\left(\frac{1}{G_0}(G_2-G_1+1)+1-\cos\beta\right)^2+\eta^2\left(\frac{G_3}{G_0}+\sin\beta\right)^2\right\}^{-1/2} \tag{6.140}$$

with

$$G_3 = -\frac{\mu}{\eta}(2-2\cos\beta-\sin^2\beta)+\beta-\sin\beta\cos\beta$$

Also, at a type II hinge, the third condition in (6.126) requires

$$\left.\frac{\mathrm{d}^2\Psi}{\mathrm{d}\phi^2}\right|_{\phi=\beta} = 2\tilde{m}_{\mathrm{H}}\left(\frac{\mathrm{d}^2\tilde{m}_{\mathrm{H}}}{\mathrm{d}\phi^2}+\frac{1}{\eta^2}\tilde{m}_{\mathrm{H}}\right) = 2\tilde{m}_{\mathrm{H}}^2\left(\frac{1}{\eta^2}-1\right) \le 0 \tag{6.141}$$

where Eq. (6.121) has been used together with $\phi=\beta$. Inequality (6.141) holds if and only if $\eta\ge1$. This implies that if $\eta\ge1$, the yield function Ψ has a local maximum at the type II hinge H, while in case of $\eta<1$, the yield function Ψ has a local minimum at the type II hinge H. An examination of the yield function at an arbitrary section in arc segment HB indeed leads to a similar conclusion, see Yu, Hua and Johnson [1985a] for details. Therefore, a mechanism with a single type II hinge provides a complete solution if $\eta\ge1$. A single hinge cannot provide

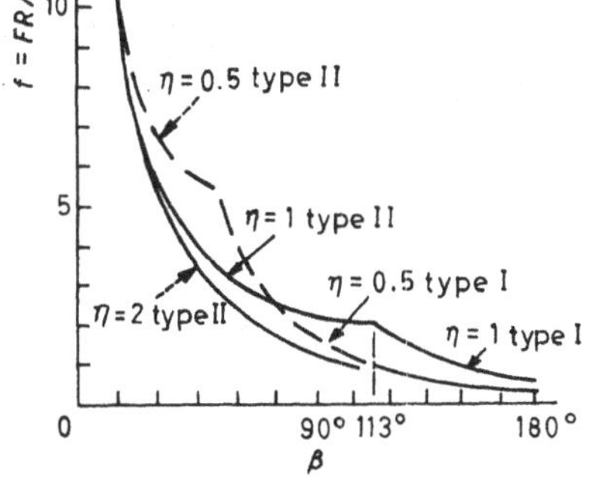

Fig. 6.29. Relationship between nondimensional step force f and hinge location β for cross-sections with $\eta=T_p/M_p=0.5$, 1.0 and 2.0.

a complete solution if $\eta < 1$, since the yield criterion is violated at some other sections. Table 6.2 shows that the ratio of fully plastic torque and bending moment satisfies $\eta \geq 1$ for only unusually elongated cross-sections; i.e. for combined bending and torsion the single hinge solution is exceptional.

Numerical Examples Calculated values relating the magnitude of the nondimensional step force and the location of the generalized plastic hinge (i.e. the f–β curves) for $\eta \equiv T_p/M_p = 0.5$, 1.0 and 2.0 are shown in Fig. 6.29. These results lead to the following observations:

1. A cantilever with circular plan form and a cross-section with $\eta < 1$ can have a plastic hinge of type II if f is very large; if f is somewhat smaller it can have a plastic hinge of type I. Neither provides a complete solution however, since the yield criterion is violated somewhere in the curved cantilever. Only when f is small and the hinge is of type I located at $\beta \geq 113°$, does the single hinge mechanism provide a complete solution.
2. For a cantilever having circular plan form and a cross-section with $\eta \geq 1$, the single-hinge mechanism with a hinge located at any β always leads to a complete solution; this hinge is type I if f is small and type II if f is large.

6.6.3 Discussion

1. The deformation mechanism with a single stationary generalized plastic hinge is *kinematically admissible* for the out-of-plane step force on a curved cantilever (Fig. 6.27), whether or not the yield criterion is satisfied in the entire cantilever. The above analysis indicates that the yield function is a maximum or extreme at the hinge if the yield condition is satisfied in the entire cantilever. In this problem, the single-hinge mechanism leads to a complete solution; in some other problems, however, the single hinge solution is not complete since the

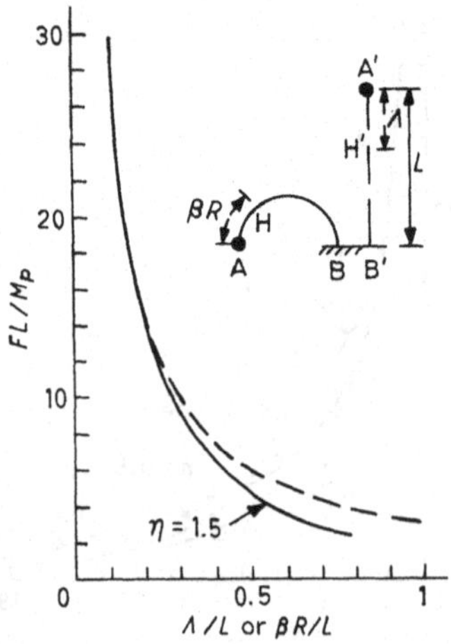

Fig. 6.30. Comparison of hinge locations in straight (- -) and curved (—) cantilevers subjected to step force F. Hinge position is given by arc length $\Lambda = \beta R$ measured from tip.

yield criterion is violated somewhere between the hinge and the root. For the latter problems, a complete solution requires mechanisms with more than one hinge. Nevertheless, the single-hinge mechanism is a kinematically admissible field so it gives an *upper bound for the final displacement obtained from any pulse load.*

2. To compare the present results for an out-of-plane step force on a cantilever of circular plan form with those for a step force on a straight cantilever (Sect. 4.1), let both cantilevers have the same length and cross-sectional properties. The relationship between force magnitude and plastic hinge location for these two cantilevers is shown in Fig. 6.30, where factor η is chosen to be 1.5. It is seen that when the plastic hinge is near the tip of these cantilevers, the hinge locations are almost the same. The effect of beam curvature on hinge location $\lambda = \beta R / L$ increases as the applied force f decreases since the hinge is further away from the tip.
3. Both Hodge [1959] and Boulton and Boonsukha [1959] observed that the yield function (6.115) is a *lower* bound for the interaction yield curve in the case of combined bending and torsion. Boulton and Boonsukha [1959] further pointed out that the values of $(\tilde{m}^2 + \tilde{t}^2)^{1/2}$ for the actual yield surface exceed unity by less than 15% for a square solid or hollow cross-section. This is unlikely to affect the previous discussion about whether or not a particular solution for the present problem violates the yield criterion.

6.7 Stepped or Bent Cantilever Subjected to Step Force

6.7.1 General Considerations

If a cantilever with circular plan form is subjected to a step force in the direction normal to the plane of its centroid, as shown in Sect. 6.6, a single-hinge deformation mechanism provides a complete solution only if the parameter $\eta \equiv T_p / M_p$ as well as the position of the hinge along the curved member satisfy limiting conditions. After examining the dynamic solution for a sharply bent cantilever subjected to a step force normal to the plane of its centroid, Hua, Yu and Johnson [1985] reached a conclusion similar to that for a circular arc cantilever; that is, a single-hinge deformation mechanism provides a complete solution that satisfies the yield criterion within the entire bent cantilever only in special cases. Otherwise, a single-hinge deformation mechanism results in a solution that violates yield in some sections of a bent cantilever. Here again, completeness of the solution for a single-hinge mechanism depends on the parameter $\eta \equiv T_p / M_p$ and the magnitude of the step force. Hence it is worthwhile to examine other possible deformation mechanisms when the single-hinge mechanism fails.

Before developing a double-hinge deformation mechanism for either stepped or sharply bent cantilevers subjected to a step force, we first consider a straight stepped cantilever as an example. This will illustrate why in some cases, a single-hinge deformation mechanism cannot lead to a complete solution. Figure 6.31a shows a straight stepped cantilever subjected to a transverse step force F at its

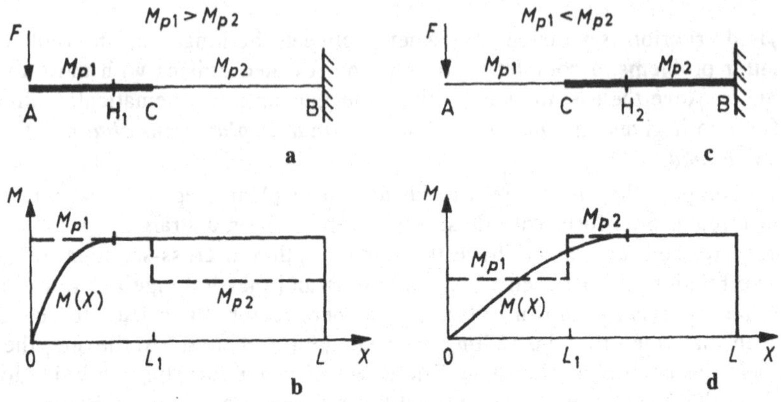

Fig. 6.31. (a) Stepped cantilever with $M_{p1} > M_{p2}$ subjected to a step force F at the tip; (b) bending moment diagram (solid curve) with hinge H_1 located in segment AC; dashed lines denote the fully plastic bending moment; (c) stepped cantilever with $M_{p1} < M_{p2}$ subjected to a step force F at the tip; (d) bending moment diagram (solid curve) with hinge H_2 located in segment CB; dashed lines denote the fully plastic bending moment.

tip. Let M_{p1} and M_{p2} be the fully plastic bending moment in segment AC and CB, respectively, of cantilever AB. If a single plastic hinge forms to the left of the discontinuous section C and $M_{p1} > M_{p2}$, then the bending moment distribution shown by the solid curve in Fig. 6.31b must violate the yield criterion (shown by the dashed lines) in segment CB. On the other hand, if a straight stepped cantilever satisfies $M_{p1} < M_{p2}$ as shown in Fig. 6.31c, the yield criterion will be violated in some portion of segment AC when a single plastic hinge H_2 forms at any section within segment CB, as indicated by Fig. 6.31d.

To overcome these difficulties Hua, Yu and Reid [1988] proposed a double-hinge deformation mechanism, in which two plastic hinges occur simultaneously in a stepped or bent cantilever. The following is a brief summary of their study on stepped cantilevers; then an analogy is established between the behavior of stepped cantilevers and bent cantilevers.

For these analyses, some of the assumptions employed in previous sections of this chapter are employed. Thus, rigid–perfectly plastic and time-independent materials are considered. The effect of shear force on yielding is neglected and the deflections are small in comparison with the length of the cantilever. Also, each segment of the stepped cantilever is considered to have uniform cross-section and density.

The dynamic response of a stepped or bent cantilever subjected to a step force at the tip, can be obtained from a five step analysis:

1. suppose that either one or two hinges form somewhere in the cantilever (for single or double-hinge mechanisms, respectively);
2. obtain two equations of motion for each rigid segment; these equations relate the external force to shear force, bending moment and acceleration distributions in each segment;
3. equations of motion obtained in (2) are nondimensionalized;
4. obtain an expression relating f (nondimensional step force) and λ (nondimensional coordinate of hinge position) by eliminating terms containing angular acceleration in the equations;

5. the following restrictions are checked: (a) yield criterion at each segment is not violated; (b) tip acceleration is in the same direction as that of step force; and (c) hinge location is inside the segment concerned.

Some of these restrictions limit the range of force f for a particular single or double hinge mechanism, or they impose a restriction on the geometry of the member. Hence, combining steps (4) and (5) above leads to a complete solution for some range of force f.

6.7.2 Stepped Cantilever

Case A: Stepped Cantilever with $M_{p1} > M_{p2}$ Consider a stepped cantilever with $M_{p1} > M_{p2}$, see Fig. 6.32a, and let L_1 and L_2 be the length of segments AC and CB, respectively, while ρ_1 and ρ_2 are the mass per unit length in segments AC and CB, respectively. By defining

$$f \equiv FL_1/M_{p1}, \quad \eta_M \equiv M_{p2}/M_{p1}, \quad \eta_\rho \equiv \rho_2/\rho_1, \quad \eta_L \equiv L_2/L_1 \tag{6.142}$$

it can be shown that plastic collapse initiates when the bending moment at the root B is equal to the yield moment. This gives a static collapse load f_{A0} for the cantilever,

$$f_{A0} = \eta_M/(1+\eta_L) \tag{6.143}$$

where subscript A pertains to Case A.

If $f > f_{A0}$, one or two hinges form somewhere in the cantilever. Following the method described in Section 6.7.1, various dynamic deformation mechanisms can be examined and finally *five* mechanisms are identified. These are summarized in Table 6.3.

In Table 6.3, H_1 denotes a hinge located in segment AC, and H_2 a hinge located in segment CB. The values of f, at which the dynamic deformation transfers from one mechanism to another are given by (6.143) and

$$f_{A1} = \frac{3\eta_M(\eta_\rho\eta_L^2 + 2\eta_L + 1)}{\eta_\rho\eta_L^3 + 3\eta_\rho\eta_L^2 + 3\eta_L + 1} \tag{6.144}$$

$$f_{A2} = 3\eta_M \tag{6.145}$$

$$f_{A3} = \frac{3(2-\lambda)}{\lambda(3-2\lambda)} \tag{6.146}$$

Table 6.3 Deformation mechanisms in stepped cantilever where $\eta_M = M_{p2}/M_{p1} < 1$

Mechanism	Plastic Hinge	Range of f
1	No hinge	$0 \le f < f_{A0}$
2	Single-hinge B	$f_{A0} \le f \le f_{A1}$
3	Single-hinge H_2	$f_{A1} < f < f_{A2}$
4	Single-hinge C	$f_{A2} \le f \le f_{A3}$
5	Double-hinge H_1-C	$f_{A3} < f$

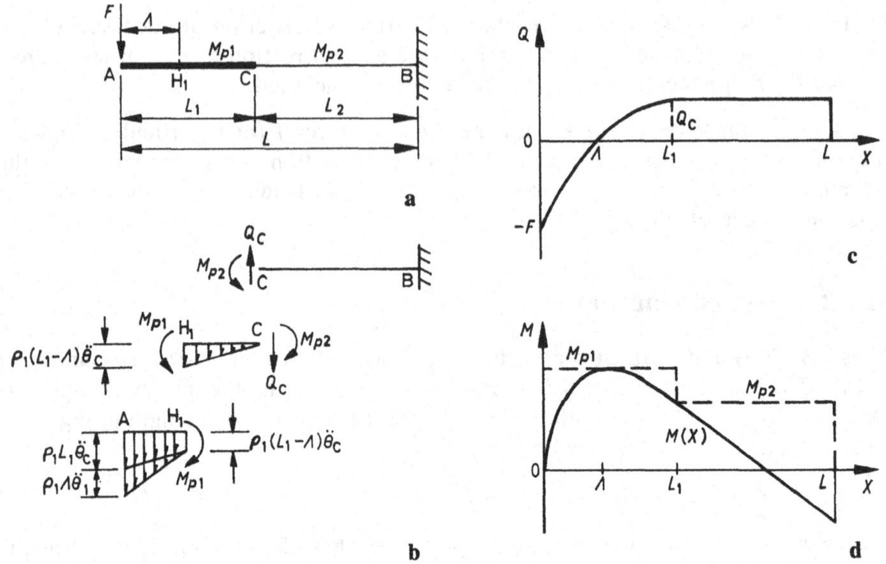

Fig. 6.32. (a) Case A: stepped cantilever with $M_{p1} > M_{p2}$ subjected to step force F at tip; a double–hinge deformation mechanism H_1–C (refer to Mechanism 5 in Table 6.3) is shown; (b) free–body diagram for the H_1–C double–hinge mechanism; (c) shear force distribution for the H_1–C double–hinge mechanism; (d) bending moment distribution for the H_1–C double–hinge mechanism; the dashed lines denote the fully plastic bending moment.

where λ is the nondimensional coordinate of hinge H_1 that can be obtained as a positive root of the cubic equation,

$$2(1-\lambda)^3 - (1-\eta_M)(3-2\lambda)\lambda^2 = 0 \qquad (6.147)$$

It is seen from Table 6.3 that a double-hinge mechanism takes place when the step force is sufficiently large. In this case, plastic hinges form at sections H_1 and C that are shown in Fig. 6.32a. Figure 6.32b is the free-body diagram for a two hinge mechanism, with $\ddot{\theta}_c$ and $\ddot{\theta}_1$ denoting the angular accelerations of segment H_1C and of AH_1 relative to H_1C respectively. Using these free-body diagrams and equations of motion, the distributions of shear force and bending moment along the cantilever are as shown in Fig. 6.32c and d, respectively. The dashed lines in Fig. 6.32d represent the fully plastic bending moment while the solid curve represents the dynamic bending moment distribution, so this figure illustrates that the solution obtained from the double-hinge mechanism satisfies the yield condition throughout the member.

The question of whether the yield function takes an extreme at a plastic hinge was investigated in Sect. 6.1; this depends on differentiability of the yield function at that hinge. In the present case, the only generalized stress is the bending moment so the yield function is

$$\Psi(X) \equiv |M(X)| - M_p(X) \qquad (6.148)$$

where

$$M_p(X) = \begin{cases} M_{p1}(X), & 0 \le X \le L_1 - 0 \\ M_{p2}(X), & L_1 + 0 \le X \le L \end{cases} \qquad (6.149)$$

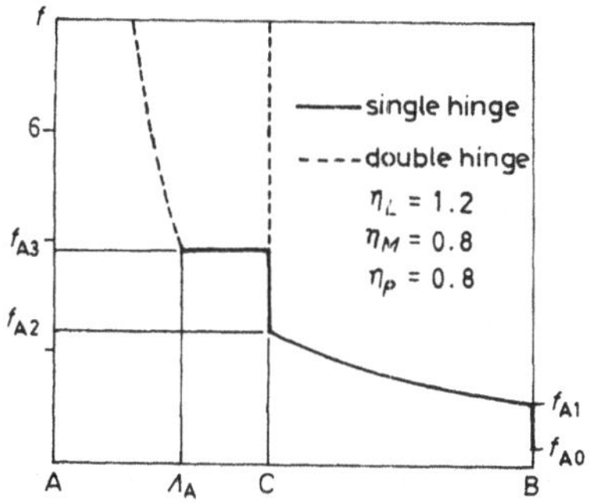

Fig. 6.33. Relationship between the nondimensional force f and hinge location λ for case A, when $\eta_L = 1.2$ and $\eta_M = \eta_\rho = 0.8$.

This yield function $\Psi(X)$ has differentiability that is sufficient to represent a travelling hinge at all sections except the stepped cross-section C where $\Psi(X)$ is discontinuous. Consequently, if a plastic hinge forms at cross-section C, then the bending moment $M(X)$ need *not* be an extreme at C. This implies that if there is a hinge at C ($\Lambda = L_1$) the shear force at that hinge can be nonzero; if a plastic hinge forms at any cross-section other than C, then the shear force must vanish at the hinge position, $dM(\Lambda)/dX = Q(\Lambda) = 0$ at $\Lambda \neq L_1$. These remarks are illustrated by Fig. 6.32c and d, where $Q_{H1} = 0$ but $Q_C \neq 0$.

Based on the analysis given by Hua, Yu and Reid [1988], the dependence of hinge position(s) on the nondimensional force magnitude f is calculated; a typical example is given in Fig. 6.33 for $\eta_L = 1.2$ and $\eta_M = \eta_\rho = 0.8$. This clearly demonstrates that with increasing force magnitude, Mechanisms 1 to 5 defined in Table 6.3 appear one after the other.

For Mechanism 5, i.e. the H_1–C double-hinge mechanism, the total energy is partly dissipated at hinge C and partly at hinge H_1. For $\eta_M = \eta_\rho = 0.8$, Fig. 6.34 shows how the part of the energy dissipated at hinge C varies with force magnitude f. The vertical distance below the solid curve represents the part of the energy

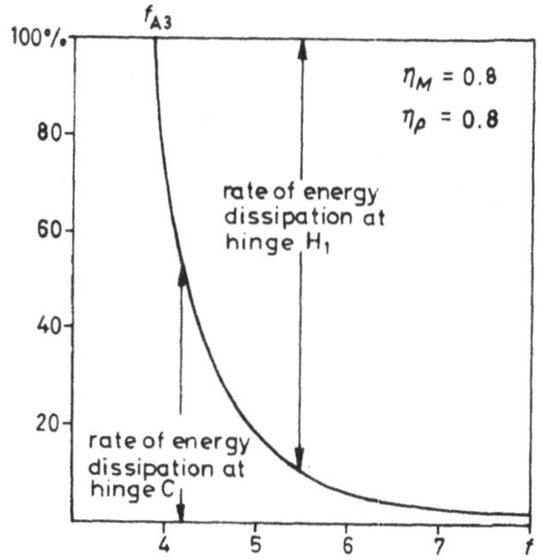

Fig. 6.34. The part of the total energy dissipation rate that occurs at hinge C depends on the force magnitude f for the H_1–C double-hinge mechanism, when $\eta_M = \eta_\rho = 0.8$ (case A).

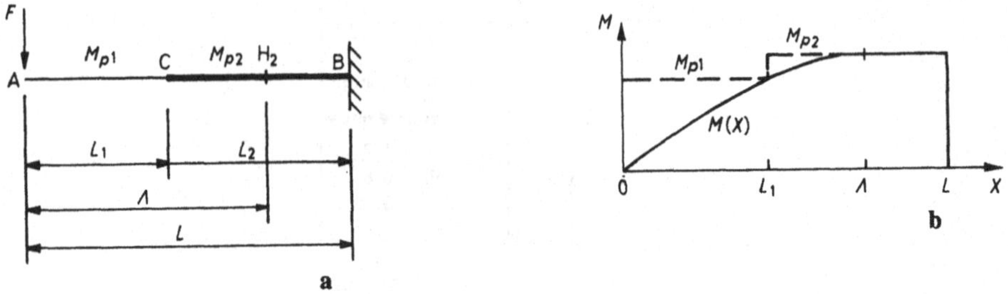

Fig. 6.35. (a) Case B: stepped cantilever with $M_{p1} < M_{p2}$ subjected to step–force F at tip. A double–hinge deformation mechanism C– H_2 is considered. (b) Bending moment distribution for the C– H_2 double–hinge mechanism; the dashed lines denote the fully plastic bending moment.

Table 6.4 Deformation mechanisms in stepped cantilever where $\eta_M = M_{p2}/M_{p1} > 1$

Mechanism	Plastic hinge	Range of f
1	No hinge	$0 \le f < f_{B0}$
2	Single-hinge B	$f_{B0} \le f \le f_{B1}$
3	Single-hinge H_2	$f_{B1} < f < f_{B2}$
4	Double-hinge C - H_2	$f_{B2} \le f < f_{B3}$
5	Double-hinge C-B	$f_{B3} \le f < f_{B4}$
6	Single-hinge C	$f_{B4} \le f \le f_{B5}$
7	Single-hinge H_1	$f_{B5} < f$

dissipation rate that occurs at hinge C; while the vertical distance above the curve represents the remaining part dissipated at hinge H_1 located somewhere between A and C. This figure indicates that the energy dissipated at the stepped cross-section C rapidly decreases at larger forces.

Case B: Stepped Cantilever with $M_{p1} < M_{p2}$ If the cantilever is stepped with $M_{p1} < M_{p2}$ (i.e. $\eta_M > 1$) as shown in Fig. 6.35a, there are *seven* dynamic deformation mechanisms for steady forces; these are summarized in Table 6.4. Figure 6.35b depicts the bending moment distribution for the C– H_2 double-hinge mechanism (i.e. Mechanism 5 in Table 6.4, where H_2 denotes a hinge position between C and B). This figure indicates that the yield criterion is not violated anywhere in the cantilever, and that the yield function is an extreme at hinge H_2 but not at hinge C.

The mechanism transition values of force f, i.e. $f_{B0}, f_{B1}, f_{B2}, f_{B3}, f_{B4}$ and f_{B5}, depend on ratios η_L, η_M, and η_ρ, see Hua et al. [1988] for details. In fact for some combinations of η_L, η_M, and η_ρ some of the mechanisms are not present. Figure 6.36 gives a typical example of Case B where $\eta_L = 2.0$ and $\eta_M = \eta_\rho = 1.2$. The distribution of energy dissipation rates at hinges C and H_2 for the C– H_2 double-hinge mechanism (i.e. Mechanism 4 in Table 6.4) are shown in Fig. 6.37.

6.7.3 Bent Cantilever

Case C: Bent Cantilever with $\eta \equiv T_p / M_p < 1$ Consider a cantilever which has been bent through an angle β with $0° \le \beta \le 90°$. A step force F acts at the tip in a

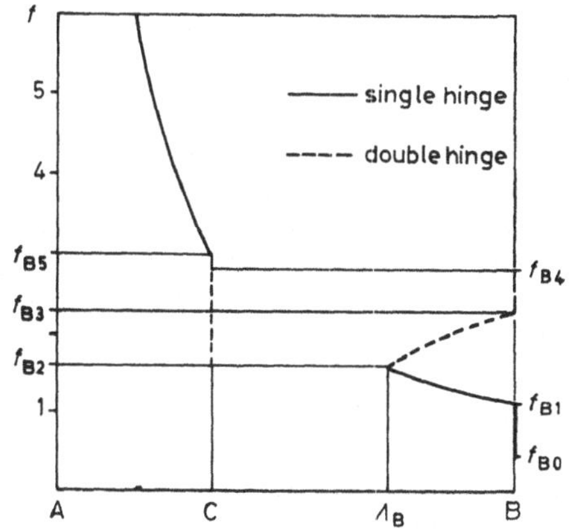

Fig. 6.36. Relationship between the nondimensional force f and hinge location λ for case B, when $\eta_L = 2.0$ and $\eta_M = \eta_\rho = 1.2$.

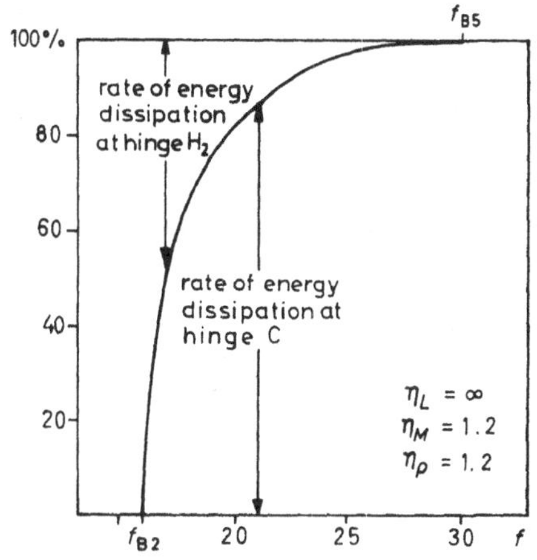

Fig. 6.37. The part of the total energy dissipation rate that occurs at hinge C depends on the force magnitude f for the C– H_2 double-hinge mechanism, when $\eta_M = \eta_\rho = 1.2$ (case B).

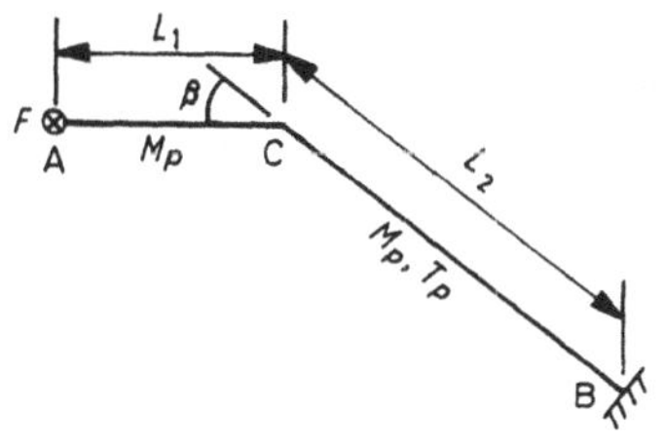

Fig. 6.38. Bent cantilever subjected to an out-of-plane step force F at tip A.

direction normal to the plane of the member, see Fig. 6.38. In addition to the assumptions made in Sect. 6.7.1, it is further assumed that the entire cantilever has a uniform cross-section and density; that is, $M_{p1} = M_{p2} \equiv M_p$, $T_{p1} = T_{p2} \equiv T_p$ and $\rho_1 = \rho_2 \equiv \rho$, where subscripts 1 and 2 pertain to segments AC and CB, respectively; M_p and T_p denote the fully plastic bending moment and fully plastic torque, respectively. Here we consider $\eta \equiv T_p / M_p < 1$; this case is representative of almost all practical cross-sections (see Table 6.2).

Within the range of small deformation, segment AC is subjected to bending while segment CB is subjected to combined bending and torsion. Hence, similar to Sect. 6.6, the yield function given by (6.115) can be written as $\Psi(\tilde{m},\tilde{t})$ where $\tilde{m}$ and $\tilde{t}$ are the nondimensional bending moment and torque at a cross-section. Table 6.1 indicates that the differentiable function $\Psi(\tilde{m},\tilde{t})$ must take an extreme value at a generalized plastic hinge located at any cross-section other than the bend C where torque is discontinuous.

By comparing the yield function for a stepped cantilever with that for a bent cantilever, an analogy can be established. If $T_p < M_p$, i.e. $\eta \equiv T_p / M_p < 1$, the beam is 'stronger' when it is subjected to merely bending rather than either torsion or combined bending and torsion. Since segment AC is subjected only to bending, it seems to be 'stronger' than segment CB, although the cross-section and density are identical for both segments. Thus, Case C is analogous to Case A that was discussed before. In the same way, Case D, which pertains to a bent cantilever with $\eta \equiv T_p / M_p > 1$, is analogous to Case B, below.

Similar to the analysis of Case A, Hua, Yu and Reid [1988] found *five* dynamic deformation mechanisms for bent cantilevers with $\eta \equiv T_p / M_p < 1$. These mechanisms can be summarized in the same way as those in Table 6.3, provided that the transition values of force f are changed from f_{A0}, f_{A1}, f_{A2}, f_{A3} to f_{C0}, f_{C1}, f_{C2}, f_{C3} accordingly. The values of f_{C0}, f_{C1}, f_{C2}, f_{C3} depend on both cross-sectional property $\eta \equiv T_p / M_p$ and the bend angle β; they can be calculated for any bend angle β,

$$f_{C0} = \eta\left\{\sin^2\beta + \eta^2(\cos\beta + \eta_L)^2\right\}^{-1/2} \tag{6.150}$$

$$f_{C2} = 3\eta_b, \qquad \eta_b \equiv \left\{\left(\frac{1}{\eta^2} - 1\right)\sin^2\beta + 1\right\}^{-1/2} \tag{6.151}$$

$$f_{C3} = \frac{3(2-\lambda)}{\lambda(3-2\lambda)} \tag{6.152}$$

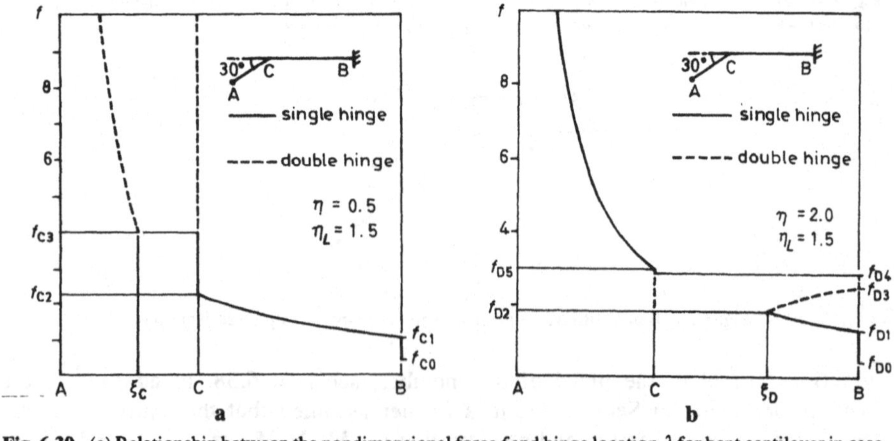

Fig. 6.39. (a) Relationship between the nondimensional force f and hinge location λ for bent cantilever in case C ($\eta \equiv T_p / M_p < 1$), when $\beta = 30°$, $\eta = 0.5$ and $\eta_L = 1.5$; (b) Relationship between the nondimensional force f and hinge location λ for bent cantilever in case D ($\eta \equiv T_p / M_p > 1$), when $\beta = 30°$, $\eta = 2.0$ and $\eta_L = 1.5$.

where λ is the nondimensional coordinate of hinge H_1. This hinge location can be obtained as a root of the equation

$$2(1-\lambda)^3 - (1-\eta_b)(3-2\lambda)\lambda^2 = 0 \tag{6.153}$$

The expression of f_{C1} is rather complicated, see Hua, Yu and Johnson [1985], so it is omitted here. Note that η_b given in (6.151) combines both the cross-sectional property $\eta \equiv T_p/M_p$ and the bend angle β as one single parameter. A comparison of (6.151) with (6.145) indicates that for bent cantilevers η_b plays the same role as $\eta_M \equiv M_{p2}/M_{p1}$ for stepped cantilevers. With the correspondence between η_M and η_b in mind, the similarities between (6.146) and (6.152) as well as between (6.147) and (6.153) are evident.

A typical example for the relationship between the magnitude of the step force f and the hinge position λ is shown in Fig. 6.39a for $\beta = 30°$, $\eta = 0.5$ and $\eta_L = 1.5$.

Case D: Bent Cantilever with $\eta \equiv T_p/M_p > 1$ This case is similar to Case B above. A similar analysis results in *seven* deformation mechanisms, which can be summarized in the same way as those in Table 6.4, provided f_{B0}, f_{B1}, f_{B2}, f_{B3}, f_{B4}, f_{B5} are changed into f_{D0}, f_{D1}, f_{D2}, f_{D3}, f_{D4}, f_{D5}. The latter values are functions of parameters β, η and η_L, see Hua, Yu and Reid [1988] for details. Also, for some combinations of these parameters, some deformation mechanisms do not occur. The relationship between the step force magnitude f and the hinge position λ for a typical example is shown in Fig. 6.39b for $\beta = 30°, \eta = 2.0$ and $\eta_L = 1.5$.

6.7.4 Discussion

1. The analyses in this Section show that the dynamic response of stepped or bent cantilevers result in double-hinge mechanisms if the applied force is sufficiently large. Generally speaking, a single-hinge mechanism fails to provide a complete solution if the force is large because the generalized stresses violate the yield criterion somewhere in the structure. In this case it is worthwhile to examine a double or multi-hinge mechanism. For example, the analysis in Sect. 6.5 shows that in some cases a single-hinge mechanism cannot lead to a complete solution for a cantilever of circular plan form that is subjected to a step force acting normal to the plane of the structure. This suggests a multi-hinge mechanism or even a mechanism with a continuous plastic zone. A solution for a particular in-plane distributed load on a curved beam gave a continuous plastic deforming region for an interaction yield condition (Cline and Jahsman [1967]).
2. An analogy between the response of stepped and bent cantilevers has been established. Application of this analogy yields a rather easy extension of the results for stepped cantilevers to those for bent ones. This analogy applies if rotary inertia of beam elements about the centroidal axis is negligible in segment CB.
3. According to this analogy, a bent cantilever with $\eta \equiv T_p/M_p = 1$ is directly analogous to a straight cantilever of uniform cross-section, so its dynamic response to step force is merely a single-hinge mechanism. The dynamic response of a bent cantilever subjected to an impulsive loading at the tip was previously

analyzed by Martin [1964]; he found that in the transient phase, a travelling plastic hinge runs continuously from the tip towards the root. At the hinge the deformation changes from bending to a combination of bending and torsion when the hinge transits the bend in the cantilever.

4. In general, if a stepped or bent cantilever with $\eta \neq 1$ is subjected to an impulsive load applied to a heavy particle at the tip, then the force at the tip can be regarded as $F = -G\dot{V}$, which decreases with time. Referring to the f–λ relations obtained in this section (see Figs 6.33, 6.36, 6.39), one can expect that the dynamic response of an impulsively loaded cantilever will undergo a series of deformation mechanisms; for instance, in Case A, Mechanisms 5, 4, 3, 2, 1 (see Table 6.3) will appear in succession during the dynamic response, and in some phases a travelling hinge will move along segments of the cantilever. If a stepped or bent cantilever is subjected to general pulse-loading, this transient phase also occurs. An analysis for this problem was given by Reid, Hua and Yang [1990]. In these cases, a double-hinge mechanism consists of a stationary hinge at section C or at the root B, together with a travelling hinge that moves along segment AC or segment CB. This travelling hinge never transits section C.
5. Extending the aforementioned studies, Reid et al. [1995a] showed that the family of deformation modes for the bent beam problem can be completed by introducing a triple-hinge mechanism. This consists of a hinge at the bend C, a hinge (H_1) in the first segment and a hinge (H_2) in the second segment. This mechanism provides a solution to the problem where the single- and double-hinge mechanisms become inadequate; these authors have shown that no further mechanism (with more than three hinges) is required. These authors also discussed cases of pulse and impact loading (Reid et al. [1995b], Wang et al. [1995]; this showed the transient behavior of the triple-hinge mechanism and the partition of the energy dissipated by plastic deformation in a multiple hinge mode of deformation.

6.8 Cantilever with an Initial Crack

6.8.1 General Considerations

The previous studies of statically and dynamically loaded cantilevers assume that any variation in section properties with axial distance is monotonic. When an initial crack or flaw exists at some cross-section this can significantly reduce the fully plastic bending moment at the cross-section and thereby alter the kinematics of deformation. If the material is brittle, fracture is likely to occur at the damaged section with little if any plastic deformation. On the other hand, in ductile materials the crack is likely to serve as the initiation site for a stationary plastic hinge that absorbs a significant portion of the input energy. Since the material in many modern structures is deliberately chosen to be ductile so that defects can be tolerated without precipitating brittle fracture, the problem of large dynamic plastic deformation of structures with initial cracks is one of current interest.

Petroski [1984a, 1984b] studied impact and the dynamic response of cracked cantilevers. First he investigated the effect of a crack at the root on dynamic deformation of a cantilever subjected to impact. An initial velocity was imparted to an attached mass at the tip. Only the modal-phase motion was considered; that is, the cracked cantilever merely rotated about the root as a rigid–body. Thus the final rotation angle at the root was easily found. This analysis served as a basis to discuss stability of the crack after impact loading. Later Petroski [1984b] considered a crack at an interior cross-section of a cantilever. The rotation angle at the cracked section after impact was estimated by employing a mechanism wherein stationary plastic hinges were located at the root and the cracked cross-section throughout the entire response. Woodward and Baxter [1986] reported their experimental study on impact bending of unnotched and notched free–free steel beams; they commented that the notched beam problem is an ideal case for application of the rigid–plastic approach because a notch localizes deformation, making it more like a hinge.

Inspired by these works and by the idea that more than one hinge could travel in a double-hinge mechanism (refer to Sect. 6.7), Yang and Yu [1991] proposed a complete analysis to an impact-loaded cantilever with an initial crack at an arbitrary interior cross-section. Both *transient* and *modal phases* were considered in the analysis. They obtained the partition of dissipated energy and a criterion for stability of the crack.

6.8.2 Impact on Cantilever with an Initial Crack

Assumptions The following assumptions are adopted in analyzing the dynamic response of a cracked cantilever:

1. The material is rigid–perfectly plastic and rate-independent.
2. The effect of shear force on yielding is neglected.
3. The deflection is small in comparison with the cantilever length L.
4. The cantilever has uniform section properties other than at an interior section where there is a crack. The initial momentum due to impact is parallel to an axis of symmetry for the cross-section. The crack spans the entire width and is located on the tensile side of the beam during deformation.
5. The crack remains stable during the dynamic response of the cantilever to impact; that is, coupling between the overall dynamic response and local crack propagation is disregarded.

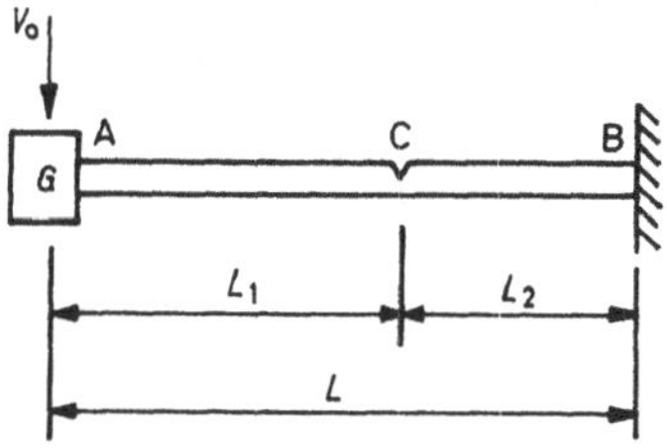

Fig. 6.40. Cantilever with crack at section C, subjected to impact at tip by a rigid mass.

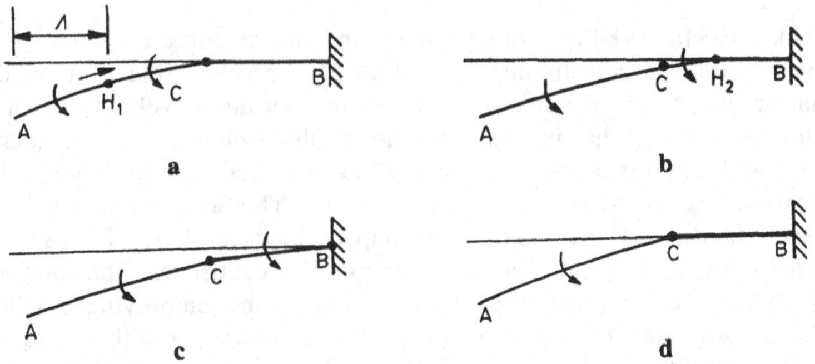

Fig. 6.41. (a) Deformation mechanism in the first phase with stationary hinge C and travelling hinge H_1. Possible deformation mechanisms in the second phase: (b) double stationary hinge C– H_2 (case I); (c) double stationary hinge C–B (case II); (d) single stationary hinge C (case III).

Transient Solution Suppose an initially straight cantilever of length L and mass per unit length ρ is struck transversely at the tip by a particle of mass G. The particle acquires an initial transverse velocity V_0 at time $t = 0$. Let a transverse crack exist at a cross-section C located a distance L_1 away from the tip, as shown in Fig. 6.40. Let the fully plastic bending moment for noncracked cross-sections be M_p, while that for the cracked cross-section C be ζM_p, where ζ, ($0 < \zeta < 1$) is a parameter depending on the size and shape of the crack, for example see Sect. 6.8.3 below.

By applying the previous analysis on stepped cantilevers (Sect. 6.7.2) to the present configuration, Yang and Yu [1991] obtained the following results.

1. The first phase of dynamic response usually consists of a stationary hinge at the cracked cross-section C and a travelling plastic hinge H_1 which moves away from the impact point at the tip, as shown in Fig. 6.41a. This double-hinge mechanism has bending moments M_p at the travelling hinge and ζM_p at the cracked section. This mechanism satisfies the yield criterion in the entire cantilever and leads to a complete solution if the crack location $\eta_L \equiv L_2 / L_1$ satisfies

$$\eta_L \equiv \frac{L_2}{L_1} < \frac{2(1+\zeta)}{3(1-\zeta)}(1 - \Lambda_1 / L_1) \tag{6.154}$$

where ζ is the crack parameter defined above and Λ_1 ($< L_1$) is the terminal position of the plastic hinge H_1 when it stops moving. For very light and very heavy colliding masses, inequality (6.154) results in

$$\eta_L < \frac{2}{3\sqrt{1-\zeta}}(\sqrt{2} - \sqrt{1-\zeta}\,), \qquad \gamma_1 \equiv G/\rho L_1 \ll 1 \tag{6.155}$$

$$\eta_L < \frac{2(1+\zeta)}{3\sqrt{1-\zeta}\;(1+\sqrt{1-\zeta}\,)}, \qquad \gamma_1 \equiv G/\rho L_1 \gg 1$$

respectively. In most cases of practical interest, the crack is small compared with the dimensions of the cross-section, so ζ is slightly less than one and inequality (6.155) is satisfied throughout most of the member. If inequality (6.155) is violated, then a third plastic hinge may be required at the root B.

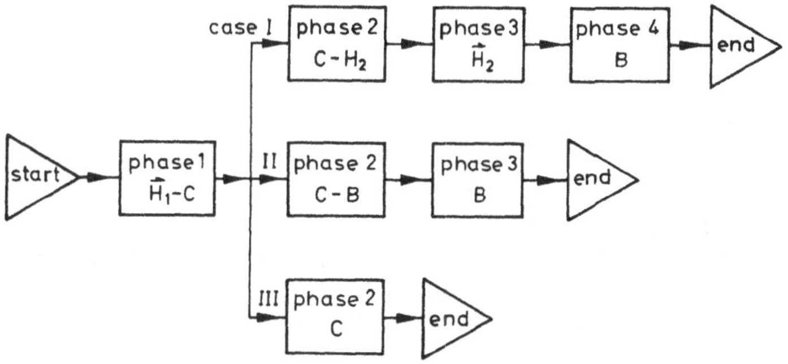

Fig. 6.42. Sequence of response phases for cracked cantilevers.

2. The first phase of dynamic response ends when the travelling hinge H_1 halts at $X=\Lambda_1$ somewhere between A and C. Thereafter, the subsequent motion follows one of *three* possible cases, depending on the range of parameters $\gamma_1 \equiv G/\rho L_1$, $\eta_L \equiv L_2/L_1$ and the crack depth ζ:

Case I: if

$$0 < \frac{1-\zeta}{\zeta} < \frac{\eta_L^2}{3\Gamma(1+\eta_L)-1} \tag{6.156}$$

where $\Gamma \equiv (\gamma_1 + 1/3)/(\gamma_1 + 1/2)$, then the second phase of the response is characterized by *two stationary plastic hinges* located at C and H_2 (i.e. C–H_2), where H_2 is in segment CB, see Fig. 6.41b.

Case II: if

$$\frac{\eta_L^2}{3\Gamma(1+\eta_L)-1} \le \frac{1-\zeta}{\zeta} < \frac{\eta_L}{\Gamma} \tag{6.157}$$

then the second phase is characterized by *two stationary plastic hinges* (C–B) located at the crack and the root, see Fig. 6.41c.

Case III: if

$$\frac{\eta_L}{\Gamma} \le \frac{1-\zeta}{\zeta} \tag{6.158}$$

then the second phase is characterized by a *single stationary plastic hinge* C located at the cracked section, see Fig. 6.41d.

3. The sequence of response phases for a cracked cantilever can be summarized by a flow chart as in Fig. 6.42. In this figure the letters designating each phase denote hinge locations; the arrow above letter H_1 and H_2 represents the direction of travel for the hinge, and the letters without a superscribed arrow pertain to stationary hinges.
4. The final rotation angle θ_{cf} at the cracked cross-section C, can be calculated from an analysis of the first and second phases during which a stationary plastic hinge forms at section C.

6.8.3 Crack Stability After Impact

Crack Stability Criteria Only very ductile materials can be accurately represented by the rigid–perfectly plastic moment–curvature relation used above, and even these materials cannot maintain stable or nonextending cracks under extreme conditions. The most useful analysis to determine the stability of cracks under conditions of large plastic strain employs the concepts of the *J-integral* and *tearing modulus* as described in fracture mechanics, e.g. in Hutchinson and Paris [1979] or Paris et al. [1979].

For perfectly plastic behavior of the cracked beam analyzed above, the value of the J-integral can be related to the deformation work done at the cracked section C. Thus for a rotation angle θ_c at the crack location,

$$J = -\frac{\partial(\zeta M_p)}{\partial A_{cr}}\theta_c \tag{6.159}$$

where A_{cr} is the crack area, which depends upon the geometry of the assumed crack. The stability of the crack is determined by a quantity known as the tearing modulus; for a perfectly plastic material it is given by

$$\overline{T} = \frac{E}{Y^2}\frac{\partial J}{\partial \overline{c}} \tag{6.160}$$

where E is Young's modulus, Y is the yield stress of the material, and $\overline{c}$ is a characteristic flaw size.

For a specified material, the material resistance curve for crack propagation can be determined by experiments. Let J_m and $\overline{T}_m$ denote the values of the J-integral and the tearing modulus, respectively, at the critical state of crack propagation. Then the crack stability criteria can be expressed by

$$\begin{aligned} &J < J_m \quad \text{and} \quad \overline{T} < \overline{T}_m, \qquad \text{for stable cracks} \\ &J \geq J_m \quad \text{or} \quad \overline{T} \geq \overline{T}_m, \qquad \text{for unstable cracks} \end{aligned} \tag{6.161}$$

Equation (6.159) indicates that the value of the J-integral is proportional to the rotation angle θ_c at the cracked section C. For the rigid–perfectly plastic idealization, rotation angle θ_c increases monotonically during the dynamic response of the cracked cantilever to impact. Hence, both J and $\overline{T}$ reach their largest values J_f and $\overline{T}_f$, respectively, when the rigid–plastic response ceases. For this final state the crack stability criteria (6.161) can be recast as

$$\begin{aligned} &J_f < J_m \quad \text{and} \quad \overline{T}_f < \overline{T}_m, \qquad \text{for stable cracks} \\ &J_f \geq J_m \quad \text{or} \quad \overline{T}_f \geq \overline{T}_m, \qquad \text{for unstable cracks} \end{aligned} \tag{6.161'}$$

with

$$J_f = -\frac{\partial(\zeta M_p)}{\partial A_{cr}}\theta_{cf} \tag{6.159'}$$

$$\overline{T}_f = \frac{E}{Y^2}\frac{\partial J_f}{\partial \overline{c}} \tag{6.160'}$$

Indeed, when the elastic contribution is neglected, conditions $J_f < J_m$ and $\overline{T}_f < \overline{T}_m$, guarantee a nonpropagating crack during dynamic response of the

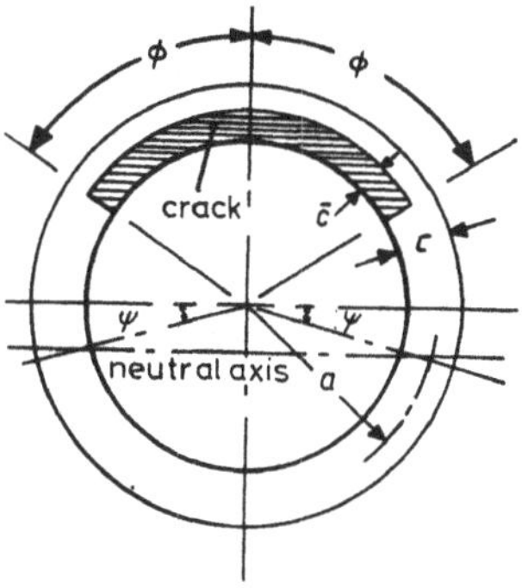

Fig. 6.43. Typical stress corrosion circumferential crack in cross-section of pipe.

cantilever. On the other hand, $J_f \geq J_m$ or $\bar{T}_f \geq \bar{T}_m$, implies that the crack extends during some part of the overall response.

Stability of Circumferential Crack in Thin-Walled Pipe Referring to Petroski [1984a], a typical stress corrosion crack in a boiling water nuclear reactor steam supply pipe is as shown by the shaded part in Fig. 6.43. Assume that this circumferential crack is located on the tensile side of the pipe during impact. This is the worst case for dynamic failure. The fully plastic bending moment for the undamaged cross-section is $M_p = 4Yca^2$, with a and c being the mean radius and the wall thickness of the pipe section, respectively. For the cracked cross-section with the geometry shown in Fig. 6.43, the fully plastic bending moment is ζM_p, with

$$\zeta = \cos\psi - \frac{\bar{c}}{2c}\sin\phi \tag{6.162}$$

where $\bar{c}$ denotes the radial depth of the circumferential crack, ϕ is the half angle subtended by the circumferential crack, and $\psi = \bar{c}\phi/2c$ is the angle subtended by the neutral axis. The crack area for this cross-section is

$$A_{cr} = 2a\bar{c}\phi \tag{6.163}$$

With the aid of (6.162) and (6.163), the J-integral can be evaluated for cases where the crack extends in either the radial or the circumferential direction. Finally, this results in

$$J_{rf} = -\frac{\partial(\zeta M_p)}{\partial\bar{c}}\frac{\partial\bar{c}}{\partial A_{cr}}\theta_{cf} = Ya\left(\sin\psi + \frac{\sin\phi}{\phi}\right)\theta_{cf} \tag{6.164}$$

$$J_{\phi f} = -\frac{\partial(\zeta M_p)}{\partial\phi}\frac{\partial\phi}{\partial A_{cr}}\theta_{cf} = Ya(\sin\psi + \cos\phi)\theta_{cf} \tag{6.165}$$

where subscripts r and ϕ represent the radial and circumferential directions, respectively, and subscript f denotes the final value after the dynamic response. Correspondingly, the final values of the tearing modulus at the cracked cross-section in the radial and circumferential directions are found as

$$\bar{T}_{rf} = \frac{E}{Y^2}\frac{\partial J_{rf}}{\partial\bar{c}} = \frac{Ea\phi}{2Yc}\left\{\theta_{cf}\cos\psi - \frac{\partial\theta_{cf}}{\partial\zeta}\left(\sin\psi + \frac{\sin\phi}{\phi}\right)^2\right\} \tag{6.166}$$

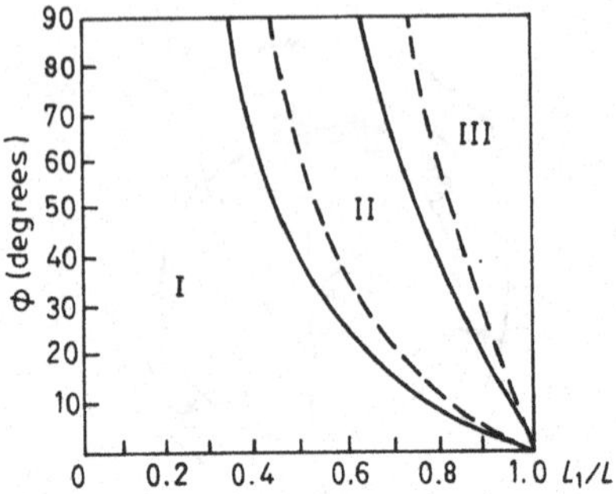

Fig. 6.44. Circumferential extent of crack ϕ and axial location L_1/L for occurrence of each of three different deformation mechanisms.

$$\overline{T}_{\phi f} = \frac{E}{Y^2}\frac{\partial J_{\phi f}}{a\partial\phi} = \frac{E\overline{c}}{2Yc}\left\{\theta_{cf}\left(\cos\psi - \frac{2c}{\overline{c}}\sin\phi\right) - \frac{\partial\theta_{cf}}{\partial\zeta}(\sin\psi+\cos\phi)^2\right\} \tag{6.167}$$

Since, $\sin\phi/\phi > \cos\phi$, Eqs (6.164) and (6.165) imply that

$$J_{rf} > J_{\phi f} \tag{6.168}$$

On the other hand, since $\partial\theta_{cf}/\partial\zeta < 0$, (6.166) and (6.167) imply that

$$\overline{T}_{rf} > \overline{T}_{\phi f} \tag{6.169}$$

provided $a\phi > \overline{c}$ or that the crack has a longer dimension in the circumferential rather than the radial direction. Since this is the case for most stress corrosion cracks, the inequalities (6.168) and (6.169) indicate that crack extension is more likely in the radial direction. This leads to a leak-before-break process that is desirable from a safety point of view. In this case most of the input energy is dissipated by crack extension in the radial direction so the pipe is unlikely to break due to circumferential fracture.

6.8.4 Numerical Example and Discussion

In the complete analysis given by Yang and Yu [1991], various combinations of parameters led to three cases with different sequences of response phases (see Fig. 6.42). For the particular crack configuration shown in Fig. 6.43 and $\overline{c}/c = 0.5$, the occurrence of these three cases can be depicted on a map of crack circumferential extent ϕ and axial location L_1/L, as shown in Fig. 6.44. In the figure, regions I, II and III correspond to Cases I, II and III in Sect. 6.8.2, respectively. The solid and dashed lines describe the boundaries between these regions for very large and very small mass ratios ($\gamma_1 \to \infty$ and $\gamma_1 \to 0$, where $\gamma_1 \equiv G/\rho L_1$ as defined in (6.155)) respectively. For $0 < \gamma_1 < \infty$, the boundaries between the regions are located somewhere between the corresponding solid line and dashed line. Since in Case III all the remaining energy after the first phase has to be absorbed by a stationary hinge at the cracked cross-section C, a crack associated with region III on the map is most hazardous with respect crack extension.

Figure 6.45a and b presents the radial and circumferential variations of final values for the J-integral as functions of the crack angle ϕ, for a crack located at

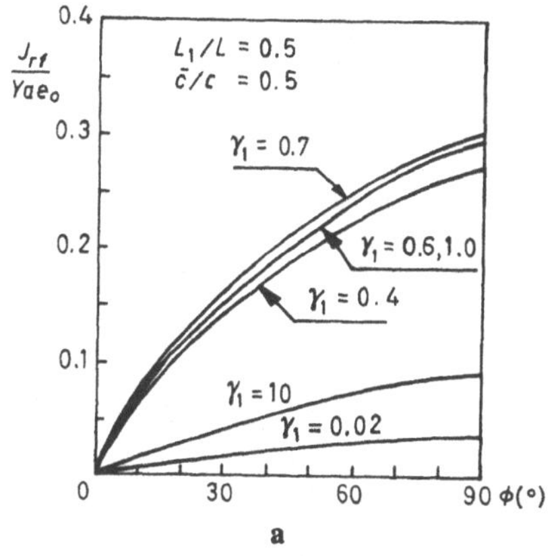

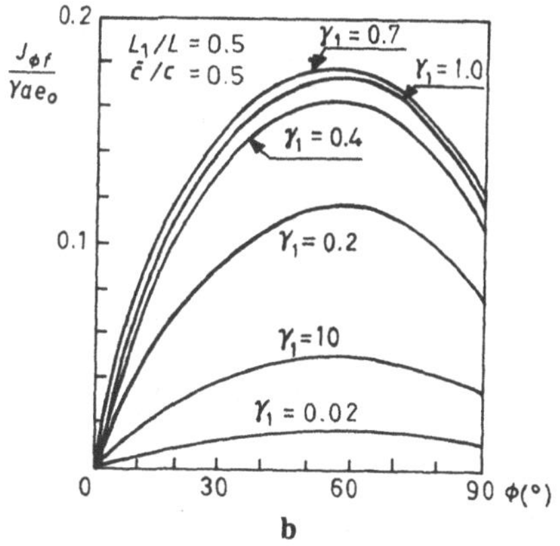

Fig. 6.45. Final values of J-integral depend on the crack geometry, for $L_1/L = 0.5$ and $\bar{c}/c = 0.5$; the final values are (a) J_{rf}; in radial direction, and (b) $J_{\phi f}$ in circumferential direction.

the half-length of the cantilever and crack depth half way through the wall thickness. Note that $e_0 \equiv K_0/M_p$ is the nondimensional initial kinetic energy as used before, so J_{rf}/Yae_0 and $J_{\phi f}/Yae_0$ are both nondimensional. In this case, both J_{rf} and $J_{\phi f}$ reach their largest magnitudes at a mass ratio $\gamma_1 \approx 0.7$. Figure 6.45a indicates that the tendency towards crack propagation in the radial direction increases with the circumferential size of the crack; while Fig. 6.45b shows that the maximum value of $J_{\phi f}$ appears for a circumferential extent of crack in the range $\phi = 50° - 70°$. This implies that even if $J_{\phi f} = J_m$, any tendency towards crack propagation in the circumferential direction is suppressed. The crack tends to propagate in the radial direction through the wall thickness and cause leakage of the pipe rather than breakage.

References

Al-Hassani, S.T.S., Johnson, W. and Vickers, G.W. [1973]. Dynamically loaded variable thickness cantilevers using a magnetomotive impulse. *Int. J. Mech. Sci.* **15**, 987–992.

Boulton, N.S. and Boonsukha, B. [1959]. Plastic collapse loads for circular-arc bow girders. *Proc. Inst. Civil Engrs.* **13**, 161–168.

Cline, G.B. and Jahsman, W.E. [1967]. Response of a rigid–plastic ring to impulsive loading. *ASME J. Appl. Mech.* **34**, 329–336.

Hodge, P.G., Jr. [1959]. *Plastic Analysis of Structures*. McGraw-Hill, New York.

Hua, Y.L., Yu, T.X. and Johnson, W. [1985]. The plastic hinge position in a bent cantilever struck normal to its plane by a steady jet applied at its tip. *Int. J. Impact Engng.* **3**, 223–241.

Hua, Y.L., Yu, T.X. and Reid, S.R. [1988]. Double-hinge modes in the dynamic response of plastic cantilever beams subjected to step loading. *Int. J. Impact Engng.* **7**, 401–413.

Hutchinson, J.W. and Paris, P.C. [1979]. Stability analysis of J-controlled crack growth. *Elastic–Plastic Fracture* (eds J. D. Landes et al.), American Society for Testing and Materials, ASTM STP668, 37–64.

Martin, J.B. [1964]. The plastic deformation of a bent cantilever, Part I: Theory. NSF GP 1115/15, division of Engineering, Brown University.

Owens, R.H. and Symonds, P.S. [1955]. Plastic deformations of a free ring under concentrated dynamic loading. *ASME J. Appl. Mech.* **22**, 523–529.

Palomby, C. and Stronge, W.J. [1988]. Evolutionary modes for large deflections of dynamically loaded rigid-plastic structures. *Mech. Struct. Mach.* **16**(1), 53–80.

Paris, P.C., Tada, H., Zahoor, A. and Ernst, H. [1979]. The theory of instability of the tearing mode of elastic-plastic crack growth. *Elastic–Plastic Fracture* (eds J. D. Landes et al.), American Society for Testing and Materials, ASTM STP668, 5-36.

Petroski, H.J. [1984a]. Stability of a crack in a cantilever beam undergoing large plastic deformation after impact. *Int. J. Pres. Ves. and Piping* **16**, 285–298.

Petroski, H.J. [1984b]. The permanent deformation of a cracked cantilever struck transversely at its tip. *J. Appl. Mech.* **51**, 329–334.

Prager, W. [1954]. Discontinuous fields of plastic stress and flow. Second U.S. National Congress of Applied Mechanics, ASME, NY, 21–32.

Reid, S.R., Hua, Y.L. and Yang, J.L. [1990]. Development of double-hinge mechanisms in a bent cantilever subjected to an out-of-plane force pulse. *Int. J. Impact Engng.* **9**, 485–502.

Reid, S.R., Wang, B. and Yu, T.X. [1995a]. Yield mechanism of a bent cantilever beam subjected to a suddenly applied constant out-of-plane tip force. *Int. J. Impact Engng.* **16**, 49–73.

Reid, S.R., Wang. B. and Hua, Y.L. [1995b]. Triple plastic hinge mechanism for a bent cantilever beam subjected to an out-of-plane tip force pulse of finite duration. *Int. J. Impact Engng.* **16**, 75–93.

Shu, D., Stronge, W.J. and Yu, T.X. [1992]. Oblique impact at tip of cantilever. *Int. J. Impact Engng.* **12**, 37–47.

Stronge, W.J., Shu, D. and Shim, V.P.W. [1990]. Dynamic modes of plastic deformation for suddenly loaded, curved beams. *Int. J. Impact Engng.* **9**, 1–18.

Symonds, P.S. [1968]. Plastic shear deformation in dynamic load problems. *Engineering Plasticity* (eds J. Heyman and F.A. Leckie) Cambridge University Press, 647–664.

Symonds, P.S. and Mentel, T.J. [1958]. Impulsive loading of plastic beams with axial restraints. *J. Mech. Phys. Solids* **6**, 186–202.

Wang, B. Reid, S.R. and Yu, T.X. [1995]. Response of a right-angled bent cantilever subjected to an out-of-plane impact and force pulse applied at its tip. Submitted to *ASME J. Appl. Mech.*

Woodward, R.L. and Baxter, B.J. [1986]. Experiments on the impact bending of continuous and notched steel beams. *Int. J. Impact Engng.* **4**, 57–68.

Yang, J.L. and Yu, T.X. [1991]. Complete solutions for dynamic response of a cracked rigid–plastic cantilever to impact and the crack unstable growth criteria (in Chinese). *Acta Scientiarum Naturalium Universitatis Pekinensis* **27**, 576–589.

Yu, T.X., Hua, Y.L. and Johnson, W. [1985a]. The plastic hinge position in a circular cantilever when struck normal to its plane by a constant jet at its tip. *Int. J. Impact Engng.* **3**, 143–154.

Yu, T.X. and Johnson, W. [1981]. The location of the plastic hinge in a quadrantal circular curved beam struck in its plane by a jet at its tip. ASME Structures and Materials Conference, Washington D.C., November, 1981. *Advances in Aerospace Structures and Materials* – 1 (eds S.S. Wang and W.J. Renton), ASME, Washington D.C., 175–180.

Yu, T.X., Symonds, P.S. and Johnson, W. [1985b]. A quadrantal circular beam subjected to radial impact in its own plane at its tip by a rigid mass. *Proc. Roy. Soc. Lond.* **A400**, 19–36.

Yu, T.X., Symonds, P.S. and Johnson, W. [1986]. A reconsideration and some new results for the circular beam impact problem. *Int. J. Impact Engng.* **4**, 221–228.

Zhang, T.G. and Yu, T.X. [1986]. The large rigid-plastic deformation of a circular cantilever beam subjected to impulsive loading. *Int. J. Impact Engng.* **4**, 229–241.

Zhang, T.G. and Yu, T.X. [1989]. Dynamic plastic response of a simply supported ring to an impact on its top (in Chinese). *Acta Mechanica Sinica*, special issue, 148–155.

Zhou, Q. and Yu, T.X. [1987]. Plastic hinge in beams of variable cross-section under intense dynamic loading (in Chinese). *Explosive and Shock* **7**, 311–318.

Chapter 7

Impact Experiments

In his seminal paper on dynamic deformation due to impact at the tip of a cantilever, Parkes [1955] showed sketches of the final profiles of some cantilevers struck by lightweight, high speed bullets, and others struck by a larger mass that collides at a relatively slow speed. For the same impact energy, these two sets of tests had deformations that were qualitatively distinct; the light mass gave a smoothly curved final shape with the largest curvature near the tip whereas the heavy mass gave almost all of the deformation concentrated near the root similar to the modal solution. This distinction on the basis of mass at the impact point in comparison with mass of the remainder of the uniform beam was convincing evidence for the utility of the rigid–perfectly plastic constitutive approximation.

7.1 Methods Of Applying Dynamic Loads

As has been mentioned, there are several methods of applying intense but short duration loads to structures. These range from impact of a colliding missile to a blast wave from a nearby explosion. For experiments designed to measure dynamic response, some methods of loading are better than others because they provide a more accurately measured input impulse or energy.

7.1.1 Missile Impact

High Speed Bullet Impact Missiles fired from guns at speeds larger than 100 m s^{-1} easily achieve contact durations of no more than 0.1 ms if both the missile and the target are composed of hard materials; i.e. if the materials have relatively large flow stresses. To obtain experimental measurements of impact phenomena, it is ordinarily best to employ spherical missiles for impact tests. This eliminates the need to consider tumbling of the missile both in free flight from the gun muzzle to the target and during impact. While it is straight- forward to measure the missile speed before impact, this method of loading suffers from difficulty in obtaining both the impulse and energy imparted to the target. The missile either rebounds from the impact with unknown velocity or it is captured in a block of soft material at the impact point. In the latter case, some soft material from the block is usually ejected backwards at high speed; also some energy is dissipated by deformation around the contact region. In both cases, the initial momentum and kinetic energy imparted to the target are not simply related to the missile impact parameters.

Drop Hammer or Powered Sledge For impact speeds of the order of 10 m s^{-1}, a gravitationally powered drop hammer can provide controlled impact conditions for heavy masses. The impact speeds of these rigs can be increased somewhat by using elastic bungee cords to increase the acceleration of the falling mass. Sometimes a higher terminal speed can be achieved by orienting the guide rails in a horizontal plane because this permits acceleration over a longer stroke; here, bungee cords may be the only source of power that drives the hammer. As the hammer (or sledge) is accelerated towards the anvil, most of these devices experience a series of small shocks applied to the hammer by the constraining guide rails and this excites large vibrations in the hammer. These vibrations are one reason why it is not ordinarily a good idea to mount slender compliant test specimens on a hammer head that is struck against a fixed anvil; the vibrations result in substantial uncertainty about initial conditions for the test specimen at impact. The duration and shape of the force pulse applied by the hammer to the anvil can be somewhat controlled by shaping the hammer head; e.g. a conical hammer head composed of a low strength material gives a force that first increases and subsequently decreases as a function of time. The hammer sketched in Fig. 7.1 illustrates some of these features of drop hammer design.

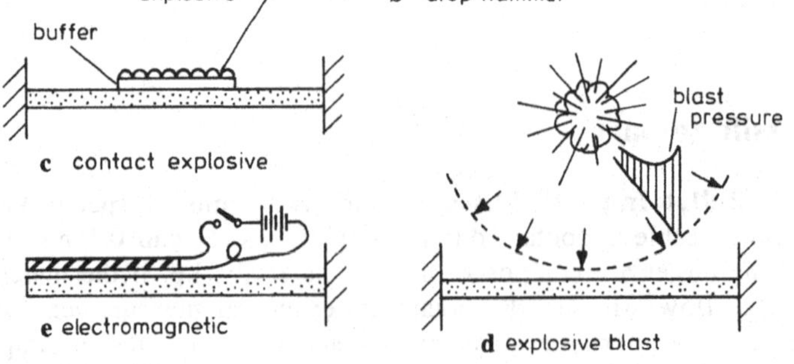

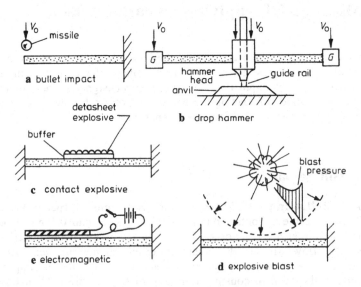

Fig. 7.1. Various methods of applying impulsive or short duration pulse loads to test specimens.

7.1.2 Explosive Loading

Contact Explosives Sheet explosive (Detasheet) or explosive cord (Primacord) have detonation speeds of 5-8000 m s^{-1} so they can be used to provide nearly simultaneous impulsive loading over a surface of a test specimen. Because explosive pressures behind the shock front are very large, explosions in contact with a test specimen can result in rear surface spallation. Spallation is prevented by covering the loaded surface with a thin sheet of neoprene used as a standoff that somewhat diffuses the pressure pulse. Contact explosions provide a uniform distribution of initial impulse with a magnitude that is controlled by changing the explosive thickness. The impulse intensity is calibrated by measuring the post-detonation velocity of free flying specimens that have the same thickness and material properties as the test specimens. Although the detonation speed is faster than elastic wave speeds in metals or plastics, if detonation is initiated at one end, that end of the structure begins to deform before the explosion pressure wave reaches the other end of the structure. Thus, it is preferable to initiate detonation at the center or use a 'wave shaper' to obtain more nearly simultaneous detonation over the specimen (see Rinehart and Pearson [1963]).

Explosive Blast Pulses A blast wave propagated outward from an explosive detonation has a pressure that decays exponentially with distance behind the shock front. The speed of the shock front decreases with increasing radius from the explosion while simultaneously, the shock pressure at the front decreases. When the blast wave impinges on the surface of a test specimen, the wave is at least partially reflected and this increases the pressure acting on the specimen in comparison with the pressure behind the free running pulse. Unless the test specimen is extremely close to the explosion, blast waves have a duration for the pressure pulse that is not insignificant in comparison with the structural response period so they should be classified as pulse loads. Blast pressure pulses of rather large amplitude can be measured directly by Manganin piezoelectric pressure gauges. Details of the decay of shock pressure and speed with increasing radius from an explosion are given by Kinney [1962] and Glasstone [1959].

7.1.3 Magnetomotive Loading

Magnetomotive loading is caused by an interaction between two currents; this is produced by a current passed through two parallel conductors or by the magnetic field from one conductor that induces eddy currents in a secondary conductor. These fields give a force that tends to drive the conductors apart; this force is proportional to the square of the current passing through the conductor. Force pulses of less than 10 ms are generated by the current from a rapid–discharge capacitor bank (see Hall et al. [1971]). The impulse imparted to the test specimen must be measured indirectly; e.g. by mounting the test specimen on a *ballistic pendulum.* Then if loading is applied to the test specimen only, the initial momentum of the system equals the impulse imparted to the specimen. This momentum can be calculated from the amplitude of swing of the pendulum.

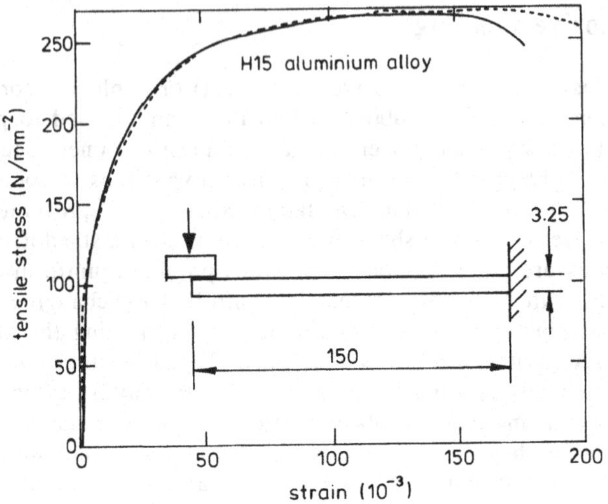

Fig. 7.2. Specimen configuration and stress–strain relation for blast loaded aluminum alloy (H15) cantilevers; $h = 3.25$ mm, $b = 6$ mm, $L = 150$ mm .

7.2 Travelling Hinges — Fiction or Fact?

Dynamic analyses of deformation processes in plastically deforming beams, plates or shells separate into a transient or moving hinge phase and a modal phase of motion. Discrete hinges that travel away from the impact point are a consequence of the rigid–perfectly plastic structural idealization. In practice, many materials show a significant range of elastic deformation and also strain hardening. These effects tend to diffuse the concentrated bending moment at a 'hinge'. Numerical investigations of dynamic response of nonlinear structures suggest nevertheless that a concentrated region of large flexural deformation does travel away from the impact point in the first phase of structural response, Reid and Gui [1987]. The length of the deforming region is larger than the thickness of the member but short in comparison with the length of the beam.

To investigate the development of plastic deformation, Shu et al. [1992] reported strain gauge measurements made on aluminum alloy cantilevers of length L = 150 mm that were hit transversely at the tip by an explosive blast pulse. The blast strikes a light block at the tip, $\gamma = G/\rho L = 0.39$. The specimen dimensions, strain gauge locations and material properties are described in Fig. 7.2. These tests showed that in the half-length adjacent to the tip, all of the plastic curvature developed in the first millisecond during the initial transient or travelling hinge phase of motion. The curvature at a section 40 mm from the tip is shown as a continuous line in Fig. 7.3a. (The curvatures in this figure have been translated from the difference between strains on top and bottom surfaces of the beam.)

Near the root the curvature developed somewhat later as indicated by the dashed line in Fig. 7.3a. Moreover, the slower rise time at the latter location showed that the *deforming segment increased in length* as it travelled. At location 2 near the root the pair of gages measured a travelling speed of the deforming segment that had slowed to roughly 50 m s^{-1} while the length of this segment had increased to $L_h/h \approx 10$. A few milliseconds after the travelling pulse passed this second pair

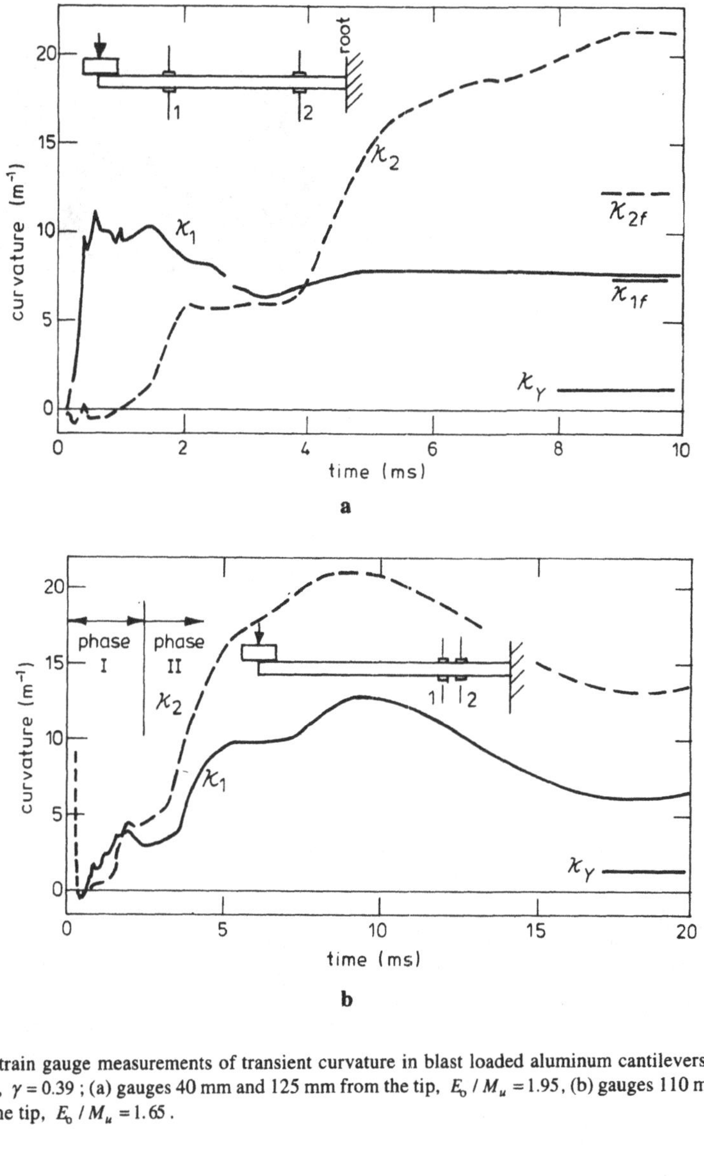

Fig. 7.3. Strain gauge measurements of transient curvature in blast loaded aluminum cantilevers with light block at tip, $\gamma = 0.39$; (a) gauges 40 mm and 125 mm from the tip, $E_0 / M_u = 1.95$, (b) gauges 110 mm and 125 mm from the tip, $E_0 / M_u = 1.65$.

of gauges, location 2 experienced another increase in curvature. This second stage of bending was more diffuse; it was associated with the region of *modal deformation* that spread from the root of this strain hardening, aluminum alloy beam.

For a similar test with somewhat less input energy, Fig. 7.3b shows the development of curvature at two pairs of strain gauges located a distance 15 mm apart at sections about $3L/4$ from the tip. The deforming segment arrived at these sections approximately 10^{-3} s after impact; at this time the average speed of the travelling flexural pulse had slowed to $35\ \mathrm{m\ s^{-1}}$. Notice that the first stage of bending began in the section that is closer to the tip at $X = 0.73L$. A few milliseconds after the flexural pulse passed each gauge location, a second phase

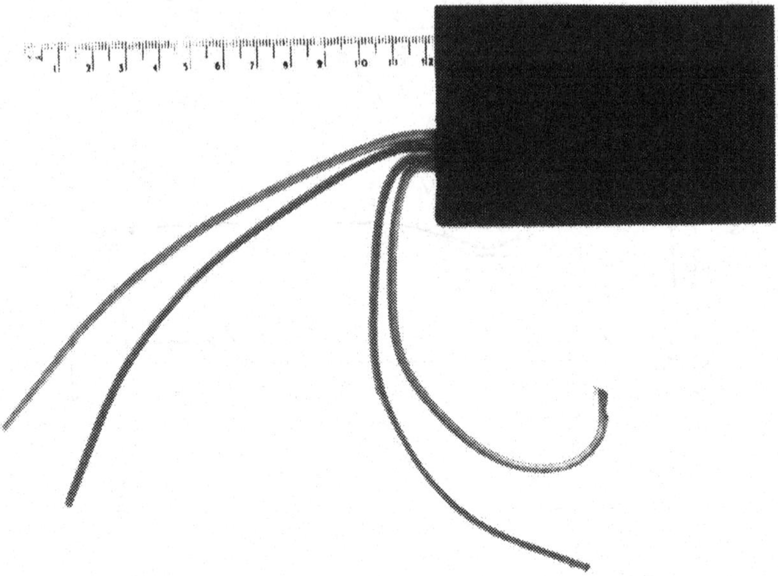

Fig. 7.4. Photograph of final deflection of aluminum alloy (H15) cantilevers with light tip mass $\gamma = 0.39$ that is hit transversely at tip by explosive blast pulse. Input energies were $E_{in} / M_u = 0.97$, 1.65 and 1.95.

of modal bending began. The modal deformation began first at the gauge located at $X = 0.84L$, close to the root. In this test with a large input energy $E_{\text{in}}/M_u = 1.65$, most of the final curvature at these sections near the root developed during the second or modal phase of deformation.

These aluminum cantilevers with substantial strain hardening show that if the input energy is large and the colliding mass is lightweight, then far away from points of displacement constraint most plastic deformation develops within a short bending segment that travels away from the impact point. The final configurations of the deformed cantilevers are shown in Fig. 7.4. In comparison with these experiments, rigid–plastic calculations of the final tip rotation underestimate the rotation if the fully plastic moment M_p is estimated by the ultimate stress σ_u and overestimate the rotation if the fully plastic moment is obtained from the yield stress Y; i.e. $M_Y < M_p < M_u$. These cantilevers with lightweight tip masses $\gamma = 0.39$ have a final deformed shape with curvature that decreases with distance from the tip. Bullet impact and explosive loading on an aluminum alloy cantilever (BS 1470 SIC) by Hall et al. [1971] showed a similar distribution of curvature. These results are not in agreement with the theory for strain hardening materials. By requiring the stress distribution to satisfy yield over the full length of the member ahead of the travelling hinge, that theory neglects limitations introduced by elastic deformations and thereby exaggerates the effect of strain hardening for the transient phase of motion. In the initial phase of motion of an elastic-plastic beam, the *effective length* of the beam is limited to the span behind the elastic wavefront that moves outward from the region of load application.

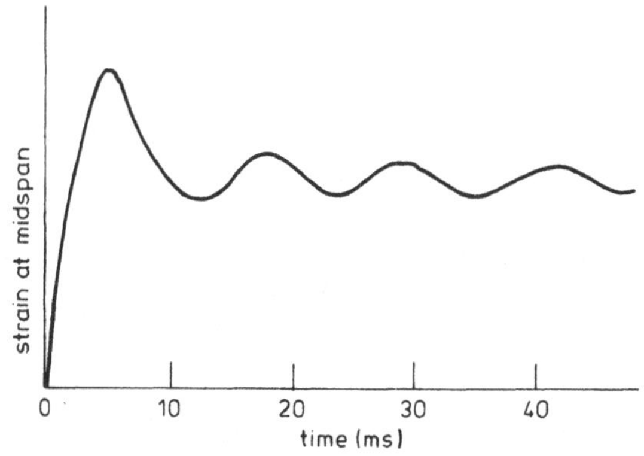

Fig. 7.5. Surface strain at midspan of magnetomotively loaded mild steel beam as a function of time. Impulse intensity $P = 750$ Pa s which is five times the minimum intensity to initiate plastic deformation.

7.3 Elastic Effects on Plastic Deformation

Slender members can experience rather large elastic deflections preceding any plastic deformation. The most obvious effect of elastic behavior at small strains is springback from the maximum deformation that occurs at each section; this effect which occurs during unloading, is most noticeable if the energy ratio is not large; e.g. $R < 10$. For simply supported beams subjected to uniformly distributed

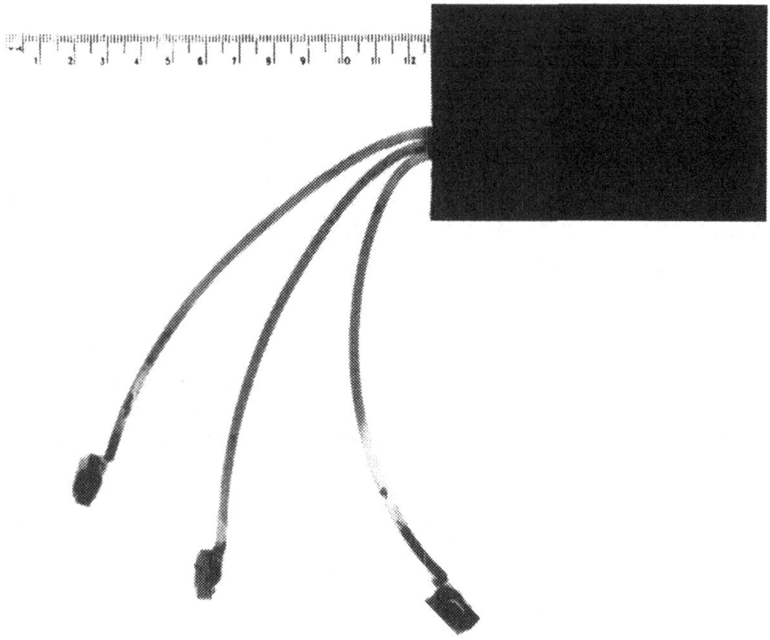

Fig. 7.6. Photograph of final deflection of aluminum alloy (H15) cantilevers hit transversely at tip by explosive blast pulse.

magnetomotive pulses of short duration, strain gauge measurements of transient response by Forrestal and Wesenberg [1976, 1977] have shown elastic springback that is almost independent of the intensity of the applied impulse (see Fig. 7.5). For slender beams, elastic deflections can exceed the depth of the beam.

In addition to widespread elastic springback, some cantilevers and free–free (unsupported) beams subjected to transverse impact near an end experience a bending moment distribution that causes the short, plastically deforming segment to hesitate and sometimes halt near midlength of the beam. This leaves a large curvature or kink that is not anticipated by the rigid–plastic theory — some examples are shown in the photographs of Fig. 7.6 . Final curvature of the cantilever in the segment between the central kink and the root is smaller than that expected from the rigid–perfectly plastic theory. The analysis of a cantilever with an elastic support at the root (Sect. 5.5) and numerical studies (Symonds and Fleming [1984], Reid and Gui [1987]) indicate that the central kink is due to elastic variations in the bending moment ahead of the travelling deforming segment. The effect of these elastic variations gives a central kink if the energy ratio R is moderately large $5 < R < 50$. In this range there is large plastic curvature but if the mass ratio γ is small enough, more energy is dissipated along the beam rather than at the root. The effect is most prevalent with small mass ratios $\gamma < 0.25$ as expected from Fig. 5.44. A similar concentrated region of large curvature appears in experiments where one end of a free–free beam is subjected to transverse impact. A numerical analysis by Woodward [1984] has shown that this is also caused by elastic variations in the bending moment ahead of a short plastically deforming segment which travels away from the impact point.

7.4 Strain Hardening and Strain-Rate Effects

Experiments on impulsively loaded structures show that calculations which use the rigid–perfectly plastic approximation exaggerate the magnitude of terminal deflections. This is due partly to neglect of the elastic strain energy stored in the deforming structure. Two other possible causes are neglect of strain hardening and strain-rate effects. Most materials exhibit strain hardening at strains $\varepsilon > \varepsilon_Y$ and many have a yield condition that is more or less rate sensitive. These effects increase the yield stress and thereby decrease the predicted deformation; they also modify the distribution of deformation. *Strain hardening and rate effects disperse the plastically deforming region* so at any instant deformation is no longer concentrated at a plastic hinge; instead, with these effects, the deformation develops in a short segment of a structure and the segment expands in length with increasing local deformation. These effects cause preliminary deformation that increases the yield moment ahead of a primary travelling deforming segment. This tends to shift the distribution of structural deformation towards the modal configuration which develops naturally as the final stage of dynamic response to impulsive loads.

7.4.1 Mode Approximations

Approximations for the stress amplification in a short plastically deforming zone were described in Sect. 5.1.2. The simplest approximation entirely neglects the transient phase of response and assumes that all of the initial input energy e_{in} is

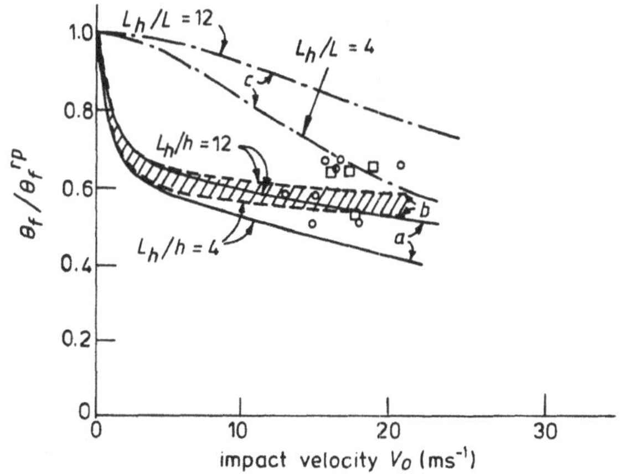

Fig. 7.7. Comparison of strain hardening and strain-rate modal solutions with test results for impact at tip of mild steel cantilever. (a) —— strain hardening and rate sensitivity; (b) - - - strain-rate sensitivity only, $\dot{\varepsilon}_0 = 40\ s^{-1}$ and $r = 5$; (c) — - - — strain hardening only, $E_t/Y = 6$ (after Symonds [1965]).

dissipated during phase II in the mode configuration. This is valid if the mass ratio γ of a rigid body attached to the structure is moderately large so deformation in phase I is negligible. To estimate the strain rate in the stationary pattern of deformation given by a mode, the deforming segment is assumed to be spread over a length L_h where $1 < L_h/h < 12$. This gives an approximate curvature $\kappa = 2\theta/L_h$ and rate of curvature $\dot{\kappa} = 2\dot{\theta}/L_h$ where the factor 2 is an approximation that accounts for a linear variation of curvature in the deforming segment.

With these approximations for curvature and rate of curvature, the bending moment M^d at a deforming section can be related to the initial fully plastic moment M_p and the rate independent moment $M_p^s(\kappa)$ that depends on strain hardening in accord with Eq. (1.18),

$$\frac{M^d}{M_p} = \frac{M_p^s}{M_p} + \left(\frac{\dot{\kappa}}{\dot{\kappa}_{or}}\right)^{1/r}, \qquad \dot{\kappa} > 0 \tag{7.1}$$

$$\dot{\kappa}_{or} = \left(\frac{2\dot{\varepsilon}_o}{h}\right)\left(\frac{2r+1}{2r}\right)^r$$

where the rate independent fully plastic moment at the current strain $M_p^s(\kappa)$ is given by

$$M_p^s(\kappa) = M_p + \alpha_m \kappa \tag{7.2}$$

$$\alpha_m = E_t I_o \tag{7.3}$$

and the last equation for the hardening coefficient α_m applies to a beam with rectangular cross-section.

The separate effects of strain hardening and strain-rate together with the combination of these two effects on the final deflections of an impulsively loaded cantilever with a moderately heavy tip mass are illustrated in Figs 7.7 and 7.8. The graphs compare calculations of the final rotation at the tip θ_{Af}^d of a hardening and rate dependent cantilever with similar calculations for the final tip rotation θ_{Af}^{rp} obtained using the elementary rate independent and perfectly plastic material approximation. The material parameters that were used to calculate the curves

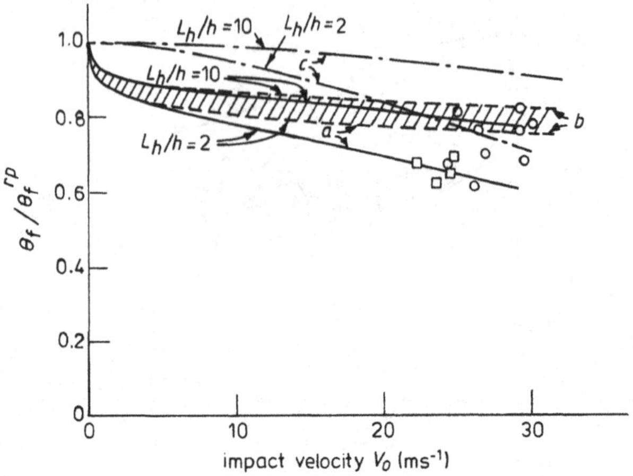

Fig. 7.8. Comparison of strain hardening and strain rate modal solutions with test results for impact at tip of 6061-T6 aluminum alloy cantilever. (a) —— strain hardening and rate sensitivity; (b) - - - strain-rate sensitivity only with $\dot{\varepsilon}_0=6500\ s^{-1}$ and $r=4$; (c) —·— strain hardening only, $E_t/Y=2$, (after Symonds [1965]).

represent mild steel and 6061-T6 aluminum alloy, respectively. To distinguish between rate independent and rate dependent effects, the graphs also contain experimental points from Bodner and Symonds [1962]; these demonstrate that the perfectly plastic material approximation gives excessively large deformations. In these tests, the ratio of incident energy to maximum elastic strain energy was large $R > 10$ and the mass at the tip was moderately heavy $\gamma = 2$ so the mode is a reasonably good approximation for the distribution of deformation in the beam. These cantilevers were loaded by a small explosive charge detonated on one side of the mass at the tip.

Comparisons of the various material approximations with the experimental data indicate that rate dependence estimated with the Cowper–Symonds parameters is more important than strain hardening effects in representing these tests. The rate effect becomes significant at a rather small energy ratio R (or initial rate of change of curvature $\dot{\kappa}$) and flattens out as strain-rate increases at larger values of R. The calculated curves show that the length of the plastically deforming segment L_h has some influence as a result of hardening but very little as a result of strain-rate.

A different modal approximation method for deformations of beams composed of rate sensitive materials is due to Forrestal and Wesenberg [1976, 1977]. Rather than assuming a priori some length for the deforming region, they calculated the initial *elastic* response of a rate-insensitive beam and obtain the strain-rate in the section where curvature is largest. This strain-rate is used in the Cowper–Symonds relation to calculate a dynamic yield stress Y_d. Once the maximum elastic stress reaches the dynamic yield stress, it is assumed that the structure is rigid–perfectly plastic with a fully plastic moment obtained from the yield stress Y_d and an initial momentum for the mode that is equal to the transition momentum of the elastic phase of response. Thus the problem has been reduced to a rate-insensitive analysis that separates into an initial elastic and a subsequent perfectly plastic phase of response. The transition between these two phases occurs when the elastic stresses satisfy a rate sensitive yield condition; the subsequent plastic

Table 7.1 Yield stress amplification due to strain-rate in mild steel beam
(Structural parameters: $L = 0.21$ m, $M_p = 107$ Nm, $T_o = 14.8$ ms)

Impulse intensity (Pa s)	Nondim. total impulse p_f	Input energy ratio K_o/M_p	Elastic period T_1 (ms)	Nondim. elastic period T_1/T_o	Yield stress amplification Y_d/Y
200	0.283	0.04	2.495	0.169	1.32
300	0.414	0.09	1.448	0.098	1.46
450	0.616	0.20	0.942	0.064	1.50
600	0.848	0.36	0.710	0.048	1.56
750	1.060	0.57	0.573	0.039	1.60

deformation depends upon the rate sensitive yield stress Y_d. The basis of this approximate method is an observation by Perrone [1965] that rate sensitivity of the yield stress is largest at small strains so the dynamic yield stress depends most on the strain-rate when yield initiates. This approximate method of incorporating rate effects was successfully employed to calculate dynamic response of stainless steel beams by Forrestal and Sagartz [1978].

Table 7.1 gives the amplification factor for yield stress calculated by Forrestal and Wesenberg [1977] for some tests on simply supported mild steel beams. These correction factors for the rate-effect gave a very accurate estimate of the maximum deflection measured in their experiments on magnetomotively loaded mild steel beams.

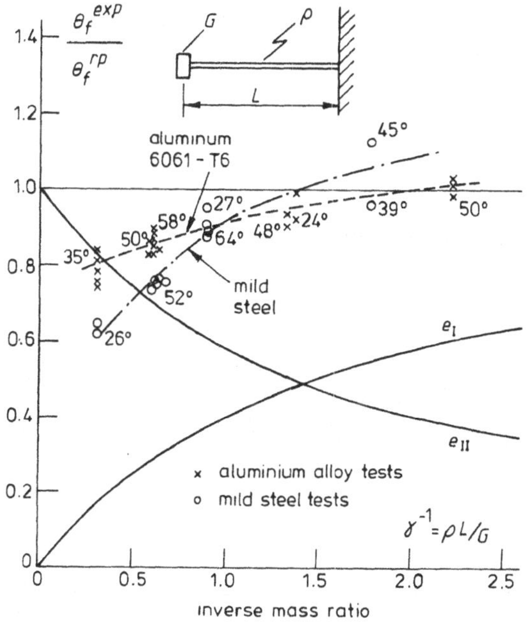

Fig. 7.9. Final tip rotations from bullet impact tests in comparison with elementary rigid–plastic theory (perfectly plastic theory gives final rotation proportional to input energy). Dashed curves are a best fit to data for aluminum alloy and mild steel beams. Solid curves are the part of the input energy dissipated in transient (e_I) and modal (e_{II}) phases according to rate independent rigid–plastic theory (after Bodner and Symonds [1962]).

7.4.2 Transient Analysis Including Rate Effects

An analysis for the transient response of an impulsively loaded, strain-rate dependent cantilever was presented in Sect. 5.1.1; instead of two separate phases, the deformation develops in a single continuous motion. This analysis assumed that the plastically deforming region initially extended throughout the entire length of the cantilever but that a rigid (nondeforming) segment of increasing length spread outward from the impact point at the tip. This analysis neglected any inertial effects in the deforming region. According to Ting [1964] the approximation has little effect on calculations of the final deflection. This form of transient rate-dependent analysis was used by Bodner and Symonds [1962] to investigate rate effects in mild steel and 6061-T6 aluminum alloy cantilevers. They performed explosive loading and bullet impact tests on specimens with a lightweight tip mass $0.3 < \gamma < 1.8$. Almost all the tests on mild steel cantilevers had input energy ratios $R > 10$, while the aluminum alloy cantilevers were hit with energy ratios $R > 3$. Consequently the cantilevers in these tests exhibited fairly large final deflections.

Results of these experiments are compared with calculations from a rigid–perfectly plastic structural analysis in the data points plotted on Fig. 7.9; there the ratio of experimental to calculated final tip rotation is given as a function of the mass ratio γ. The experimental values given on the graph indicate that the final tip rotations are in the range $24^{\circ} \le \theta_{Af} \le 64^{\circ}$. In comparison with the predictions of the rate independent, perfectly plastic analysis, the deformation in most tests is reduced, especially at large mass ratios. This is attributed to strain-rate dependence of the flow stress; this effect is largest at large mass ratios where a modal pattern accounts for most deformation. The curves labelled e_{I} and e_{II} indicate the part of the input energy dissipated in phases I and II of the rate independent perfectly plastic analysis, respectively.

7.5 Dynamic Rupture

Rupture occurs when a crack penetrates through the full thickness of a member. In ductile members the deformation processes that culminate in rupture almost always involve shear or stretching in addition to bending. Thus the elementary rigid–perfectly plastic approximation is of little help in analysing or predicting rupture. To consider effects of shear and stretching superposed on bending, it is necessary to include these other components of deformation and the conjugate stresses in the structural response analysis.

7.5.1 Location and Mechanism of Rupture

Rupture of a ductile member initiates when either the strain or the density of dissipated energy at a section exceeds some characteristic value. The pattern of deformation determines where energy is dissipated at an instant of response; at any time the distribution of accumulated deformation throughout a structure can

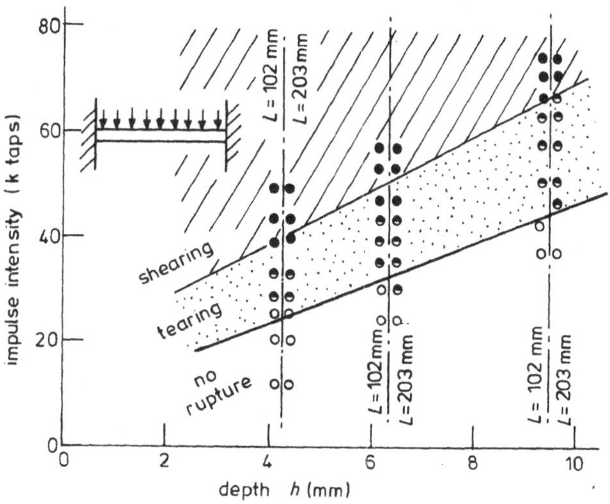

Fig. 7.10. Impulse intensity at rupture of explosively loaded 6061-T6 aluminum alloy beams clamped at the ends where tap = dyne s cm^{-2} (after Menkes and Opat [1973]).

be obtained by integrating the rate-of-deformation from the initial to the current time.

Distributed Dynamic Loading The distribution and relative magnitude of various components of deformation or generalized strain are most evident for steady pulse loading on a rate-independent structure. While a constant pressure is applied to some part of a structure, the pattern of deformation is stationary and each component of deformation increases linearly with time. Symonds [1968] used a square yield surface to approximate the interaction between flexure and shear in an analysis of deformation in a simply supported beam subjected to a uniform pressure pulse. This showed that more energy can be dissipated by shear adjacent to the supports rather than flexure if the length L/h is small and the applied force $F/F_c >> 1$.

Experiments on impulsively loaded thin aluminum alloy (6061-T6) beams by Menkes and Opat [1973] have shown that with both ends clamped to prevent any displacement or rotation, the deformation that culminates in rupture is even more complicated; this is because clamped ends constrain axial displacement and induce stretching. These beams were driven transversely by a uniform explosive load and those that ruptured always broke adjacent to the root. The fractures initiated on the convex surface of the beam where bending, stretching and shear combine to give the largest effective strain. With increasing intensity of the applied impulse, Menkes and Opat identified three patterns of deformation among which the latter two mechanisms culminate in rupture:

I. Large flexural deformation
II. Large flexural deformation with tensile tearing at ends
III. Small flexural deformation with shear deformation and shear rupture at ends

Figure 7.10 illustrates that the behavior changed from one of these mechanisms of rupture to another at the same impulse intensity for beams of 102 mm and 203 mm length. At least for shear rupture in mechanism III, this meant that the input energy was large enough so that shear sliding at the generalized hinges adjacent to the supports persisted until rupture. For impulsive loading, the shear sliding mechanism occurs at discontinuities in the initial velocity field.

These experiments were analyzed by Shen and Jones [1992]. They attributed the initiation of tearing rupture at the transition to mechanism II to a critical strain on the convex surface that develops when the central deflection is large enough to add significant stretching to the flexural strain at a clamped end. For impulses larger than that for transition to mechanism II, shear sliding at the hinge at each end of the beam reduces the effective thickness and hence decreases the stretch integrated over the total length that is required to rupture the cross section of the beam. This integrated stretch or increase in length is directly related to the central deflection. For the aluminum alloy beams in Fig. 7.10, they calculated that the mechanism II–III transition occurred when roughly 45% of the energy dissipated at end hinges was due to shear sliding.

Concentrated Dynamic Loading If a transverse impulsive load is applied to a concentrated mass on a slender structure or a heavy missile strikes a light beam, the pattern of deformation and hence the distribution of energy dissipation at any time are different from those obtained from a uniform impulsive load. Again shear sliding occurs at the velocity discontinuities but now these are located adjacent to the colliding missile as described in Sect. 5.3.3. In this case the plastic hinges are more diffuse than was found with uniform impulsive loading where the hinges were located adjacent to displacement constraints; consequently more energy can be dissipated at each hinge before rupture.

Liu and Jones [1987] performed drop weight impact tests on aluminum alloy and mild steel beams of half-length $L = 51$ mm with clamped end conditions. The colliding missile had a nose radius roughly equal to the depth of the beam and was much heavier than the beam so the mass ratio $\gamma > 250$. This missile struck either the center or an off-center position along the clamped beam at a transverse impact speed in the range $2 < V_0 < 12\ \mathrm{m\,s^{-1}}$. Although the initial kinetic energy of the colliding mass was very large in comparison with the capacity of the beam, the impact speed was slow so both aluminum and mild steel beams ruptured by predominantly shear sliding only if the impact point was immediately adjacent to a clamped end. Otherwise, rupture occurred at the impact point after stretching of the beam had become significant due to large deflections. These ruptures initiated on the tensile side of the beam. These results for impact by a heavy striker have been correlated with results for other loading conditions by Jones [1989a].

7.5.2 Measurements of Generalized Strain at Rupture

Rupture initiates at the section with the largest tensile strain when the largest principal strain equals a rupture or ultimate strain ε_r. In general the rupture strain depends on triaxiality of the strain state; for unaxial stress however, it is a material property. Values of the rupture strain measured in uniaxial tension tests are given in Table 7.2.

The tensile strain in the outer fiber at a plastic hinge is the sum of the largest flexural strain $\max(\varepsilon_m)$ and the strain due to stretching ε_n. The criterion for

Table 7.2 Material properties for aluminum alloy and steel specimens

Material	Thickness (mm)	Yield Stress (N mm^{-2})	Ultimate Stress (N mm^{-2})	Rupture Strain
Al. alloy plate (BS1470-HS15)[a]	–	182	318	0.19
Al. alloy bar[a]	3.81	354	475	0.19
	5.08	354	475	0.19
	6.35	354	475	0.19
	7.62	412	553	0.15
Al. alloy sheet (H15)[b]	3.25	103	269	0.18
Al. alloy sheet (6061-T6)[c]	<20.	280	336	0.10
Steel plate (BS4360-43A)[a]	–	324	501	0.31
Steel bar[a]	3.81	337	464	0.39
	5.08	337	464	0.39
	6.35	302	444	0.40
	7.62	302	444	0.40

[a] Liu and Jones [1987]
[b] Shu et al. [1992]
[c] Mil. Hdbk. 5 [1959]

tensile or tearing rupture is that the largest tensile strain equals the rupture strain ε_r; i.e.

$$\max(\varepsilon_m) + \varepsilon_n = \varepsilon_r \tag{7.4}$$

In any particular problem, the strains are related to displacements through kinematic constraints; they also depend on the hinge length L_h (i.e. the extent of a plastically deforming region).

For example, the fundamental mode of deformation for a rigid–perfectly plastic uniform beam of length $2L$ is illustrated in Fig. 7.11. If the ends of the beam are clamped, there is a plastic hinge at the center and one at either end. These hinges are short in comparison with the length of the beam, $L_h/L \ll 1$. Assuming that stretching due to moderately large deflections is distributed in proportion to the curvature at each hinge, we can estimate the separate components of strain and hence the smallest central deflection for tensile rupture at a hinge,

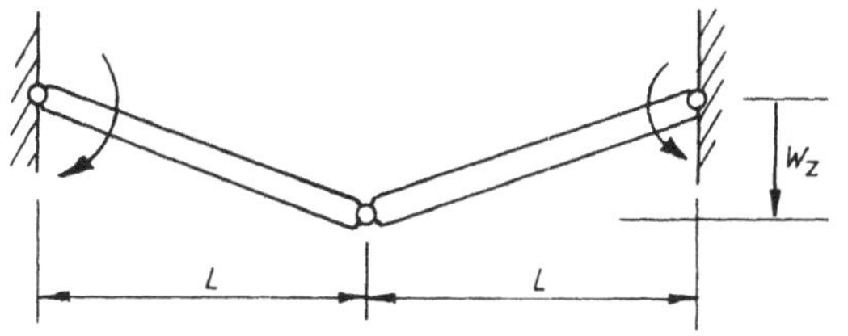

Fig. 7.11. Fundamental dynamic mode of deformation for uniform beam clamped at the ends.

$$\max(\varepsilon_m) = h\kappa/2 = hW_z/2LL_h$$

$$\varepsilon_n = \left\{\left(1 + W_z^2/L^2\right)^{1/2} - 1\right\}\left(L/2L_h\right)$$

Here it is clear that hinge length L_h is a crucial but ill-defined parameter for obtaining predictions of rupture. The length L_h is typically specified in the range $1 < L_h/h < 12$; this length certainly increases due to dispersion in strain hardening and strain-rate dependent materials. Most likely, hinge length depends also on the shear deformation at a hinge. In analysing experimental results for transition from one mechanism of rupture to another, Shen and Jones [1992] observed that the hinge length decreased as the dissipation due to shear D_q became a larger part of the total dissipation D at any particular hinge. Their empirical equation for the data of Menkes and Opat was as follows:

$$L_h/h = 1.3 - 1.2D_q/D \tag{7.5}$$

As hinge length decreases the tensile strain increases where the strain is compatible with deflection of an axially constrained beam; thus, large impulses that give significant shear deformation at initial velocity discontinuities in axially constrained beams also reduce the central deflection at which rupture initiates.

Rupture by Shear Sliding Shear deformation at a generalized hinge causes transverse relative displacement; the extent of this relative displacement at rupture is strongly affected by the length of the plastically deforming region; i.e. L_h. For Menkes and Opat's tests of impulsively loaded clamped beams of 6061-T6 aluminum, Shen and Jones [1992] found that at rupture in mechanism III, $\Delta_f/h \approx 0.6$. In contrast, blunt bullet impact against 6.35 mm thick 2024-T3 aluminum alloy plates gave $\Delta_f/h \approx 0.75$ for a bullet diameter of 7.82 mm (Liss, Goldsmith and Kelly [1983] or Jones [1989b]).

Rate Effect on Rupture Strain In rate dependent materials, large strain rates increase the flow stress for plastic deformation but decrease the ultimate or rupture strain. The dynamic rupture strain ε_r^d can be related to the static rupture strain ε_r^s by an expression similar to the Cowper–Symonds equation:

$$\frac{\varepsilon_r^d}{\varepsilon_r^s} = \left[1 + \left(\frac{\dot{\varepsilon}}{\dot{\varepsilon}_0}\right)^{1/r}\right]^{-1} \tag{7.6}$$

where $\dot{\varepsilon}_0$ and r are the usual Cowper–Symonds material constants. This expression gives a fracture strain that is inversely proportional to dynamic stress so the dissipation density at fracture is independent of strain-rate (if hinge length is rate independent).

References

Bodner, S.R. and Symonds, P.S. [1962]. Experimental and theoretical investigation of the plastic deformation of cantilever beams subjected to impulsive loading. *J. Appl. Mech.* **29**, 719–727.

Forrestal, M.J. and Wesenberg, D.L. [1976]. Elastic–plastic response of 6061-T6 aluminum beams to impulse loads. *J. Appl. Mech.* **43**, 259–262.

Forrestal, M.J. and Wesenberg, D.L. [1977]. Elastic–plastic response of simply supported 1018 steel beams to impulse loads. *J. Appl. Mech.* **44**, 779–780.

Forrestal, M.J. and Sagartz, M.J. [1978]. Elastic–plastic response of 304 stainless steel beams to impulse loads. *J. Appl. Mech.* **45**, 685–686.

Glasstone, S. [1959]. *Effects of Nuclear Weapons.* McGraw-Hill, New York.

Hall, R.G., Al-Hassani, S.T.S. and Johnson, W. [1971]. The impulsive loading of cantilevers. *Int. J. Mech. Sci.* **13**, 415–430.

Jones, N. [1989a]. On the dynamic inelastic failure of beams. *Structural Failure* (ed. T. Wierzbicki and N. Jones). John Wiley & Sons, 133–159.

Jones, N. [1989b]. *Structural Impact.* Cambridge University Press, 274.

Kinney, G.F. [1962]. *Explosive Shocks in Air.* Macmillan, New York.

Liss, J., Goldsmith, W. and Kelly, J.M. [1983]. A phenomenological penetration model of plates. *Int. J. Impact Engng.* **1**, 321–341.

Liu, J. and Jones, N. [1987]. Experimental investigation of clamped beams struck transversely by a mass. *Int. J. Impact Engng.* **6**, 303–335.

MIL-HDBK-5 [1959]. *Strength of metal aircraft elements.* Armed Forces Support Center, Washington, D.C.

Menkes, S.B. and Opat, H.J. [1973]. Broken beams — tearing and shear failures in explosively loaded clamped beams. *Experimental Mechanics* **13**, 480–486.

Parkes, E.W. [1955]. The permanent deformation of a cantilever struck transversely at its tip. *Proc. Roy. Soc. Lond.* **A228**, 462–476.

Perrone, N. [1965]. On a simplified method for solving impulsively loaded structures of rate-sensitive materials. *J. Appl. Mech.* **32**, 489–492.

Reid, S.R. and Gui, X.G. [1987]. On the elastic-plastic deformation of cantilever beams subjected to tip impact. *Int. J. Impact Engng.* **6**, 109–127.

Rinehart, J. and Pearson, J. [1963]. *Explosive Working of Metals.* Macmillan, New York.

Shen, W.Q. and Jones, N. [1992]. A failure criterion for beams under impulsive loading. *Int. J. Impact Engng.* **12**, 101–121.

Shu, D., Stronge, W.J. and Yu, T.X. [1992]. Oblique impact at the tip of a cantilever.*Int. J. Impact Engng.* **12**, 37–47.

Symonds, P.S. [1965]. Viscoplastic behavior in response of structures to dynamic loading. *Behavior of Materials Under Dynamic Loading* (ed. N.J. Huffington, Jr.). ASME, New York, 106–124.

Symonds, P.S. [1968]. Plastic shear deformations in dynamic load problems. *Engineering Plasticity* (ed. J. Heyman and F.A. Leckie), Cambridge University Press, 647–664.

Symonds, P.S. and Fleming, W.T. [1984]. Parkes revisited; on rigid–plastic and elastic–plastic dynamic structural analysis. *Int. J. Impact Engng.* **2**, 1–36.

Ting, T.C.T. [1964]. The plastic deformation of a cantilever beam with strain rate sensitivity under impulsive loading. *J. Appl. Mech.* **31**, 38–42.

Woodward, R.L. [1984]. Transverse projectile impacts on beams. *J. Appl. Mech.* **51**, 437–438.

Index

9 783540 760139